Extrasolar Planets and Astrobiology

Extrasolar Planets and Astrobiology

CALEB A. SCHARF

COLUMBIA ASTROBIOLOGY CENTER
COLUMBIA UNIVERSITY

University Science Books
Sausalito, California

University Science Books
www.uscibooks.com

Production Manager: Side by Side Studios
Manuscript Editor: Lee Young
Design: Paul C. Anagnostopoulos
Illustrator: Lineworks
Compositor: Windfall Software, using ZzTEX
Printer & Binder: Maple Vail Press

Cover design and cover composite by Mark Ong, Side By Side Studios. Saturn rising image courtesy of NASA and the NSSDC. Beach photograph © 2008 JupiterImages Corp.

This book is printed on acid-free paper. ∞

Library of Congress Cataloging-in-Publication Data

Scharf, Caleb A., 1968–
    Extrasolar planets and astrobiology / Caleb A. Scharf.
       p.   cm.
    Includes bibliographical references and index.
    ISBN 978-1-891389-55-9 (alk. paper)
    1. Extrasolar planets.   2. Exobiology.   3. Life on other planets.   4. Space biology.   I. Title.
QB820.S28 2008
576.8′39—dc22
                                                                    2008020138

Printed in the United States of America
10 9 8 7 6 5 4 3 2 1

Abbreviated Contents

Contents

Foreword

This book by Caleb Scharf offers a formal exposition about an exploding development throughout the world: the merging of multiple scientific disciplines to form a compelling new field, namely astrobiology. The most profound questions posed by ancient civilizations largely remain unanswered, left for religious leaders, philosophers, and the curious among us to ponder throughout the ages. Ancient Greek philosophers intensely debated the uniqueness of our Earth and the possibility of life elsewhere, but they and their academic successors made little progress for 2400 years. Suddenly and startlingly, we stand at the brink of answering those old questions.

The answers are arriving from two directions: progress on the physical prerequisites for life, and hints of the first habitable worlds. Astronomers have found that the fundamental equations of gravity, electricity, and quantum physics (some yet to be fully understood) are the same everywhere throughout the spacetime of the universe. The atoms and molecules of which life is composed are also ubiquitous throughout the universe. The 92 naturally occurring atoms are seen in stars and galaxies by spectroscopic analysis of their light, filling the periodic table for most of the 13.7 billion year age of the universe. Those atoms combine into complex organic molecules such as alcohols and amino acids, which are found in comets, moons, and interstellar clouds. The LEGOs™ of life are everywhere. Moreover, water, the great chemical cocktail mixer, is among the most abundant of substances in the universe, found on planets, moons, and comets. Water mobilizes, destroys, and recombines the organic molecules, to create uncountable molecular permutations of great size and complexity. Few can doubt that proteins and nucleotides will form given enough time.

The energy required for life comes in many forms, including starlight, geothermal, tidal, and radioactive, offering the complex organics myriad ways to power further reactions. Chemical replication will surely occur by molecular precursors to RNA and DNA, leading to a competition for both the valuable molecular building blocks and for the free energy. The successful precursors will multiply and outcompete their neighbors, populating their environment. The blurry but inexorable transition from "prebiotic" to "living" clusters of organic molecules reveals life as a natural phenomenon, forged in the furnaces of stars.

Meanwhile, astronomers have developed techniques to detect the first habitable abodes in the universe. The subsurface water in Mars, the oceans inside Europa and Ganymede, and the geysers of Enceladus all constitute rich chemistry labs, each one percolating for billions of years. The exploration of these new worlds seems as compelling as the most courageous transoceanic voyages in human history. Should robotic probes discover the first alien lifeforms, the ubiquity of life in the universe will be established and the diversity of life sampled.

Beyond our Solar System, over 300 planets have now been discovered orbiting other stars. So far, only large ones similar in size to Jupiter, Saturn, and Neptune have been found. But new techniques are now in hand to detect terrestrial-sized planets. The Kepler mission will search for stars that dim periodically due to earths that cross in front, blocking only 1/10,000 of the starlight. Astronomers will also make precise Doppler measurements of stars to detect their reflex motion in response to the gravitational tug by terrestrial planets. Some clever astronomers are surveying the smallest stars, the red dwarfs, to detect their dimming and reflex motion, to reveal earth-sized planets.

For the future, NASA and the Jet Propulsion Laboratory have developed the Space Interferometry Mission that will use the interference of light waves gathered by a spaceborne pair of telescopes to detect earth-like planets, and measure their masses, around nearby stars. Just over the horizon are plans for a spaceborne telescope that blocks the glare of nearby stars, allowing us to take images of Earth-like planets and to determine their chemical composition from their spectra. Any worlds having oxygen atmospheres and surface oceans will smell fishy from 40 light years. This census of habitable earths will fill GoogleGalaxy with ports-of-call for our grandchildren who will send robotic probes and later themselves, at least those with extreme daring and patience. The urge to explore these new worlds comes from our anthropological roots at Olduvai Gorge two million years ago. What sets us apart from the stones and the stars is our insatiable desire to understand our kinship with both.

Geoffrey W. Marcy
University of California, Berkeley

Preface

The questions that surround the quest for an understanding of the nature of life in the universe are perhaps as diverse as life itself. Is life on Earth unique? How did life on Earth begin? Are there other planets like the Earth? Are there other systems like the solar system? Do these systems harbor life? The prospect of writing any kind of text that can be used to teach the science that is necessary to begin to address such questions is, to say the least, daunting.

I will therefore begin with some excuses as to why this book does not attempt to deal with all such questions. The original idea behind the material was to offer the analytically literate person an introduction to the emerging "interdiscipline" of astrobiology. It is absolutely *not* meant to be either complete, or definitive. In fact, I think that anyone claiming to produce a complete or definitive text (especially one for teaching) on astrobiology at this stage is probably an alien—so watch out. What this book *does* try to do is lay some of the groundwork for a few of the methods and physical principles that serve as tools for understanding questions of star and planet formation, astronomical observation, and chemical and biological modeling. As the book's title suggests, the study of planets other than our own is a central part of the material. The primary audience is therefore either a student or researcher in astronomy or physics, or possibly someone from the geophysical, chemical, or biological sciences, looking for a deeper understanding of the "astro" in astrobiology.

As the book developed, more and more purely descriptive material crept in - culminating in some of the speculative material in the final chapter. I think this speculation is entirely valid, since so much of astrobiology is (at present) speculation - albeit increasingly based on real information. I also hope that this can demonstrate to a student how speculation with some analytic basis can serve a useful purpose by stimulating further discussion and thought.

A lot of emphasis is placed in this book on the importance of not being blinded by an "Earth-centric" view of astrobiology. Earth is the best, and only, template that we have available in the present quest for

Extrasolar Planets and Astrobiology

the description of the extraordinary advances and discoveries that have driven the growth of astrobiology largely to their appropriate places in the following chapters.

There is nonetheless an underlying philosophical theme to much of what we will discuss. Put simply, we work with the hypothesis that life emerges as a natural consequence of the physical laws in this Universe. It might therefore be treated as a fundamental property of the Universe much like stars and planets. In this sense "life" is merely one way that matter becomes organized on certain scales and in certain circumstances. We do not deal with this in a particularly deep way. Nuances of quantum mechanics and subatomic physics are left to other works; as are discussions of any merit on the philosophical underpinnings of life, consciousness, metaphysics, and the human experience. The intent is not to trivialize the discussion of life and its origins, but rather to avoid any tendency to imbue it with any "special" significance. Life is certainly special to us, but whether it is special or mundane to the Universe remains an open question—one that we might hope to answer through the science of astrobiology.

On the face of it one might then expect this to be a rather dry exploration; a lot of it is about solving equations or slotting physics into answering basic questions. However, one cannot fully appreciate the excitement and beauty of the science without experiencing the deeper, quantitative, side of things. From this knowledge springs the ability to connect seemingly unrelated phenomena and to make genuine progress. The incredible power of even the simplest piece of quantitative reasoning is a very exciting thing, especially when applied in new ways to the ancient questions posed by astrobiology. It is probably also fair to say woe betide the mechanic who forgoes the toolbox when trying to disassemble and reassemble an engine.

Thus, while a fluency with basic scientific technique and knowledge is generally assumed throughout; for the sake of self-containment, should the reader become stranded on some tropical planet, a brief discussion of the broader context is given below.

1.1.1 The Current Domain of Known Life

The most abundant and diverse life we know of (and currently the *only* life we know of) is that which exists on Earth, the prototype for terrestrial planets (Figure 1.1). Earth is a rocky world, rich in heavy elements,

Figure 1.1. The full Earth photographed from Apollo 17, en route to the Moon on December 7th, 1972, for the final lunar landing of the Apollo era. The continent of Africa is seen to the upper left, Antarctica is seen at the bottom (NASA).

orbiting a G-type dwarf star at a distance where the mean solar insolation (radiation flux) is some 1300 watts per square meter (1.3×10^6 erg cm^{-2}). It harbors a nitrogen–oxygen surface atmosphere, and exhibits a seven-tenths surface coverage of liquid water—with commensurate local surface temperatures ranging from a record minimum of 184 K [2] to a record maximum of 331 K [3] and a global mean of approximately 287 K.

---

2. Recorded at Vostok station in Antarctica
3. Recorded at El Azizia, Libya

Life (as defined by our knowledge of terrestrial biochemistry) appears to exist in a relatively narrow shell on Earth, which we might (rather arbitrarily) consider to be defined by the atmosphere up to and including the stratosphere 50 km above the planet's surface, to as much as ~ 5 km below the mean sea level (i.e., in and below both the deep ocean and below the solid dry-land/continental surface). This represents less than some 3% of the total volume of the Earth (Figure 1.2).

Although the major thrust of astrobiology is to understand life in the cosmos, and in particular the processes that give rise to its potential habitats and chemical and energetic resources, one of the ultimate outcomes must be a deeper understanding of the Earth, its origins, evolution, and future. In that sense, while Earth and our solar system serve as prototypes, the effort to place them in true physical context must be considered as equally important.

The first step along this path takes place at the very beginning of the currently observable Universe.

1.1.2   A (Very) Brief History of Normal Matter

The Universe is currently some 13.8 billion years old and is still expanding and cooling with a present mean temperature of about 2.78 K—as defined by the photon field of the cosmic microwave background; the remnant of the hot Universe at an age of three hundred thousand years. In terms of primary composition, the mass content of the Universe appears to be dominated by a form of "dark" matter, which has a negligible cross section of interaction with normal matter, or indeed with itself, and whose presence is felt only through gravity. By contrast, normal matter (by which we mean particles such as protons, neutrons, electrons), comprises only about ten percent of the total gravitating matter in the Universe.

By the time the Universe was approximately three hundred thousand years old, the normal matter component consisted of gaseous, neutral, atoms of Hydrogen (74%) and, as a result of **primordial nucleosynthesis**, helium (25%), with trace amounts of the light elements deuterium (1%), lithium, and beryllium. Within only a few hundred million years of this time, gravity was able to draw together the first condensations of normal matter (aided by the dominant dark matter component), and from there the first generation of stellar objects. We will return to the general question of star formation in Chapter 2. Once stars were formed,

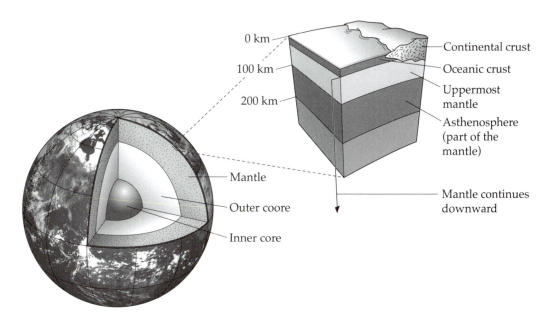

Figure 1.2. A cutaway schematic of the Earth. The atmosphere and outer crust represent a tiny fraction of the total volume of the planet. The oceanic crust is different than that of the continents, and consists of basalt some 5–10 km thick. The continental crust is more varied in composition and is less dense, but is typically some 20–70 km deep. Both are less dense than the material in the mantle, and effectively float on top of it (see Chapter 5). The mantle extends downwards to some 2500 km in depth, and varies in temperature from 100 °C at its upper levels to more than 4000 °C at the boundary with the outer core. At greater depths it is primarily solid due to the enormous pressure from overlaying material. The outer core is thought to be primarily liquid, and extends from a radius of some 1200 km to about 3500 km; the inner core is thought to be solid. Both are largely composed of iron and nickel (as well as some heavier and some lighter elements). During the early stages of the Earth's formation there would have been strong **differentiation** of material (see Chapter 3), with heavier/denser elements finding their way down through the gravitational well of the planet. Intriguingly the inner core is likely too hot to sustain a permanent magnetic field, which must be generated by the outer liquid core. Furthermore, the inner core may actually rotate slightly faster than the rest of the planet, by some 0.3–0.5 degrees per year (Zhang et al. 2005).

stellar nucleosynthesis began to build all heavy elements, or in astronomical terms: **metals** (meaning anything heavier than helium), as well as disperse them throughout the cosmos via stellar outbursts and explosions (nova and supernova). Thus began a cosmic cycle of element production and dispersal, creating the potential for normal matter to

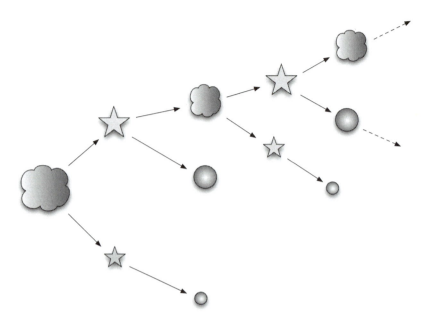

Figure 1.3. A schematic of the cyclical nature of element production and dispersal by stars in the universe. On the left, a molecular cloud collapses and forms both high-mass stars (upper arrow), and low-mass stars (lower arrow). The lower mass stars go through their Main Sequence lifetime and ultimately produce a low-mass white-dwarf stellar remnant. The high-mass stars can end their Main Sequence lifetime in a supernova explosion, producing both a nebula of dispersed elements and a high-mass stellar remnant, either a neutron star or a black hole. The dispersed elements are eventually incorporated into a new collapsing molecular cloud (see Chapter 2), which in turn can produce both high- and low-mass stars. Over time the elemental abundance of metals in stars therefore increases, as does the amount of normal matter sequestered into stellar remnants (and, as a smaller fraction, into planetary bodies).

condense into forms such as dust and rocky bodies, as well as remaining as (or being recycled into) gas and plasma (Figure 1.3).

1.1.3 Stars

What we typically refer to as stars are objects actively powered by nuclear fusion processes. This distinguishes them from the more general term **stellar object**, which may refer to a variety of things, including

stellar remnants, such as white dwarfs, neutron stars, and black holes. Stars span a wide range of masses, from a few hundredths of a **solar mass** (one solar mass is 1.99×10^{33} g, or $1M_\odot$) to as much as 200 solar masses—which has recently been observed to be an apparent upper limit to the mass of present-day stars. Stars are extraordinary objects—they quite literally represent the balance between gravity (which acts to contract objects) and thermodynamics (and in some instances even quantum mechanics). Although often considered in popular terms to be "giant balls of hot gas," the internal makeup of stars is really that of a plasma (an ionized gas). Central (core) densities can be a million times that of the Earth's atmosphere at sea level, with temperatures and pressures of the order of millions to billions of times (respectively) those of the Earth's atmosphere. Stars therefore represent conditions radically different to those we are familiar with on Earth. Most stars are prevented from collapse from their own weight by a mixture of gas and radiation pressure (they can, to first order, be considered as in **hydrostatic equilibrium**, Chapter 2); these are in turn sustained by the energy production of nuclear fusion processes in the stellar core. The sustained fusion of elements occurs inside stars because the temperatures and densities are such that atomic nuclei are able to overcome their mutual electrostatic repulsion. Aided by quantum mechanics (the uncertainty principle and the particle–wave duality of matter) the nuclei can then find themselves close enough that the nuclear force (a nucleon–nucleon force, not to be confused with the strong nuclear force mediated by gluons and operating between quarks within the nucleons themselves) takes over and binds the protons and neutrons together. If the new nuclear configuration represents a lower overall energy state then the surplus energy is released as highly energetic photons (gamma rays). Thus, a new, heavier element is formed, and energy is liberated.

Through a variety of reaction chains of different rates or probabilities, heavier and heavier elements can be formed. This occurs as long as the core plasma has sufficient temperature to allow successive nuclei to overcome their electrostatic repulsion due to the increasing baggage of nuclear protons. Specifically, the two forces at play are the nuclear force that decreases exponentially with physical separation, and the electromagnetic force between protons which decreases as the inverse of the separation squared (i.e. $1/r^2$). As the number of nucleons increases up to the elements iron and nickel (Fe^{56}, Fe^{58}, and Ni^{62}) the nuclear force dominates and produces ever more tightly bound nuclei. Eventually though, the longer range electromagnetic force begins to offset the gain in binding energy of adding more nucleons. For heavier elements than iron and nickel the nuclear configurations become sufficiently loose that the fusion reactions become effectively **endothermic**. All such elements are formed via secondary processes—such as during stellar explosions (supernova) where intense particle fluxes enable the construction of more massive nuclei—and subsequent radioactive decay then fills out the periodic table beyond iron.

For our purposes, one of the key features of stars is that the majority of their lifetime (i.e., that period during which fusion processes are ongoing in their cores) is spent fusing, or "burning" their core hydrogen. The reasons for this are that hydrogen fusion is initiated at the relatively low temperature of $\sim 10^7$ K, there is more hydrogen than anything else in a new star (even if many generations beyond those in the early universe), and the energy production *rate* due to hydrogen fusion is *less* temperature dependent than that of heavier elements. For example, the energy production rate in the triple-alpha process, in which helium is burnt to heavier elements such as carbon, nitrogen, and oxygen, is a factor $\sim 10^{38}$ times more sensitive to the stellar core temperature than the simplest proton–proton burning chain of hydrogen fusion. During the period of a star's life in which it is predominantly burning hydrogen, we refer to it as a **Main Sequence** star —for historical reasons based on the position such stars take in the parameter space of stellar luminosity versus temperature (the **Hertzsprung–Russell diagram**, Figure 1.4).

However, the absolute length of time spent as a Main Sequence star is a function of the mass of the star. The more massive the star, the shorter its Main Sequence lifetime, and indeed the shorter its overall lifetime. The reason is simple: as gravity contracts a more massive star, the interior temperature and density must reach higher values in order to halt the contraction—as dictated by hydrostatic equilibrium (see Chapter 2). The higher the core temperature the faster the nuclear reaction rates, and the more rapidly material is burnt. For these reasons too, more massive stars are both hotter and more luminous than lower mass stars (see H-R diagram). Once a star has used up all the hydrogen in its core it is no longer a Main Sequence star. Its subsequent evolution depends critically

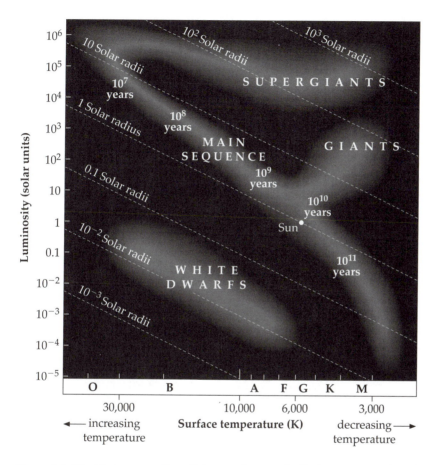

Figure 1.4. The Hertzsprung–Russell diagram for stars. Stellar luminosity is plotted against effective temperature (i.e. measured from the observed photosphere)—increasing to the left. The Main Sequence is illustrated as the thick locus running from the upper left to lower right. The lifetime of stars on the Main Sequence is indicated, and correlates with the stellar masses. Lines of constant stellar radii are overlaid. Outside the Main Sequence the populations of supergiant and giant stars, as well as white dwarfs, are also shown.

on its mass. For lower mass stars (less than approximately three solar masses) the end point is often a white dwarf, an object of **electron degenerate** matter. In such objects it is the quantum mechanical properties of electrons (specifically the Pauli exclusion principle) in dense matter that enable them to resist further compression and thereby support the mass of the star even though there is no further energy generation. For higher mass stars, the late stages of their evolution can be complex, heavier elements can be produced, and once these fusion pathways are depleted the star may undergo violent death as a supernova, leaving remnants such as neutron stars or even black holes.

During a lower mass star's Main Sequence lifetime there is also modest evolution of its external properties. As hydrogen is burnt to helium, the mass per particle of the core plasma increases. Consequently the classical gas pressure decreases, and the entire core must shrink and raise its temperature to ensure that hydrostatic equilibrium is maintained. Since the core temperature is slightly elevated, the energy production rate must also increase, and to the external observer the star will become slightly hotter and slightly more luminous. Thus, over a Main Sequence lifetime a star will gradually become more luminous, and the peak of its radiation output (since the star is to first order a blackbody) will shift towards higher energies. Such evolution may well be of vital importance in the consideration of potential habitats for life surrounding a star (Chapters 6 & 9).

1.1.4 Planets

What is a planet? This seemingly innocent question is actually quite hard to answer in astrophysical terms. In part this is due to genuine subtleties in the physical classification of celestial objects, but it is also due to a current lack of consensus on what is the sensible thing to do (for a review see Basri and Brown 2006).

There are two ends to this particular serpent. At the high-mass end is the distinction between a planet and a "stellar object," and at the low-mass end is the distinction between a planet and smaller bodies such as asteroids.

On the face of it, the problem at the high-mass end appears to have what is perhaps a better, physically motivated, solution. We have described how a star is distinct from other stellar-type objects by virtue of its internal energy production. Since the core temperature of a star is determined largely by its overall mass in hydrostatic equilibrium, this informs us that *below* a certain mass the core temperature will never become high enough to initiate sustained fusion reactions. Indeed, the temperature required for the fusion of deuterium ($^2_1$H) is less than that of pure hydrogen ($\sim 10^7$ K), due to a larger cross section of interaction. This suggests a natural lower limit to the mass of what we would term a star, since deuterium is present in all primordial material in the universe. Such an object has a mass of $\simeq 0.01 M_\odot$, or, switching to units of Jupiter masses (1.899×10^{30} g $= 1 M_J$) approximately $13 M_J$. This is not to suggest that *sustained* deuterium fusion can occur—that probably happens at masses above $\sim 0.01 M_\odot$—or indeed that even sporadic fusion has ever occurred in such an object, but it could have. This clearly leaves a gap between our stated lower mass for stars in the previous section (which, more precisely is $\sim 0.08 M_\odot$) and planets. Objects indeed exist in this mass range, and are known as **Brown Dwarfs**. These are a fascinating class of "sub-stellar" objects, most likely supported in their cores by the same electron degeneracy pressure that supports white dwarfs (see above). Their internal structure may be such that they are entirely **convective**, which is to say that energy transport occurs by the up- and downwelling of material, from core to surface. This mixes their contents very thoroughly. Their relatively low temperatures (less than 1000 K) also allow quite fragile molecular species to exist in their outer atmospheres.

The upper-limit mass scale of $13 M_J$ for planets is the definition, or convention, that we will stick to in this book. It is worth noting however that a planet of mass just below $13 M_J$ is still quite an exotic object compared to most of the planets we are familiar with in our solar system. Such an object will be largely gaseous and a strong source of infrared photons, since during its formation out of a collapsing cloud of gas (Chapter 2) gravitational potential energy will have been converted to thermal kinetic energy. The cooling for such an object is slow, since as it radiates away energy it will continue to contract (ignoring additional stellar radiation input from any parent star) and maintain its temperature (see Chapter 2). From a purely emotional point of view such an object is perhaps more a failed star or brown dwarf than a planet. Indeed, current theories of planet formation suggest that making planets more massive than about $10 M_J$ may be tricky—*if* they are to be made by the same mechanisms as lower mass planets (Chapter 3). Thus, while the definition of a planet based on a mass of $13 M_J$ is well motivated, it does *not* actually tell us how, or where, such objects should form.

One way to complete this definition for a massive planet is to combine it with the requirement that the object be in stable orbit around a larger, stellar-mass, body—thereby relegating free-floating objects to yet another category known sometimes (and confusingly) as "sub-brown dwarfs."

We can now move on to consider the low-mass end of the planet definition debate. This can often be a rather emotional subject. People of the twentieth century formed a strong attachment to the notion that our solar system is occupied by nine planets—where the outermost, and last to be discovered, is of course Pluto, whose largest companion is Charon (note the avoidance of the word "moon"). The problem with considering Pluto to be the ninth, and somehow final, planet is that in recent years additional, equivalent, if not greater bodies have been discovered in the outer solar system. This suggests that Pluto is really just one of a potential myriad of outer "worlds" that straddle the orbital region between the truly massive, thick atmosphere–bearing, outer planets (i.e. Uranus and Neptune) and the Kuiper belt of material left in a relatively pristine state after the formation of the solar system (see below). Indeed, some of these bodies may even originate from the Oort cloud (see below). To be a little more precise—Pluto has a mass of $0.0021 M_\oplus$ (and Charon is about 1/8th of the mass of Pluto), compared to $0.055 M_\oplus$ for Mercury, the smallest inner planet, and $14.5 M_\oplus$ and $17.1 M_\oplus$ for Uranus and Neptune. With limited data we know that the diameter of three of the recently discovered outer system bodies may be as much as ~ 104–130% that of Pluto (Eris, née "Xena"), $\sim 90\%$ (Sedna), and $\sim 60\%$ (Quaoar). Poor little Pluto does indeed seem to pale a little by comparison, and certainly loses some of its luster as the outermost "planet" (Figure 1.5, and see for example Brown et al. 2005).

It seems therefore that there are likely to be other significant outer system bodies, just waiting to be discovered by our ever-improving techniques—which indeed have only now begun to usurp Pluto, which is a great testament to the skills of Clyde Tombaugh, who first identified it in 1930.

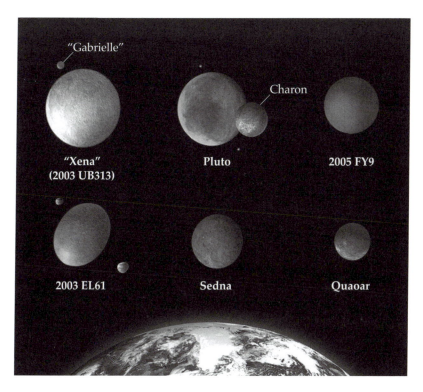

Figure 1.5. An illustration of the relative sizes of the Earth and the estimated sizes of the Kuiper belt, or "Trans-Neptunian" objects Eris (née "Xena"), Sedna, Quaoar, 2005 FY9 and 2003 EL61 (Adapted from NASA/ESA and A. Feild (STScI)) images).

What then to call such bodies—are they planets, or are they asteroids (which means *minor planet*)? There does not seem to be a particularly convenient measuring rod in this case. One potential, physically motivated, definition is that a body should be considered a planet only if it is large enough for its self-gravity to have forced it into a closely spherical shape (occuring roughly at a few hundred kilometers diameter), and that it is in independent orbit about the parent star (to differentiate it from a moon). This too has ambiguities—such as the dependency of when sphericity is reached on the material composition (e.g., ice versus rock). It also allows for the largest asteroids (such as Ceres, between

Mars and Jupiter) to be classed as planets. An additional caveat would then need to be that the object must have also cleaned out small material in the vicinity of its orbit in order to be considered a planet rather than just more detritus. At the time of writing there is considerable debate about where to "draw the line" for planets at this low-mass end. Irrespective of any astronomical definitions that are made official, it is likely that this debate will continue.

Perhaps the distinction is best dealt with in terms of how particular objects participated in the original formation of the solar system, and how their orbital configuration matches that of the majority of massive bodies in the system (for example, eccentricity and orbital inclination). There may be no universal, physically motivated, boundary between planet and less-than-a-planet that holds in *every* planetary system we encounter. What ultimately matters is not the semantic definition, but the "family" to which a given object belongs, which in turn describes its origins. We therefore leave this open, and try to be very specific in describing objects, in order to distinguish them.

1.1.5   Our Solar System

Until quite recently our *only* point of reference for the study of planetary systems and the potential for life has been the Solar System (Figure 1.6).

In addition to the major planets (from which we will exclude Pluto), six of which also play host to moon and satellite systems—ranging from a single moon (Earth) to over 60 moons and satellites (Jupiter) [4] —there are tens of thousands of other known objects in our system. These range from substantial asteroids, in a variety of orbital configurations (from system-crossing orbits, to those in the asteroid belt and trojan points throughout the system), to comets and small rocky bodies in streams and groups (e.g., Figure 1.7a). The existence of vast fields of material beyond the orbits of Uranus and Neptune is also strongly suspected and/or indicated.

Indeed, as we step outwards from Neptune we first encounter the Kuiper Belt, predicted in 1950 by Gerard Kuiper. This is a region between ~ 30 AU from the Sun (i.e. the orbit of Neptune) to at least 50–60

---

4. The distinction between moons and satellites is, much as for planets and asteroids, somewhat fuzzy.

Figure 1.6. The Sun and planets of the solar system (and their major moons) to approximate scale. Note that while we include Pluto out of historical respect, it is perhaps better considered as a different class of object (adapted from NASA/JPL).

AU where at least ~ 1000 significant objects (ranging from the sizes of Pluto and Eris to smaller asteroidal bodies) exist in a disk-like structure, extending the territory of the outer Solar System (Brown et al. 2005). These objects are left over, we think, from the early development of the Solar System, and may have much to do with the migration (Chapter 3) of the outer planets (Neptune in particular) and have likely been scattered outwards due to interaction with these larger worlds in this early period. Occasionally Kuiper Belt objects may be perturbed and fall inwards as short-period comets.

Beyond this ragged edge to our familiar solar system is the postulated Oort cloud (Figure 1.7b). In a shell some 50,000 to 100,000 AU (~ 1 light year) from the Sun the Oort cloud may comprise the remnant material from the collapsing nebula, or molecular cloud, which was the progenitor of our Solar System (Chapter 2). It was originally proposed as the source of the majority of (long-period) comets entering the solar system, owing to the estimated aphelion distances of these objects. Containing some 10^{12} bodies (with a total mass of as much as $\sim 100 M_{\oplus}$) the

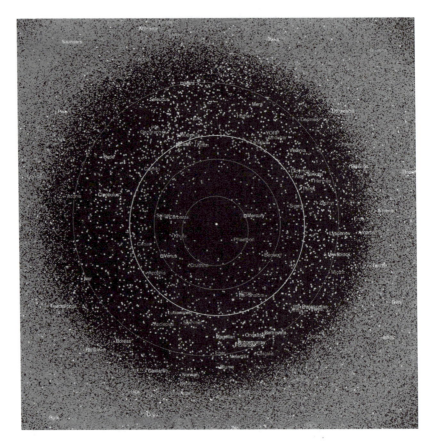

Figure 1.7. (a) An illustration of the distribution of small, asteroidal, bodies in the inner solar system. The orbits of Mercury, Venus, Earth, and Mars are shown. Thousands of asteroids are known, including those on so-called Earth-crossing orbits—with the potential for eventual collision. The majority of objects in this figure occupy the asteroid belt, between the orbits of Mars and Jupiter (Scott Manley/Armagh Observatory).

Oort cloud may have been formed as condensing material much closer to the proto-Sun was flung outwards by the gravitational action of the forming gas giant planets, such as Jupiter. While it has yet to be confirmed around our Sun, or indeed around other stars, it is an appealing picture, and certainly comets are coming from *somewhere* well beyond the orbit of Neptune.

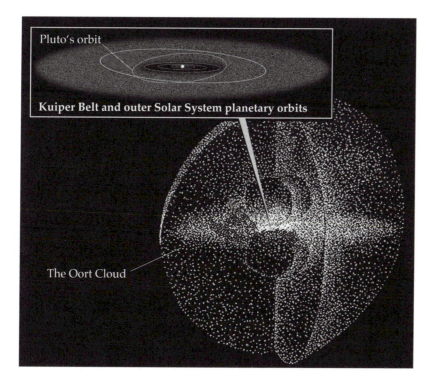

Pluto's orbit

Kuiper Belt and outer Solar System planetary orbits

The Oort Cloud

Figure 1.7 *(continued).* (b) An illustration of the location and geometry of the Kuiper belt, extending from approximately the orbit of Neptune to some 50–60 AU from the Sun, and of the postulated Oort cloud (NASA, A. Feild (STScI)).

Beyond the postulated Oort cloud is true interstellar or intra-galactic space. The closest star *system* is Alpha Centauri (actually a triple star system) at a distance of 4.37 light years—or ∼4–5 times farther than the proposed edge of the Oort cloud.

1.1.6 Our Sun

It is often easy to forget that the Sun is very much a participant in the daily phenomena of the Solar System. In addition to providing stellar insolation (flux) ranging from a scorching 6000–14,000 watts per square meter (these SI units are used in this particular example because watts are more familiar day-to-day units) at the orbit of Mercury, to a feeble 1 watt per square meter at the orbit of Neptune, the Sun floods our system

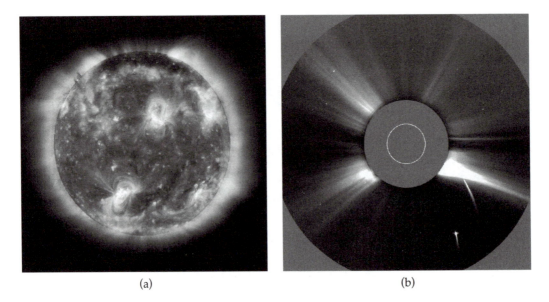

(a)   (b)

Figure 1.8. (a) An extreme ultraviolet image (at 171 Å) of the Sun taken by the EIT instrument onboard NASA's Solar and Heliospheric Observatory (SOHO) spacecraft. Unlike an optical waveband image, this UV picture shows dramatically some of the higher energy coronal structures of the Sun, including plasma that is trapped and accelerated along magnetic field lines (SOHO (ESA & NASA)). (b) An extreme ultraviolet image of the Sun taken by the LASCO instrument onboard NASA's SOHO spacecraft. In this picture the disk of the Sun has been blocked out by a coronagraph to enable the contrast to be enhanced to show the highly extended coronal structures emanating from the stellar surface. These structures contain accelerated particles (such as protons and oxygen ions) that stream out across the solar system. To the mid-lower right of the image a powerful solar prominence is seen. Close to this are images of a pair of *sun-grazing comets*, whose material is being sublimated off their surfaces by solar radiation, producing classical cometary tails. In this case the comets did not reappear on the other side of the Sun, suggesting that they were destroyed, or at least depleted of volatiles (SOHO (ESA & NASA)).

with high-energy particles (e.g., fast-moving electrons and protons). This flux of high-energy particles originates in the coronosphere—the outer atmosphere, above the visible "surface" photosphere—where the intense magnetic field structures generated by the Sun can accelerate and heat material to solar escape velocities (Figure 1.8a and 1.8b). The most intense outflows of radiation occur during solar flares or prominences, but even when quiescent the Sun continues to output a particle flux.

The precise impact of this high-energy radiation on the present-day solar system is not fully understood; however, it clearly influences the magnetic fields and atmospheres of planets, as seen in the aurora on Earth, Mars, Jupiter, and Saturn. It also contributes to the radiation environment of the entire system (that already includes the cosmic ray background originating from the rest of the Galaxy, and possibly further afield). Radiation modifies the chemistry of a system - for example the surface composition of asteroids and comets, and is generally destructive for the complex chemistry of life as we know it. Particle radiation can also play a major role in the loss, or erosion, of planetary atmospheres (see Chapter 6). This is particularly interesting when the Sun is compared to stars of different masses. For example, M-dwarfs (mass 0.1–$0.5 M_\odot$) are far more numerous than G-dwarf stars (such as our Sun) but are also prone to far more flaring and consequently create a much higher local particle flux—which would almost certainly impact complex chemistry in those systems.

It also seems that the particle and high-energy photon flux of *young* stars (i.e. during formation as protostars or shortly after hydrogen ignition) plays a major role in sculpting the chemistry and matter aggregation of surrounding proto-planetary material (Chapter 7).

1.1.7 Exoplanets

Since it was understood that stars were members of our Sun's family, humans have speculated that there must also be other planetary systems. Interestingly, this has been the case even in the absence of a plausible theory for the formation of planets—it has just seemed natural that stars and planets go together. Until the last few years of the twentieth century there was however no direct, verifiable, evidence for worlds around distant stars.

We now have increasingly extensive and sophisticated information on the existence and characteristics of extra-solar planets, or exoplanets. In Chapter 4 we go into some detail on the present methods by which planets are detected and investigated. In writing a text such as this it is always a little risky to describe in specific terms the breadth (or narrowness) of current knowledge, since by next week it will almost certainly be out of date. In light of this we restrict ourselves to a somewhat historical discussion, concentrating on those facts which are unlikely to change.

The first generation of planets to be discovered have, by the nature of the observational techniques used, tended to be massive ($> 1M_J$) and in relatively close orbits to their parent stars (see Chapter 4 for a detailed discussion). Not only that, but the great majority of these planets have substantially elliptical, or eccentric, orbits—by comparison our solar system is a paragon of circularity. Here then are giant (presumed gaseous) planets existing close to their stellar parents (far closer than Jupiter is to the Sun), and often with significantly eccentric orbits— which is *not* a minimum energy configuration.

Thus, our first glimpses into the flora and fauna of exoplanets have provided a set of systems that bear almost *no* resemblance to our own Solar System. Although it is premature to say that our Solar System is therefore the odd one out, it would be correct to say that if we are to understand planetary systems we must step well beyond the norm established by our own.

That being said, there are many things about this first generation of exoplanets that make good physical sense. For example, there is a seemingly good correlation between the metal (heavy element) abundances of stars (which reflects the initial composition of the material from which star and planet are assumed to have formed, Chapter 2) and the likelihood of detecting massive planets. One would expect it to be easier to form massive planets from proto-planetary/proto-stellar material with a higher abundance of silicates and iron—there is simply more raw material to get things going with (although it should be noted that this statement assumes a particular mechanism for giant planet formation, which may be an oversimplification, see Chapter 3). Furthermore, there appear (with some hindsight) to be ready mechanisms for producing massive planets close in to parent stars, and for producing eccentric orbits (Chapter 3).

It is not only exoplanets which have become part of the armory of astrophysical data—the observation of *proto-planetary* (or proto-stellar) disks (Chapters 2, 3 and 7) is now an ever improving and growing source of information. Indeed, many such disks exhibit the clear signatures of the presence of otherwise unseen massive planets, which are sweeping and perturbing the dusty disk material—leaving their ghostly mark.

There are of course many more questions than answers with this first generation of observational knowledge. There are also many examples of situations which don't even fit into these new pictures—giant planet

Table 1.1. **The number and biomass of prokaryotes (Chapter 5) on Earth. The total represents at least half of the total biomass of the planet, which includes macroscopic organisms. Adapted from Whitman et al. 1998.**

Environment	Number of prokaryotic (microbial) cells	Mass of carbon (grams)
Aquatic habitats	1.2×10^{29}	2.2×10^{15}
Oceanic subsurface	3.6×10^{30}	3.0×10^{17}
Soil	2.6×10^{29}	2.6×10^{16}
Terrestrial subsurface	$2.5 \times 10^{29} - 2.5 \times 10^{30}$	$2.2 \times 10^{16} - 2.2 \times 10^{17}$
Total	$4.2 \times 10^{30} - 6.4 \times 10^{30}$	$3.5 \times 10^{17} - 5.5 \times 10^{17}$

atmospheres which are more extended than expected (Chapter 4), apparent giant exoplanets hundreds of times *farther* from their parent star than the newly established norm, and so on.

1.1.8 Life and Habitats

Much as with exoplanets, what we once thought was the norm for life on Earth has turned out to perhaps be the exception. Humans (or apes) do not rule the planet, nor do birds, fish, insects, or even plants. In fact the "humble" microbe rules supreme—if sheer biomass, genetic diversity, and length of time they have existed on the planet are the criteria to judge by. The *total* living biomass of the planet Earth is estimated to be several 10^{19} grams. This figure is however constantly being revised upwards, for the reason that ever more microbiota are being discovered. Already microbes account for $\sim 10^{18}$ grams of carbon biomass, corresponding to (including other elements) > 50% of the total terrestrial biomass. Much of this life exists in environments previously considered uninhabitable—in dark, chemically harsh places of extreme temperature (both hot and cold, Chapter 5), and even high in the atmosphere. The overwhelming *majority* of microbial life exists in the Earth's *subsurface*—in the first kilometer or so beneath the dry land surface and to an even greater extent in the oceanic sedimentary subsurface. Table 1.1 adapted from Whitman et al. (1998) summarizes some of these distributions of microbial life on Earth.

These "non-classical" biota do not necessarily consist of highly active life; indeed many so-called *extreme* organisms live a decidedly low-energy existence—for good reasons of conservation. Nonetheless, it appears that subsurface microbial populations have a genetic diversity greater than that of surface, macroscopic, organisms—and have likely been in place for hundreds of millions, if not a billion or more, years.

It also seems that a very clear distinction needs to be made between "simple" and "complex" life—especially in the context of the central questions of astrobiology. If the ground truth on Earth is a fair example (and it may or may not be) then it seems reasonable to say that while "simple" life (i.e. microbial life) may be ubiquitous, "complex" (i.e. multicellular, macroscopic life) may not. There are many arguments that can convince one that the conditions necessary for the existence of complex life like ourselves on a planet are so numerous, and need to be so finely tuned, that such life must be very unusual in the Universe (see also Chapter 10). The fact of the matter is that we just don't know. Having already had our faces rubbed in the dirt somewhat over our preconceived notions of what a "normal" planetary system should look like (see above), we should probably proceed with as open a mind as possible until *proven* otherwise. This is especially true in light of the fact that we do *not* understand the origins and evolution of life on Earth, and we do *not* even know whether the biochemistry of terrestrial life is indeed the only, or most efficient, that the Universe allows.

That being said, in order to make progress we need to have at our disposal at least a working hypothesis. This is immediately apparent when we consider questions of **habitability**. Habitability in the astrobiological sense means the capacity to sustain life—which may be a function of time, space, and physical and chemical conditions. The most commonly used central tenet of habitability—and indeed a fine working hypothesis—is that *liquid water* is essential for life, be it simple or complex. As we will discuss in some greater detail in Chapter 9, liquid water is the most versatile and chemically important solvent we know of in biochemistry. It provides the molecular mobility that terrestrial biochemistry requires to function, and the nature of the electronic structure of the water molecule is such that it lends itself to an extraordinary range of roles in chemistry and molecular structures (Figure 1.9).

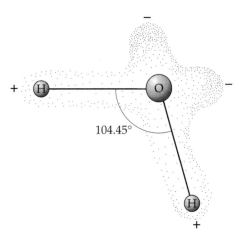

Figure 1.9. An illustration of the geometry and electron orbital structure of the water molecule, H_2O. The mean electron density is approximately 10 times greater around the oxygen atom than the two hydrogens, giving rise to the strong polar nature of the molecule.

From the astrophysical point of view water should certainly be plentiful in the Universe, since hydrogen is primordial, and oxygen is readily produced in moderately massive stars. In terms of habitability the question is usually phrased as whether or not water exists in a *liquid* state on or below the surface of a rocky planet. There are many assumptions involved here—our immediate default to a terrestrial type world for one. It is important to be aware of these assumptions, for while they may be very well founded (terrestrial type worlds for example may also offer atmospheres and tectonic activity critical for all forms of life), it behooves our scientific minds to always allow for the seemingly improbable, since that too may be the truth.

Classically, or at least in the last half of the twentieth century, one criterion for the habitability of a terrestrial world has therefore been whether it orbits its parent star at a distance where the stellar insolation allows for liquid surface water. The range of orbital radii where a planetary surface can be sustained between the freezing and boiling temperature of water (273–373 K at one Earth's atmosphere pressure) is known as the circumstellar habitable zone (CHZ, Chapter 9). This zone depends on the reflectivity (or **albedo**) of the planet, the stellar luminosity, the atmospheric composition of the planet, its orbital characteristics (e.g., eccentricity), its spin rate and spin axis orientation (**obliquity**), and many other factors—one of the most intriguing of which is whether or not life exists and itself modifies the planet characteristics. Obviously then, if one is trying to use an estimate of the CHZ as an indicator of the likelihood of life one has to proceed cautiously, and not become entangled in circular reasoning.

It is also possible that environments with liquid water exist entirely cut off from stellar insolation. Subsurface conditions might exist where either chemistry or local heating (for example tectonic or radiogenic sources) allow for either bodies of liquid water or microscopic pockets or thin films where biochemistry can occur. It is also possible that entire oceans of liquid water can be generated by the effects of tidal heating. The Jovian moon Europa has been the most discussed example of this, although other intriguing objects such as Saturn's Enceladus exist (Chapter 10). Europa is caught in a Laplacian orbital resonance with the moons Io and Ganymede such that the ratios of the orbital periods of Io, Europa, and Ganymede are 1:2:4. The effect of this resonance is that Europa's orbit is consistently tugged by the other moons so that rather than being circular it is slightly eccentric. This slight eccentricity results in Europa being stretched and torqued by Jupiter. The same thing is happening to the other moons in the resonance, Io in particular is being tidally stretched, or heated, by Jupiter to such an extent that it is a mess of molten sulphur and silicates, with volcanoes and local temperatures as high as 1800 K. The surface temperature due to solar insolation at the orbit of Jupiter is only some 100 K. Europa is subject to tidal heating some fifty times less than that of Io, however this may be sufficient for the strain and friction forces to warm its interior to around 273 K. Europa has a water ice surface. However this surface exhibits a remarkable topology of cracks and undulations, entirely reminiscent of the surface of a terrestrial polar ocean, where ice has cracked, rolled, and re-frozen above liquid water (Figure 1.10a and 1.10b). As discussed more in Chapter 10, Europa also appears to possess *induced* magnetic fields, pointing

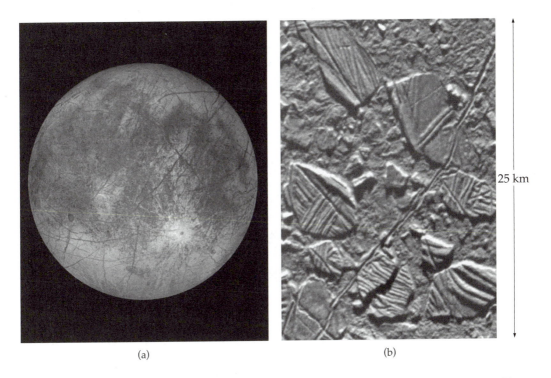

(a)   (b)

Figure 1.10. (a) An image of the water-ice encrusted Jovian moon Europa taken by NASA's Galileo spacecraft. The surface is relatively free of recent cratering (with the exception of the bright crater structure to the lower right) and exhibits evidence of both extensive cracks or rifts and variations in surface color/composition (see Chapter 10). (b) Close-up image of the surface of Europa taken by NASA's Galileo spacecraft. The image is some 25 km in the vertical dimension and clearly shows a remarkable surface topology akin to broken, tumbled, and re-frozen ice sheets over a liquid ocean (NASA/NSSDC).

towards the existence of a subsurface conductive medium—compatible with a subsurface salty water ocean. It appears possible then that Europa does in fact possess a subsurface liquid ocean, powered not by the Sun, but by gravitational and rotational energy dissipation in the Jovian system.

There is also a question of what we truly mean by "habitat." As noted above, in the astronomical sense it is generally used to mean a niche which can sustain life—however, sustaining life can mean anything from enabling the proliferation of biota, to simply not destroying life. At the latter end of things, the question of *dormant* life arises. Bacteria, for example, can exist in dormant states known as **spores** for extremely long periods in physically harsh conditions (Chapter 5). There are even claims of dormant bacteria surviving incarceration within fossil tree resin (amber) for tens of millions of years. There are certainly promising cases of bacterial spores apparently being revived from dry or icy conditions thousands of years after taking this form, and bacteria appear to have survived exposure to the lunar surface for several years (Chapter 5). Such characteristics have led to speculation not only of the ability of life to lie in wait for local conditions to return to past glories (as might be the case with Mars), but also for life to be dispersed through the cosmos—either as naked bacterial spores, or as spores embedded in rocky or icy material flung throughout, or even beyond, a planetary system by impacts and gravity. The notion of such **panspermia** has had a long and checkered career (the promotion of the idea by Fred Hoyle and colleagues did not do so well, although history may yet yield them the upper hand). However, our more recent appreciation of the tenacity of life on Earth, and questions surrounding the origins of life on Earth, have led to such ideas being reconsidered.

It is certainly an appealing notion—that life can disperse like seeds throughout the cosmos, or at very least throughout a given planetary system. However, it is *not* a full solution. It does not tell us where life truly originates (except to say that particularly fertile habitats will provide the seeds for lesser ones). It is also entirely predicated on the idea that bacterial spores, as robust as they might be, can survive the extreme radiation environment of intra- and inter-planetary space for millions if not billions of years. It is not clear how many pieces of genetic code can be wrecked by the passage of high-energy particle radiation before an organism is incapable of revival—or indeed whether there would be anything left other than the raw amino acid ingredients (which we know exist both in meteorites and in interstellar space).

Finally, there are many other factors that must come into play when we consider other aspects of habitability as defined above. One of these is the *longevity* of a particular environment. A terrestrial type world that experiences constant, and profound, variations in its environment may not bode well for the lengthy processes of evolution as we know them.

A popular example is that of cometary or asteroidal impact (Chapter 8). A planet being constantly subject to tens or hundreds of megatons of explosive impact events may present just too variable a surface environment for life to gain a hold—and most certainly seems unlikely to allow the development of complex life. Similarly, a planet in a dense stellar environment may experience the devastating radiation of too many nearby supernova events over its history to allow for life. Conversely, life as we see it on Earth arguably benefits from a certain level of adversity. Without environmental pressure, then what we understand about natural selection indicates that life might "stagnate," both in terms of diversity (and hence ability to exploit resources and niches) and in terms of the single life-time adaptability of an organism.

1.2   Preview

Much of the above is really a skeletal outline of the following chapters. What we now aim to do is to zoom in to study the astrophysical, and to a lesser degree biological and geophysical, tools employed to investigate the fundamental questions of astrobiology. There will be many equations, however, the level of mathematics is kept well within that familiar to a physical sciences undergraduate. There is clear beauty in many of the problems we will tackle, and this beauty really cannot be fully appreciated without doing the hard work of writing Greek letters—nor can a practical knowledge be acquired without the ability to compute answers.

References

General Astrobiology Readings

Chyba, C. F., & Hand, K. P. (2005). ASTROBIOLOGY: The study of the living universe, *Annual Review of Astronomy and Astrophysics*, **43**, 31.

Gilmour, I., & Sephton, M. A., eds. (2004). *An Introduction to Astrobiology*, The Open University, Cambridge University Press.

Goldsmith, D., & Owen, T. (2002). *The Search for Life in the Universe*, 3rd ed., University Science Books.

Lunine, J. I. (2006). *Astrobiology: A Multidisciplinary Approach*, Addison Wesley.

Specific references for this chapter

Basri, G., & Brown, M. E. (2006). Planetesimals to brown dwarfs: What is a planet? *Annual Review of Earth and Planetary Sciences*, **34**, 193.

Brown, M. E., Trujillo, C. A., & Rabinowitz, D. L. (2005). Discovery of a planet sized object in the scattered Kuiper belt, *Astrophysical Journal*, **635**, L97–L100.

Burbidge, E. M., Burbidge, G. R., Fowler, W. A., & Hoyle, F. (1957). Synthesis of the Elements in Stars, *Reviews of Modern Physics*, **29**, 547.

Dole, S. (1964). "Habitable Planets for Man," RAND corporation report R-414-PR, available at http://www.rand.org/pubs/reports/2005/R414.pdf.

Udry, S., & Santos, N. C. (2007), Statistical properties of exoplanets, *Annual Review of Astronomy and Astrophysics*, **45**, 397.

Whitman, W. B., Coleman, D. C., & Wiebe, W. J., (1998). Prokaryotes: The unseen majority, *Proceedings of the National Academy of Sciences*, **95**, 6578–6583.

Zhang, J., et al. (2005). Inner core differential motion confirmed by earthquake waveform doublets, *Science*, **309**, 1357–1360.

Problems

1.1   Describe the physical processes responsible for the production and dispersal of heavy elements in the Universe. You may wish to refer to the 1957 article by Burbidge et al.

1.2   Read the review article by Basri & Brown (2006) on planet definitions. Discuss the physics and physical phenomena that help determine the characteristics of planetary objects of increasing mass (from small to large). Summarize the overall conclusions of this paper on the definition of a "planet."

1.3   Using the review article by Udry and Santos (2007) on the statistical properties of exoplanets describe the types of known exoplanets, the classes of orbital phenomena seen (e.g. eccentricities, resonances), and the relationships between planet populations and the properties of host stars.

1.4   Read the article by Whitman, Coleman and Wiebe (1998) on prokaryotic/microbial life. Describe the methodology that they employ to evaluate the global inventory of "life" on Earth. Discuss how they approach the issue of genetic diversity in microbial populations and how they argue for a large diversity in microbes.

1.5 In 1964 Stephen Dole wrote an extraordinary document entitled (with some hubris) "Habitable Planets for Man" for the RAND corporation (see References). Much of what is in this document is incredibly prescient for our current study of astrobiology. Read Chapter 2 of Dole's work and discuss each of his subsections on the issue of habitability. Which stand up to more modern standards and which appear less relevant today? (Consider what we now know about life on Earth.) You may wish to refer ahead in this present book to later chapters.

Proto-stellar Collapse and Star Formation

2.1 Introduction

In this chapter we will investigate some of the basic physical concepts needed to understand why, and how, stellar systems form out of murky nebula (molecular clouds). Why do we care about this if we want to study planets and astrobiology? We care because the formation of planets and the properties of planetary systems are intimately linked to the formation of stars. Planets and planetary systems appear to offer the most likely harbors for life as we know it, and, as we will investigate in later chapters (e.g., Chapters 5, 6, 9, 10), are probably also intimately linked to the origins of life.

We can furthermore make what may seem to be a bold, if not provocative, statement—namely, that the fundamental reason why planets form systems around stars (although not discounting the possibility of other formation routes, discussed at the end of this chapter) is due simply to the form of the gravitational force and Nature's urge to minimize energy and conserve angular momentum. If we formulate things in this way then the production of planets around stars becomes a common natural phenomenon, an inevitable consequence of physics. While we might not have ever doubted this to be the case, it is important to make the distinction between a phenomenon that is unusual, or very finely tuned, and one which is generic.

Later we will discuss how life, at least here on Earth, seems to have emerged in fairly short order following the formation of the Solar System. If this too is a generic property of places in the cosmos that can

31

spawn or harbor organisms, then the readiness with which planets are formed, and the diversity of their properties, may be deeply connected to the origins life.

2.2   A Brief Cosmography

Before embarking on an investigation of the physics behind star formation it is useful to form a mental picture of exactly where the star formation regions are in our own neighborhood and in the Galaxy as a whole. This cosmography will also serve as a reference point for when we discuss the practical issues of searches for exoplanets.

If we begin with our stellar locale we find that the Sun is within a loose collection of nearby stars. In Figure 2.1 a simple map of the local stellar turf is shown—representing the volume of space some 24 light years across, centered on the Sun, and indicating the brighter nearby stars. If we zoom out a little farther, then Figure 2.2 contains a similar map, now some 500 light years across, again indicating a selection of brighter star positions. Going further, Figure 2.3 encompasses a region some 9000 light years across, and we now see the positions of some of the larger, best known, nebulae or molecular clouds. These include the Orion nebula, which is a highly active region of star formation, containing many proto-stars and pre-main-sequence stars (which we will discuss further below). The distribution of these clouds is correlated with the grand spiral-arm structure of our Galaxy—the Milky Way—and in the two panels of Figure 2.4 the Galaxy is seen from a vantage point above its plane (see Chapter 9 for more on Galactic structure).

In Figure 2.5 two optical images of molecular clouds illustrate their general appearance: the Orion nebula and part of the Eagle nebula—sometimes referred to in rather optimistic terms as the "pillars of creation." As stately as they appear, these majestic structures are nonetheless dynamic. Indeed, as shown in Figure 2.6, we can detect significant motions in the gaseous material—they are dynamically complex and possibly even rotating on scales as large as $\sim$0.5–1 pc ($\sim$1–3 light years). The fact that they are not static has implications for our later discussions of planet formation and angular momentum evolution.

The largest of these structures are known as giant molecular clouds (GMCs), and span a mass range between approximately 10^5 and a few

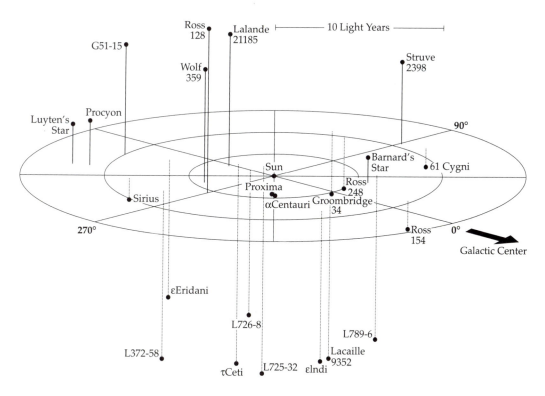

Figure 2.1. The local stellar neighborhood. The positions of bright stars within approximately 12 light years of the Sun are shown in an isometric projection. The plot is centered on the Sun and oriented to the plane of the Galaxy (shown by concentric circles and Galactic longitude markings)—the Galactic center is some 28,000 light years distant. The closest bright stars, Proxima and Alpha Centauri, are almost directly in the plane of the Galaxy at a longitude of some 315°. (Adapted from Richard Powell, www.atlasoftheuniverse.com.)

$10^6 M_\odot$, and a spatial scale of several tens of parsecs. As suggested by Figure 2.5, GMCs have significant structure, and they are really best considered as **cloud complexes**, consisting of smaller "clumps" of mass $\sim 10^3$–$10^4 M_\odot$ and sizes in the range of 2–5 parsecs. Within these clumps themselves there can be denser structures, or "cores"—which may have masses of the order of $\sim 1 M_\odot$ and sizes of the order of ~ 0.1 parsecs. Temperatures may be as low as ~ 10 K. Very often, proto-stars are seen associated with these structures. It seems likely that GMCs are really

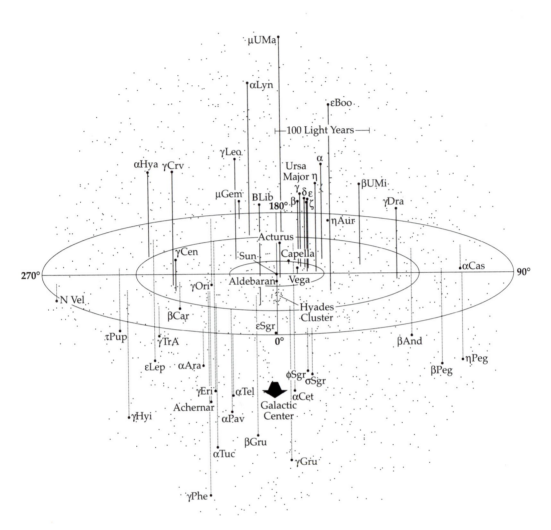

Figure 2.2. A larger volume of space centered on the Sun. The brightest stars within 250 light years of the Solar System are shown. A notable feature is the inclusion of the Hyades cluster—an open cluster of stars numbering some 218 members in total. These stars almost certainly formed from the same original molecular cloud (see main text). (Adapted from Richard Powell, www.atlasoftheuniverse.com.)

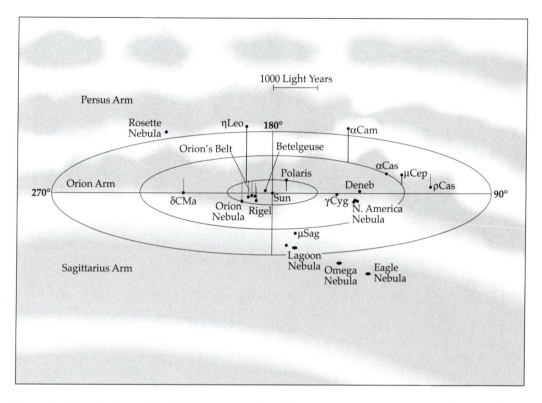

Figure 2.3. A selection of the brightest stars and well known nebular structures are shown within approximately 4500 light years of the Sun. The major spiral arm structures of the Milky Way are now clearly seen (illustrated as grey shading). The Sun resides within the Orion arm, which lies between the inner Sagittarius and outer Perseus arms. (Adapted from Richard Powell, www.atlasoftheuniverse.com.)

transient in nature, due to the shifting galactic arm structure (spiral density waves), the effect of supernova explosions, and could even be considered to be temporary collections of smaller dwarf molecular clouds (DMCs). DMCs are seen throughout the Galaxy and tend to be more evenly distributed than GMCs, not necessarily tracing the spiral arms of the Milky Way. Much like the clumps within GMCs the DMCs have masses $10^3 - 10^4 M_\odot$ and it appears that a major fraction of the gaseous molecular material in the Galaxy exists in these clouds.

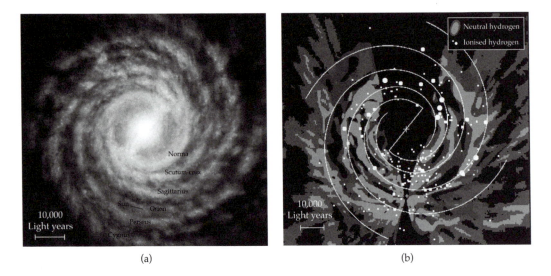

(a)   (b)

Figure 2.4. The grand spiral arm structure of the Milky Way galaxy is shown for an external observer at the Galactic north pole. In the first panel the distribution of stars and nebula is presented in an "artist's impression." In the second panel the currently observed distribution of neutral and ionized hydrogen gas is plotted in the same coordinate system. Both neutral and ionized gas follow the visual spiral arm structure. (Adapted from Richard Powell, www.atlasoftheuniverse.com.)

This is the very brief cosmographic context of star formation in our Galaxy. There are very particular regions where stars appear to be forming, and there are very particular regions where we can access formed systems with our astronomical instruments.

2.3 Observed Proto-stellar Structures

The subject of star formation generally falls into two categories: one is the study of the *populations* of forming stars and their environments, the other is the study of the details of individual systems or classes of systems. This latter area might be best termed as **stellar embryology**. It is a challenging astronomical task since proto-stellar systems tend to be cloaked in dense molecular gas and dust. However, high-energy photons (e.g., X-rays) and low-energy photons (e.g., infrared) emitted

(a)   (b)

Figure 2.5. Two examples of nebular or molecular clouds; the Orion nebula (left panel) and part of the Eagle nebula (right panel). These images have been taken using multiple optical filters and enhanced to emphasize the cloud structures. In the upper left of the Orion nebula image clear bands of dark, dust-rich material are seen obscuring the inner regions of the cloud. The glow in the center of Orion comes from the presence of newly formed stars (NASA/ESA and M. Robberto). In the Eagle nebula some of the densest dust and gas structures are shown, orientated as "pillars." The sculpted appearance of these structures is primarily due to photo-evaporation from UV radiation winds from young stars in the nebula that are stripping the less dense material away, leaving the densest cores. The smallest visible "knots" of material at the uppermost part of the left hand pillar may contain proto-stellar systems (NASA/ESA and J. Hester, P. Scowen).

by proto-stellar objects can penetrate these shrouds and reveal the inner workings of such systems.

Star formation appears to seldom occur in isolation. In star-forming (nebular) regions we see clusters and groupings of nascent systems. As we will discuss below, this is likely due to the fact that the conditions required to trigger the conversion of molecular clouds into stellar objects often occur, or act, in concert across entire structures.

Objects that match our expectations for proto-stars fall on the high-luminosity side of the main sequence in the HertzsprungRussell (H–R)

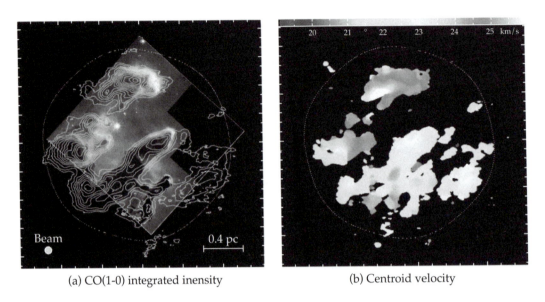

(a) CO(1-0) integrated inensity   (b) Centroid velocity

Figure 2.6. An example of how detailed molecular line emission can be used to measure the dynamics of molecular clouds. In the left panel the intensity of carbon monoxide (CO) seen via the $J = 1 - 0$ rotational transition (a millimeter wavelength emission line) in the Eagle Nebula is plotted as contours overlaying the Hubble Space Telescope optical image. The emission clearly traces the denser nebula structures. In the right-hand plot the Doppler map of the CO $(1 - 0)$ derived line-of-sight velocity is plotted. This clearly demonstrates that the nebula is not static, but has substantial dynamics—which may include large scale rotation. Similar results are obtained for other molecular clouds (Figure by Marc W. Pound, University of Maryland. (CO $(1 - 0)$ data obtained with the Berkeley-Illinois-Maryland Association interferometer (see Pound (1998), ApJ, 493, L113).

Diagram (Figure 1.4). This high luminosity is thought to be a consequence of the greater physical size of a proto-star. Treating a proto-star to first order as a blackbody, the luminosity L scales with surface area: $L \propto R^2$, where R is the radius of the object. So the bigger the object (at fixed temperature) the more luminous it is. Herein lies part of the puzzle of where stars actually come from, and what we actually mean by *proto-star*.

These objects are really just clouds of gas collapsing or contracting in on themselves, which, as we will prove below, must heat up as they shrink. In the early stages of collapse, however, they will not heat up particularly quickly. During this period a solar-mass proto-star will tend to gradually diminish in luminosity as it physically shrinks, while staying at a similar temperature. Indeed, in the H–R diagram it will trace out over time an almost vertical track, known as the **Hayashi track**, down towards the main sequence. Proto-stellar systems are also seen to almost invariably produce extraordinary **bipolar outflows** or **jets**, in symmetric pairs leading directly out along the presumed spin axis of the systems (Figure 2.7). Below and in Chapter 3 we will discuss some of these observational details a little more. Suffice to say here that the precise mechanism of these jets is not fully understood, but that they clearly represent material being ejected from the system from very close proximity to the central proto-star itself. These outflows also likely play a major role in regulating the infall of material in the system and in helping remove or redistribute angular momentum. In Figure 2.8 a summary of the observed stages of star and planet formation is presented—while we have not yet demonstrated this, stars and planets do indeed seem to be inexorably entwined from birth.

One class of proto-stellar object is known as a T Tauri type. T Tauri objects often exhibit jets, as well as a thick, dusty, disk of material enshrouding the central proto-star and extending over some 100 AU in diameter. It is often assumed that a T Tauri object represents a particular stage of the star formation process, prior to the clearing of the surrounding disk material, and lasting between 100,000 and 3 million years. Systems which appear to be at a slightly later stage of evolution are often termed **pre-main-sequence stars**. These are close to the main sequence on the H–R diagram, and are likely undergoing sporadic fusion in their centers, but have not yet fully "switched on" sustained core hydrogen burning (Chapter 1). Often these systems show evidence for extensive dusty disks surrounding the central stellar object. These disks are no longer as massive as for proto-stars, but they are nonetheless substantial, and some even show evidence for the presence of unseen planet-sized bodies, interacting with, and distorting the disk material.

From the observation of these many stages it is possible to deduce their approximate timescales. For example, for material in a molecular cloud to contract into something that would be identified as a progenitor of a solar-mass system probably takes of the order of 2–3 million years. The time it takes this proto-stellar object to switch on as a zero-age main sequence star may take an additional 30–40 million years. This latter timescale is highly dependent on the mass of the final star. For a star

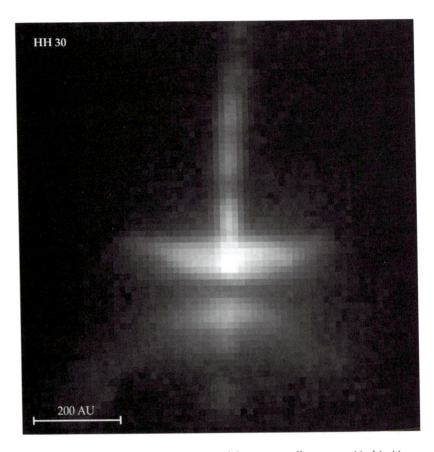

Figure 2.7. A Hubble Telescope image of the proto-stellar system Herbig-Haro 30. A clear bipolar outflow, or jet, is seen in the vertical direction—both above and below a horizontal structure. The dark horizontal band dividing the central hourglass-like shape is a thick gas and dust disk surrounding the central proto-stellar object. The top and bottom surfaces of this disk are brightly illuminated, and heated, by the central proto-star. See §2.3.1 for classifications of proto-stellar systems (NASA/ESA/STScI).

ten times as massive as the Sun the timescale is measured in *hundreds of thousands* of years, while for a star a tenth of the mass of the Sun the timescale stretches to a few 100 *million* years. Overall though, compared to the main sequence life-span of a solar-mass star ($\geq 10\,\mathrm{Gyr}$), things happen pretty quickly in the formation process.

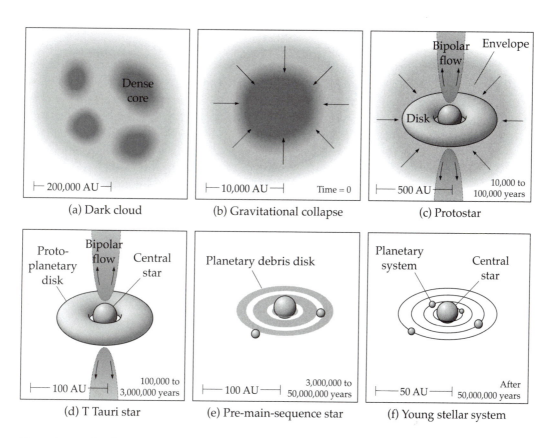

Figure 2.8. A sketch summarizing the overall observed and theorized sequence of star formation. In this picture the formation of planets is also indicated: we will discuss this in much more detail in Chapter 3. Approximate scales and times are indicated assuming a solar-system like mass and composition. Copyright *American Scientist* 89: 316–325, July–August 2001, all rights reserved.

2.3.1 The Four Stages of Star Formation

Before we actually look at some of the basic theory behind star formation it is useful to further summarize the current best-bet picture, derived from both theoretical and observational investigations—as described in the previous section. This will help provide the context and chronology of subsequent discussion. Shu (1987) put forward a scheme that involves four distinct phases—each of which also has distinct observational characteristics. This is illustrated in Figure 2.9. The dusty nature of proto-stellar systems, combined with the pre-main sequence temperatures of proto-stars, results in a strong infrared emission spectrum. The shape of this spectrum (often given in the form of a **spectral energy distribution** or SED, which is simply frequency times flux) changes between these different phases (Figure 2.9) and can be used to observationally classify these systems. In the first phase, or **CLASS 0**, a central proto-star and disk has formed deep within a still infalling envelope of dust and gas. The SED of this phase is peaked in the far-infrared and millimeter wavelengths. In the second phase, **CLASS I**, the proto-star accretes matter *through the disk*, and a strong stellar wind (radiation and particles) can escape along the rotational axis of the system (effectively breaking out above and below the disk)—creating a **bi-polar outflow** as described previously. Now the SED is peaked to somewhat higher frequencies, and deep spectral absorption features are seen as the disk material absorbs, rather than entirely blocks, the proto-stellar emission. In **CLASS II** the surrounding nebular material above and below the disk has been blown away, leaving just the flattened disk, which can now be considered as a proto-planetary disk, with planet formation in full swing. Again, the SED shifts to higher frequencies as more of the increasingly hot proto-star's radiation can escape. In **CLASS III**, the central object is essentially a **pre-main sequence** star - almost ready to initiate its hydrogen burning phase. The gaseous component of the disk has largely dissipated through accretion onto the proto-star, photo-evaporation, stellar winds, and accretion into planetary objects. Substantial dust remains, but is really a **debris disk** of material from the collisions and "grinding down" of planetesimals and rocky proto-planets (Chapter 3). The SED is now dominated by the central stellar object, with a tail of infrared emission from the remaining dust.

2.4 Proto-stellar Collapse

How (and why!) does a cloud of cool gas and dust turn into stars? This is the central question we need to answer to understand the nature of star formation in molecular clouds. In the previous section we outlined the observational picture, here we begin to look into the theory of the underlying physical mechanisms. This is in truth a highly complex problem, and at the time of writing it would be fair to say that the jury is still out on any definitive answer. However, as will often be the case in

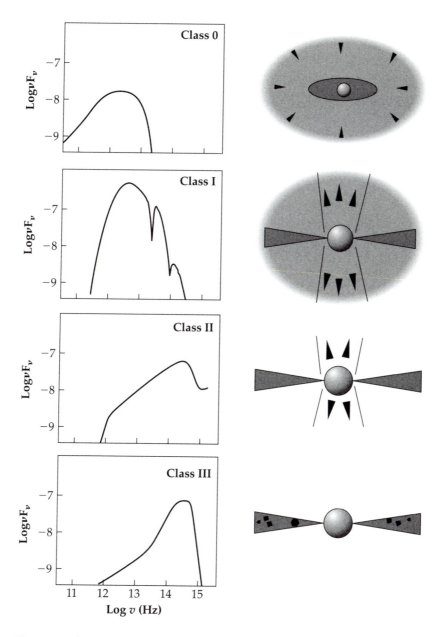

Figure 2.9. Illustration of the four stages or classes of proto-stellar systems commonly considered. These are observationally distinguished by clear differences in their emitted radiation, shown by the left-hand plots of spectral energy distribution (SED), which is simply frequency times flux. As described in the text, as the infalling and disk materials are cleared by the central proto-star's activity, the SED changes dramatically as the central object is increasingly revealed.

this exploration, we can nonetheless gain insight from simple physical arguments. We need to first investigate the balances at play between *gravity* and *kinetic energy*, or *gas pressure*, since this is at the heart of the question.

2.4.1 The Virial Theorem

For a spherical cloud of gas in **hydrostatic equilibrium**, gravitational forces are balanced by pressure forces. Specifically, for a spherical shell of gas at radius r, thickness dr and area $4\pi r^2$ the balance of pressure and gravitational forces is

$$-dP\ 4\pi r^2 = \frac{GM}{r^2}4\pi r^2 \rho\ dr,\qquad (2.1)$$

where $M = M(<r)$, the total mass contained within r. P is the (uniform) gas pressure, and ρ is the (uniform) gas density (Figure 2.10).

Thus,

$$\frac{dP}{dr} = -\frac{GM\rho}{r^2}.\qquad (2.2)$$

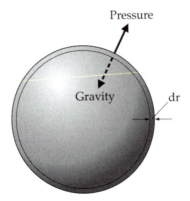

Figure 2.10. The simple spherical gas cloud model used to derive the Virial Theorem. Zero rotation is assumed, and a given shell of gas of thickness *dr* is held in equilibrium by competing gravitational and gas pressure forces.

With the benefit of hindsight we'd like to write this expression in terms of potential and kinetic energies. Multiplying Equation 2.2 by $4\pi r^3$,

$$4\pi r^3 dP = -GM4\pi r\rho \, dr, \tag{2.3}$$

we can eliminate dr since $dM_{\text{shell}} = 4\pi r^2 \rho dr$, yielding

$$4\pi r^3 dP = -\frac{GM}{r} dM. \tag{2.4}$$

Now, the right-hand side of this expression is looking like the gravitational potential energy, and we still need to work on the left. We need to integrate over volume (V) and mass (M_{total}) to sum things up over the spherical gas cloud:

$$3 \int_{P_{\text{center}}}^{P_{\text{surface}}} V dP = - \int_0^{M_{\text{total}}} \frac{GM}{r} dM. \tag{2.5}$$

Doing this by parts, we have

$$3 [PV]_{\text{center}}^{\text{surface}} - 3 \int_0^{V_{\text{total}}} P dV = - \int_0^{M_{\text{total}}} \frac{GM}{r} dM. \tag{2.6}$$

Now, the term $[PV]_{\text{center}}^{\text{surface}} = 0$ if we assume that $P_{\text{surface}} = 0$ and $V_{\text{center}} = 0$, which seems a reasonable first step. Furthermore, we now recognize that the right-hand side of 2.6 is indeed just the total gravitational potential energy for a uniform density sphere, and we denote this as $-\Omega$. We still need to get something that looks like a total kinetic energy on the left-hand side, so we can substitute $dM = \rho dV$, and obtain

$$3 \int \frac{P}{\rho} dM + \Omega = 0. \tag{2.7}$$

It's almost there; we know that for an ideal gas $P = \rho kT$ where k is the Boltzmann constant and T is the gas temperature. We also know that the thermal (kinetic) energy *per unit mass* is just $u = \frac{3}{2}kT$. In other words, $P = \rho \frac{2}{3} u$ and so $\frac{3P}{\rho} = 2u$. Now we can see that $3 \int \frac{P}{\rho} dM$ is simply $2U$, where U is the *total* thermal (kinetic) energy of the gas sphere. Then Equation 2.7 becomes

$$2U + \Omega = 0, \tag{2.8}$$

which is known as the **Virial Theorem** (so named for the "virial of Clausius," where a viral is a summation). The Virial Theorem is *also* true for purely gravitational, *time averaged*, systems, (i.e., without gas pressure) such as satellites, planets orbiting stars, star clusters, galaxies, and clusters of galaxies. Alternative derivations include those that explicitly treat a system statistically, averaged over time. The importance of this theorem cannot be overstated. If someone stops you in the street and asks what the most magical thing in astrophysics is you may well find yourself answering "the Virial Theorem," it is truly a thing of beauty. As a side note, the precise form of Equation 2.8 is clearly a consequence of the form of the gravitational force law (i.e., "$1/R^2$") as well as the number of internal degrees of freedom in the gas (assumed here to be monatomic and ideal). Since the electrostatic force is also a "$1/R^2$" law, the Virial Theorem also has application in chemistry to the understanding of molecular structures and interactions.

We can now obtain our first physical insight regarding the nature of collapsing clouds by considering the above example of a non-collapsing, equilibrium cloud. The total energy E of the cloud is simply

$$E = U + \Omega. \tag{2.9}$$

Applying the Virial Theorem we see that

$$E = -U = \frac{\Omega}{2}. \tag{2.10}$$

Now, suppose that such a cloud *radiates* energy away, thereby decreasing E. This implies that Ω decreases, *but that U must increase*! We immediately learn that it must be hard for a hydrostatically supported cloud to cool down. This is tremendously important. It means that a cooling (radiating) gas cloud will shrink (Ω decreases), but as it does so its internal temperature will be pushed up—and so it will initially get hotter as a result of radiating away energy! Now, if the cloud can be treated as a blackbody we know that the radiative energy loss goes as $L \propto T^4$. Therefore a hotter cloud will radiate energy away at a much higher rate—so it should be able to continue to shrink, but it will keep getting hotter as it does so, which is counterintuitive for a cooling object. This is nonetheless the situation for an object such as a proto-star where gravity and gas pressure are in competition. In what follows we will see some of the consequences of this in more explicit detail.

2.4.2 Collapse

We now have a basic tool with which to examine the conditions where a self-gravitating cloud of gaseous material might be willing to collapse. It is important to emphasize that what follows is true only for a cloud or object of *uniform* density. Nature seldom provides such objects, and indeed the real key to a complete understanding of cloud collapse and stellar formation will come from investigating the "lumpiness" of nebulae as well as other factors in addition to classical gas pressure. Nonetheless, in order to possess the mathematical and physical language to understand these phenomena we can use this idealized example, which still has application in certain circumstances.

If for some reason $2U < |\Omega|$ then pressure forces are overwhelmed by gravity and a cloud must begin collapsing. As discussed above, we will continue to assume that our spherical cloud has uniform density. Let us further suppose, in at least any initial stages of collapse, that this density remains approximately *constant*. If the initial collapse is slow then this is not an entirely crazy thing to do.

We know that the gravitational potential energy within a radius R is

$$\Omega = -4\pi G \int_0^R M(r)\rho r \, dr. \tag{2.11}$$

Since $M(r) = \frac{4}{3}\pi r^3 \rho$ we can, by substituting M for ρ, integrate 2.11, substitute ρ for M and obtain

$$\Omega = -\frac{3}{5}\frac{GM^2}{R}. \tag{2.12}$$

We now seek an explicit expression for U in order to apply the Virial Theorem condition for collapse and to see what this requires in terms of actual cloud properties. We know that the total thermal (kinetic) energy is $U = \frac{3}{2}kTN$ where N is the total number of gas particles in the cloud.

At this point it is convenient to introduce the concept of **mean molecular weight**. In reality a cloud will have both a mixture of different elements and molecular species, and neutral atoms/molecules together with ions and free electrons. Thus, the *mean* mass of the gas particles will depend on the precise particle content of a cloud. We can define a mean molecular weight as the ratio of mean gas particle mass to the mass of a hydrogen atom:

$$\mu = \frac{\bar{m}}{m_{\mathrm{H}}}. \tag{2.13}$$

The explicit calculation and explanation of μ we leave to other astrophysics texts since we will delve only a little way into this. Suffice to say that for typical (cosmic) abundances of elements in an interstellar cloud, μ ranges from about 1.3 for an entirely *neutral* gas, to about 0.6 for an entirely *ionized* gas.

We now write

$$U = \frac{3}{2}\frac{M}{\mu m_{\mathrm{H}}}kT. \tag{2.14}$$

Thus, if $2U < |\Omega|$,

$$\frac{3MkT}{\mu m_{\mathrm{H}}} < \frac{3}{5}\frac{GM^2}{R}. \tag{2.15}$$

Now, if the *initial* cloud radius is just $R = (\frac{3M}{4\pi\rho})^{1/3}$ then we immediately obtain the *minimum mass* cloud which *will* collapse:

$$M = \left(\frac{5kT}{G\mu m_H}\right)^{3/2}\left(\frac{3}{4\pi\rho}\right)^{1/2}. \tag{2.16}$$

This is known as the **Jeans Mass**, which we will refer to as M_{Jean}. We can also rewrite this expression to obtain the **Jeans Radius**, or **Length**, which is the minimum *size* cloud of this ρ, T and μ which *will* collapse.

$$R_{\mathrm{Jean}} = \left(\frac{15kT}{4\pi G\mu m_{\mathrm{H}}\rho}\right)^{1/2}. \tag{2.17}$$

What we see then is that for gas of a given composition, temperature, and density, if a cloud exists that is either more massive or larger than the Jeans scale then it cannot pressure support itself against its own gravity and it will collapse. The sense in which this operates is that hotter clouds need to be more massive in order to collapse, denser clouds can be smaller and still collapse, and an ionized cloud must be more massive than a neutral cloud in order to collapse.

Before we get too excited though we should remind ourselves that we have made some gross assumptions in the above calculation—in fact we will continue to make these for the time being, but ultimately we need to be aware of this.

As with any physical model, it is only as good as its description of nature. Let us put the above to the test. A typical diffuse giant molecular cloud has a mean $T \simeq 100$ K, $\mu \simeq 1$ and mean $\rho \simeq 10^{-24}$ g cm^{-3}. Writing 2.16 in a more convenient form:

$$M_{\text{Jean}} = (1.2 \times 10^5) M_\odot \left(\frac{T}{100 \text{ K}} \right)^{3/2} \left(\frac{\rho}{10^{-24} \text{ g cm}^{-3}} \right)^{-1/2} \mu^{-3/2} \quad (2.18)$$

we then find the typical $M_{\text{Jean}} \approx 10^5 M_\odot$. This appears to be a problem! It is telling us that "average" giant molecular clouds (e.g., the Orion nebula) should be collapsing to make $10^5 M_\odot$ objects—so how do we make stellar mass objects this way (0.08–$100 M_\odot$)? Even if we set the cloud temperature down to a minimum of 10 K we are left with the same conundrum, and if we consider denser gas then DMC scales should also be collapsing.

2.5 Outside-in Versus Inside-out Star Formation

The result of the previous section raises the question, What mechanism(s) sets the final masses of stars in our Galaxy? If $10^5 M_\odot$ GMCs and lower mass DMCs are gravitationally unstable then what happens during their collapse to make stellar mass objects? There are two basic solutions: the first is that something breaks the collapsing cloud up into stellar-mass pieces, the second is that such clouds don't really collapse en masse but rather it is the known, smaller, internal structures (e.g., clumps and cores) that suffer dynamical instability.

As described in §2.2, we know from direct observation that molecular clouds actually have significant internal structure, and star formation predominantly occurs in those regions of higher density within the cloud (**molecular cores**). The general consensus today is that, in most cases, stars form via the "inside-out" mechanism. That is to say that proto-stars are built up by the agglomeration, or **accretion** of material, thereby growing initially small mass concentrations into stellar-sized objects. The process is halted when, for example, the proto-star starts its main sequence fusion cycle and disperses the surrounding material (§2.3.1).

Nonetheless, we need to explain why this happens. In order to do this we will first consider the "outside-in" possibility. There are two reasons for spending time on a phenomenon that we don't think is an accurate description of most star formation. The first is that the physical argument used is a useful one, and while it may not describe the dominant mechanism of converting molecular clouds into stars it does almost certainly have application in certain circumstances. The second is that it teaches us about how to apply basic physical arguments to astrophysical questions.

What follows in the next section then is a conceptual argument for what might be considered the "classical" picture for turning $10^5 M_\odot$ clouds into stellar mass objects. Following this we will discuss the more likely, inside-out, route to star formation.

2.5.1 Classical Fragmentation: An Example of Technique

We begin by considering the **free-fall timescale**: t_{ff}. This is the time it would take a uniform molecular cloud to collapse down to a single point, assuming gravity is the sole force to contend with—in other words we switch the gas pressure *off* (clearly a gross and irredeemable assumption, but astronomers are crazy people). This is

$$t_{\mathrm{ff}} = \left(\frac{3\pi}{32G\rho} \right)^{1/2}. \tag{2.19}$$

For a giant molecular cloud $t_{\mathrm{ff}} \sim 10^8$ years. This is going to be a decent lower limit to the collapse, since adding gas pressure back in would act to slow things down. Now, this time is much larger than the **thermal adjustment** timescale—i.e., the time it takes the cloud to re-equilibrate by radiative cooling after gaining thermal energy due to the collapse (c.f. the Virial Theorem arguments from Equation 2.10). Therefore it is safe to assume that the cloud can remain *isothermal* during at least the early collapse stages.

As it collapses further the density ρ must increase—which means that the local Jeans Mass must *decrease* during collapse ($M_{\text{Jean}} \propto \rho^{-1/2}$). The immediate consequence of this is that it becomes possible for the cloud to *fragment* into smaller, also collapsing, sub-clouds, since the lower M_{Jean} now makes such sub-clouds unstable. Any minor density structure in the original cloud can now begin collapsing on its own. Now, it is tempting to breathe a sigh of relief at this point, since this appears to be a mechanism to break a large cloud into smaller and smaller pieces. However, therein lies the next problem—in this picture there is nothing preventing the fragmentation occurring all the way down to *sub*-stellar masses.

We must again appeal to a crude physical argument to solve this new problem. At some smaller cloud size the thermal adjustment time *does* become comparable to t_{ff}. The cloud cannot cool fast enough to maintain isothermality and becomes adiabatic. As a side note: as the cloud becomes smaller and denser the radiated photons will have an increasingly hard time transferring through the gas; this increased **opacity** helps drive the move to an adiabatic compression, since the time required to re-establish thermal equilibrium increases relative to the free-fall time.

In this case then $T \sim \rho^{2/3}$ and so $M_{\text{Jean}} \sim \rho^{1/2}$, the local Jeans Mass will start to *increase* and further fragmentation will cease.

In order to see if this solves the problem of making stellar mass fragments we need to ask, At what point does this happen? Now, $t_{\text{ff}} \sim (G\rho)^{-1/2}$ and the total gravitational energy that needs to be radiated away during the collapse is $\Omega \sim GM^2/R$. In order to maintain a constant T during collapse then the *rate* of radiation output must be to first order the total energy divided by t_{ff}:

$$\frac{dE}{dt} \sim \frac{GM^2}{R}(G\rho)^{1/2}. \qquad (2.20)$$

Putting all the numerical factors back in and replacing ρ yield

$$\frac{dE}{dt} = \left(\frac{3}{4\pi}\right)^{1/2} \frac{G^{3/2}M^{5/2}}{R^{5/2}}. \qquad (2.21)$$

Now, if the cloud is in thermal equilibrium, which certainly has a good chance of being true in the early collapse stages, it cannot radiate more than the blackbody luminosity L_{BB} (energy/unit time) for a given T, that is,

$$L_{BB} = 4\pi R^2 \sigma T^4, \tag{2.22}$$

where σ is the Stefan–Boltzmann constant.

When the cloud is isothermal $L_{BB} >> \frac{dE}{dt}$, and the transition to adiabaticity must occur only when $\frac{dE}{dt} \approx L_{BB}$. To inject slightly more realism into this discussion we should also allow for the fact that the luminosity of the cloud will more typically be a fraction f of the maximum blackbody rate, $L = f L_{BB}$ (this allows for the increasing internal opaqueness of the ever denser cloud, as described above). Equating 2.21 with 2.22 we then obtain the criterion for fragmentation to *stop*:

$$M^5 = \frac{64\pi^3 \sigma^2 f^2 T^8 R^9}{3G^3}. \tag{2.23}$$

This expression then tells us the *smallest* Jeans Mass which can be reached—the smallest structure that will collapse. By substituting for R_{Jean} in 2.23, setting $\mu = 1$, and with some rearrangement we can arrive at this smallest M_{Jean}:

$$M_{Jean}^{smallest} = 0.02 M_\odot \frac{T^{1/4}}{f^{1/2}} \tag{2.24}$$

Suppose $T = 100$ K for a molecular cloud, and setting $f = 1$, then $M_{Jean}^{smallest} \simeq 0.063 M_\odot$. Rather amazingly this is right in the brown dwarf regime—at the boundary between stars and giant planets!

In fact, it is *hard* to make $M_{Jean}^{smallest}$ much smaller. Smaller f boosts the size, and even if $T \sim 0$ K then $M_{Jean}^{smallest}$ is lowered by only a factor of three. One set of factors we have not included, except through the parameter f, are the deviations from the blackbody cooling of Equation 2.22. The presence of metals in the molecular cloud gas provides for a wider range of atomic and molecular emission lines than for just hydrogen and helium gas. Many of these emission lines kick in strongly at certain gas temperatures and can greatly enhance the cooling rate of the gas, and need to be taken into account in detailed modeling.

Even allowing for the litany of approximations we have made along the way, if this picture of fragmentation were to hold in even a general way then it would already be telling us something quite profound; objects smaller than brown dwarfs, namely, planets, probably arise in a *different* process than direct collapse from a molecular cloud.

As appealing as this picture is, we know it to be flawed. The internal structure of molecular clouds is complex, and star formation almost certainly occurs in regions where denser "cores" of gas exist. Jeans scales still describe the basic criteria for collapse, but the notion that an enormous molecular cloud will neatly fragment down to a set of stellar mass objects is highly simplistic and incomplete for present-day star formation. There is also another problem, which we discuss below, to do with the observed *rate* of star formation in the Galaxy.

2.5.2 Inside-out Star Formation

The alternative, and currently widely accepted picture, is based on much of the same physics, but operates more as an "inside-out" process. In this case proto-stellar objects indeed form from the collapse of denser molecular cores in GMCs and DMCs, but their final mass is more a function of how much material they can accrete from their surroundings. Nonetheless, the above argument on isothermal versus adiabatic collapse and fragmentation is important background. In the following we discuss some of the factors that must be incorporated into any complete model of star formation.

By using the many assumptions made in the previous sections we ignored major physical realities, some of which we summarize here in no particular order.

- The density of a molecular cloud will *not* be uniform, and clouds will *not* be spherical in general. Star formation is seen to occur in dense molecular cores within clouds.
- The molecular cloud (or clumps/cores) will not be truly isothermal at any stage.

- During collapse energy can go into ionizing the gas and altering μ, and the chemical composition of the cloud may not be uniform.
- The cloud/clumps/cores will in general have non-zero angular momentum, in which case the Virial Theorem must be revised to $2T + 2U + \Omega = 0$, where T is the rotational kinetic energy, and thus R_J is increased.
- Magnetic fields and external radiation fields will strongly impact the collapse physics.

Of these, the first item is arguably one of the most important, as it turns out is the last. To consider the consequences of *non-uniform* density structure and gravitational instability we can apply what we have already learnt about basic collapse physics.

If pressure forces were switched off in a spherical system then the gravitational free-fall time-scale was given by Equation 2.19. Consider a two-density, Jeans unstable clump (Figure 2.11) consisting of a higher density core surrounded by a lower density shell. In this case then

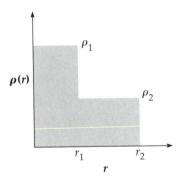

Figure 2.11. The radial density distribution of a two-density, Jeans unstable clump of gas. An inner core of uniform density ρ_1 is surrounded by an outer shell of lower density ρ_2. As described in the text, the timescale for collapse to the origin of the inner core is less than that of the outer shell. Thus, this system will collapse "inside out."

$t_{ff}(r_1) < t_{ff}(r_2)$. In fact, if we further allow for the density to increase as the clump collapses then this is an *accelerative* process as t_{ff} decreases with time. We can also see this by simply studying the equation of motion for a zero-pressure collapse:

$$\frac{d^2r}{dt^2} = -\frac{GM(r)}{r^2}.$$

(2.25)

Thus, $\frac{d^2r}{dt^2} \propto \frac{1}{r^2}$, and collapse is accelerative for a fixed M at a given r. What this is telling us is that, if $\rho(r)$ is a decreasing function of r, then the non-uniformity must grow with time (assuming that gas pressure or other forces are negligible). In Figure 2.12 this behavior is sketched out further.

As simple as this description is, it does nonetheless provide us with an important insight. As a region with an initially slightly denser core

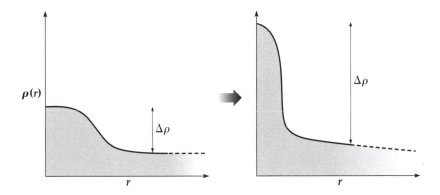

Figure 2.12. A more general example of the change with time of the radial density profile of a Jeans unstable clump of gas (spherical symmetry is again assumed). The left-hand plot shows a smoothly varying density profile with a density difference $\Delta\rho$ between the clump core and outskirts. The right hand plot shows the clump some time later. In this case we have assumed that the gas distribution has a very large extent, and therefore during collapse all densities continue to rise. In reality it will have a finite extent, or edge, in which case the outer regions will eventually drain inwards. As discussed, the collapse is accelerative for a fixed M at a given r and so the **shape** of the density distribution will change with time and, as shown, $\Delta\rho$ must increase.

collapses the core will rapidly increase in density while the outer envelope density increases more slowly. This is *all* we need to see that the
idea of a proto-stellar object, surrounded by less-dense material, is an
entirely natural consequence of gravitational collapse. Furthermore, this
implies that an initial density profile of this form will allow for the formation of a proto-stellar object that is some *fraction* of the total mass of
a collapsing region.

Although we have not yet included rotation (angular momentum,
see below) it is clear that this naturally differential collapse will lead to
a situation much like that described in §2.3.1 corresponding to a CLASS 0
proto-stellar system—a dense central object surrounded by still infalling
material.

A practical difficulty exists with the inside-out picture of star formation. If we take the observed properties (temperature, density) of dense
cores and DMCs, then based on the Jeans mass criterion, essentially all
such structures should be actively collapsing and producing stars. However, this would imply a star formation rate well in excess of that seen in
the Galaxy! So, something is stopping, or slowing the collapse of dense
cores and DMCs, as well as the larger structures within GMCs.

The most likely answer lies in magnetic field support (Shu et al. 1987).
We will not go into this in detail, suffice to say that if fields of the order
of a few tens of μG exist (and such field strengths are indeed measured)
in molecular clumps, and are "frozen in," then they provide sufficient
support (additional pressure) to significantly raise the mass required
for collapse. In fact this new critical mass should be of the order of
$M_{\text{crit}} \sim 10^3 M_\odot$ compared to the Jeans mass estimate as low as a few $M_\odot$
for the densest molecular clumps.

Specifically

$$M_{\text{crit}} \sim 10^3 M_\odot \frac{B}{30\mu G} \left(\frac{R}{2\,\text{parsec}}\right)^2. \tag{2.26}$$

Of course we *do* see star formation occurring, so something must
cause the magnetic field support to give way over time. The most likely
mechanism is known as **ambipolar diffusion**. To understand this we
must understand how the magnetic fields provide a pressure support.
Only electrically charged particles in a cloud feel a force from the mag-

netic fields. The majority of a cloud is electrically neutral and therefore feels the magnetic support indirectly through collision with the charged particles (e.g., ions, electrons)—a "frictional drag." In a cloud with relatively *low ionization* (low relative density of charged particles or plasma) ambipolar diffusion refers to the slow drift between neutral and charged particles, so that eventually the neutral matter does indeed collapse according to the Jeans criterion. In fact, as a denser clump of gas forms its ionization tends to *decrease*, since ions and electrons have a better chance of recombining, and since radiation (e.g., cosmic rays) can penetrate the gas less well and therefore produces less ionization. The process of the magnetic field "leaking out" is therefore an accelerative one. So rather than an entire DMC collapsing, the magnetic fields ensure that only parts are able to become unstable, and these are rather gently collapsed into structures like the observed molecular cores, still surrounded by more extended envelopes. At this point the inside-out collapse of a core can progress with vigor.

We also note that in the above general picture situations may exist where a cloud *is* sufficiently massive that it can overcome all support pressures, and would indeed be able to collapse according to the basic Jeans argument. In such a case some amount of classical internal fragmentation might also ensue. This may be the case in regions producing **high-mass** stars, such as in the Orion nebula.

Finally, there appears to be a causal connection between magnetic fields in molecular clouds and **turbulence** (for a definition of turbulence see §2.9.2). Turbulence could also, in principle, provide support against collapse since it introduces an additional kinetic component to the gas, above that just due to thermal motion. However, in molecular clouds turbulence should dissipate on relatively short timescales. Observations suggest that significant motions and turbulence *do* exist in clouds—and at some level contribute to the cloud support—but what maintains the turbulence? In a medium such as a molecular cloud, with a plasma component, **magnetohydrodynamic** (MHD) *waves* can propagate. These in turn pull on the plasma and can generate turbulent movement. It appears that this can happen with just the right magnitude to match both the magnetic support of the cloud and the observed turbulence. The origin of MHD waves is likely from stellar winds, explosions, and cloud collisions. However, these must occur predominantly *inside* the clouds themselves, otherwise they will act to *compress* the cloud and encourage collapse—as we discuss below.

2.6   Triggering Collapse

In general, observations of star-forming regions indicate that, for massive stars (O and B stars), formation tends to happen en masse. In at least some cases it can then be argued that an external force acts to trigger star formation. Possible mechanisms include shock, or density, waves from distant supernovae or Galactic scale phenomena. As these pass through a cloud they may either raise the ambient pressure of the lower density envelope surrounding the denser cloud cores, or directly compress the cores—leading to a drop in the local Jeans Mass and overcoming any magnetic or turbulent support. Other, perhaps more local events, such as the "switch on" of a pre-main-sequence star as it finally joins the main sequence could also raise the ambient radiation pressure, thereby changing the hydrostatic balance in the denser cloud cores. This might be particularly true for high-mass stars that form quickly and have high luminosity (see discussion in §2.11 below).

In these situations there is clearly a (potentially complex) feedback process in action between stellar birth and death.

2.7   Angular Momentum in Collapse

Returning to one of the further complications to our initially simple model of collapse and Jeans Mass, let us consider the presence of finite angular momentum. We know that stars, and our own solar system, have finite angular momentum, we also (§2.2) know that molecular clouds have dynamical structure. There are really two issues to be considered here:

- The impact of rotation on a collapsing system.
- The net angular momentum of a system, which tends to be conserved, and may be hard to shed.

Consider the first issue. In Figure 2.13 we illustrate a scheme for a particle of mass m in a collapsing, rotating spherical cloud. The particle

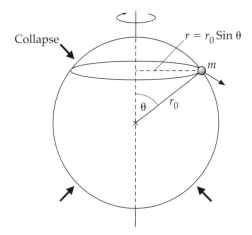

Figure 2.13. An illustration of the simple geometry considered here for rotating collapse. A particle of mass m resides in a spherical shell of initial radius r_0 in a collapsing cloud, which is also rotating. The distance of the particle from the spin axis is just $r_0 \sin \theta$.

of mass m is initially at a radius r_0 from the cloud center, and this radius subtends an angle θ with respect to the spin axis. The distance from the spin axis is then just $r = r_0 \sin \theta$.

If the rotation frequency of the particle is $f = v/2\pi r$ revolutions/unit time, where v is the rotational velocity, then the angular velocity is $\omega = 2\pi f$ and the angular momentum $L = mvr = 2\pi m f r^2 = \omega m r^2$. The *specific*, or per unit mass, angular momentum is just $h = \omega r^2 = \omega r_0^2 \sin^2 \theta$.

Therefore we can see that for a particle at $\theta = 0$, its post-collapse location will be at $r_{\text{final}} = 0$, since $h = 0$. A particle at $\theta = \pi/2$ will however have specific angular momentum h. If h is conserved, then after collapse the particle can enter an orbit. Assuming this is circular then the orbital radius (see §2.8) will obey $r = h^2/GM$ where M is the total enclosed mass. Thus, the final orbital radius of this particle will be $r_{\text{final}} = \frac{\omega^2 r_0^4}{GM}$, which is known as the *centrifugal radius*. All material between $\theta = 0$ and

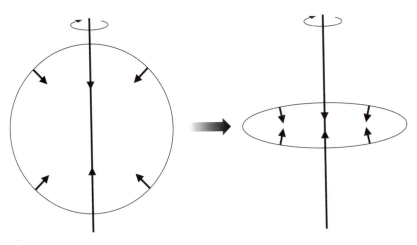

Figure 2.14. Schematic of the fundamental nature of rotating collapse. A spherical system will collapse to an increasingly flattened disk structure. The final radius of the disk (assuming no transport of angular momentum initially) will be at the centrifugal radius: $\omega^2 r_0^4 / (GM)$.

$\theta = \pi/2$ will merge onto a common plane of rotation between the centrifugal radius and the center (Figure 2.14). The net *linear* momentum (i.e., that perpendicular to the lane of rotation) of the collapse is assumed to be zero. As collapsing material reaches the plane of rotation it will of course have a vertical velocity. If there were no collisions, or dissipation, the material would simply oscillate in the vertical direction about this plane. If however the material is gaseous, or consists of finite-sized particles then radiative cooling or random collisions will, over time, bring the material into a new equilibrium about this plane. The final thickness of the disk so formed will be determined by a force balance between the vertical component of the disk self-gravity and the gas pressure, or a balance of vertical particle orbits according to the virial theorem. We will revisit this later on in discussing planet formation. It is important to realize however that the resulting disk *does* have a finite thickness.

Thus, in the presence of initial finite angular momentum, a collapsing cloud will wind up forming a disk owing to the additional rotational acceleration - this is just a natural consequence of the competition of forces. We can immediately see that the gross orbital architecture of our own solar system (and indeed the observation, or inference, of astrophysical disk structures around objects as diverse as proto-stars and black holes) would be a direct result of a collapsing, rotating, cloud.

As an historical aside, in the mid 1700s both Kant and Laplace formulated models of the formation of the solar system from a slowly rotating cloud of material collapsing due to self-gravity. They postulated that a flattened disk would form and that the planets would condense from this disk. The so-called **Kant–Laplace nebular hypothesis** appears to be well verified by both further theoretical work (as above) and direct observation of proto-planetary disks.

2.7.1 Conservation of Angular Momentum

Before going further we need to take note of another result of angular momentum conservation, one which will have far-reaching consequences for the formation of stars and planets.

The final and initial spin frequencies (f_2, f_1) of a mass conserving angular momentum but shrinking from a radius of r_1 to r_2 are related by

$$\frac{f_2}{f_1} = \left(\frac{r_1}{r_2}\right)^2.$$

(2.27)

This equation represents the old physics chesnut of an ice skater spinning on the spot, drawing his or her arms in, and therefore spinning faster. Let us use a trivial example to consider the ramifications of this for star and planet formation. Consider a 0.1 pc radius gas cloud, and suppose that it collapses to a stellar size object, say, to a size $R_\odot$. Measurements of the rotational velocities of a typical molecular clump are of the order of 10 km s^{-1}, or $f_1 = 5 \times 10^{-13}$ for our cloud. Therefore $f_2 \sim 1$, and the star should be spinning once every *second*! Not only is this vastly different from the rotation, or spin period, of our Sun (some 25–27 days depending on solar latitude), it is also unphysical in the sense that centrifugal forces would then overwhelm gravitational forces and the Sun could be disrupted.

Thus, *if* angular momentum were in fact uniformly conserved in a collapsing, rotating system then the central condensation—presumed to be the central star—would spin up to an unallowable rate. In fact, observations of proto-stellar systems suggest that the central objects are indeed spinning much faster than typical main sequence stars. Clearly then, there must be mechanisms at work which can redistribute and/or dissipate angular momentum within a collapsing, rotating, proto-stellar system. These mechanisms must operate during these early stages in order to produce the relatively slowly rotating stars we see around us. In order to begin to understand what these mechanisms might be, we need to step back a little and consider some more physical facts.

2.8   Orbital Basics

At this point it is important to introduce the basic laws of orbital dynamics. We will not derive anything, rather just define the laws we will employ. There are many sources from which a full treatment of the derivation of these laws can be obtained.

In general, bound orbits are elliptical (see Figure 2.15). The orbital radius at any point is given by

$$r = \frac{a(1 - e^2)}{1 + e \cos \theta},$$

(2.28)

where a is the *semi-major* axis of the ellipse (the average of the maximum and minimum orbital distance from, for example, a star).

Non-circular orbits are often referred to not as elliptical, but rather as **eccentric**, with eccentricity e, where $e \equiv (1 - b^2/a^2)^{1/2}$ and $2b$ is the minor axis of the ellipse. The closest point in the orbit to the mass at the focus is termed **periapsis**, and the farthest is **apoapsis**. For our own Sun we refer to peri-*helion*, and ap-*helion*, for other stars we use the terms peri-*astron*, and ap-*astron*. The quantity θ is known as the *true anomaly*, and is the angle between an orbiting object's current position and its position at periapsis. In general a system of gravitating bodies is considered in terms of its center-of-mass frame (see Figure 2.16).

The center of mass coordinates are given by

$$\mathbf{R} = \frac{m_1 \mathbf{r}_1' + m_2 \mathbf{r}_2'}{m_1 + m_2},$$

(2.29)

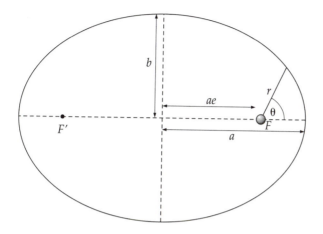

Figure 2.15. The geometry and parameterization of an elliptical orbit. The position of an object is measured relative to the principal focus F and is given by the distance r and angle θ as shown.

and the *reduced mass* is given by

$$\mu = \frac{m_1 m_2}{m_1 + m_2}.$$

(2.30)

Then in the center-of-mass frame the total energy of a gravitationally bound system is

$$E = \frac{1}{2}\mu v^2 - \frac{GM\mu}{r},$$

(2.31)

where M is the total system mass and r is the separation of two masses.

Finally, **Kepler's Laws** describing bound orbits are paraphrased here:

- An object orbits a mass in an ellipse, with the mass at one focus of the ellipse.

- A line connecting the object to the mass sweeps out equal areas in equal times, that is, $\frac{dA}{dt} = \frac{1}{2}\frac{L}{\mu}$ where L is the angular momentum.

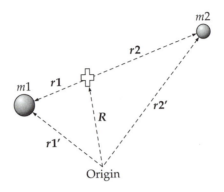

Figure 2.16. The definition of the center of mass of a system. In this example two masses, m_1 and m_2, are at position vectors $\mathbf{r}'_1$ and $\mathbf{r}'_2$ relative to the coordinate origin. The center of mass of this system (shown as a cross) is located at $\mathbf{R}$, given by Equation 2.29. In the center of mass reference frame $m_1\mathbf{r}_1 + m_2\mathbf{r}_2 = 0$.

- $P^2 \propto a^3$, where P is the orbital period and a is the semi-major axis of the object orbit, and then in general,

$$P^2 = \frac{4\pi^2 a^3}{G(m_1 + m_2)}. \tag{2.32}$$

Consequently, for a Keplerian orbit, the angular velocity of a small mass around a large mass M is $\omega = \left(\frac{GM}{r^3}\right)^{1/2}$ (where $\omega = v/r$).

2.9   Disk Evolution

From the very naive considerations in §2.7.1 above we have seen that somehow we need to lose angular momentum from a proto-star, and indeed from the inner region of a disk during the disk collapse and evolution. We will continue to assume that the net angular momentum of a system is conserved and not dissipated (by, for example, frictional heating and radiation energy loss). What follows here is just one possible way of formulating the problem which lends itself to providing insight as to what happens in a forming proto-stellar system.

2.9.1 Energy Minimization and Angular Momentum Conservation

Instead of treating the full problem of a disk of material we can look at a simplified model, consisting of what amounts to pieces of a disk (following Lynden-Bell & Pringle 1974). Consider two small bodies of mass m_1 and m_2 in Keplerian orbits around a central mass M.

The total energy of the system is then

$$E = \frac{-GM}{2}\left(\frac{m_1}{r_1} + \frac{m_2}{r_2}\right),\qquad(2.33)$$

and the total angular momentum (ignoring the spin of any of the masses)

$$L = (GM)^{1/2}(m_1 r_1^{1/2} + m_2 r_2^{1/2}).\qquad(2.34)$$

Now, suppose that the orbits are *perturbed* by a small amount by changing the orbital radii, while *preserving* angular momentum L.

We can do this by setting $\Delta L = 0$ and making the perturbations Δr_1 and Δr_2 obey

$$m_1 r_1^{-1/2}\Delta r_1 = -m_2 r_2^{-1/2}\Delta r_2.\qquad(2.35)$$

This will correspond to a change in E, in terms of body 1, of

$$\Delta E = \frac{-GMm_1\Delta r_1}{2r_1^2}\left[\left(\frac{r_1}{r_2}\right)^{3/2} - 1\right].\qquad(2.36)$$

Now, suppose that $r_1 > r_2$, then, if m_1 were perturbed to a *larger* radius then the total E will *decrease*. Similarly, if $r_1 < r_2$ then E will be decreased if m_1 moves to a smaller orbital radius.

Therefore, the total energy E can be reduced (which thermodynamics dictates is nature's preference in any system), while conserving the net angular momentum of this system, by moving the inner orbiting body farther in and the outer orbiting body farther out. In this very simple toy model, evolution driven by the minimization of energy and the conservation of angular momentum tends to *spread* the radial distribution of orbiting mass.

This indicates that a disk may evolve in the same way—it will tend to spread out over radius with time. However, thus far we have not allowed for the possibility of *mass transfer*, either between the two orbiting bodies, or indeed between zones in a disk. As we will see below (§2.9.2), there is a general physical argument for the transfer of mass between parts of a disk of particles, due simply to differential velocities. We do not go through the calculations here, but summarize—if mass transfer (between adjacent disk zones) could occur then it is found that the energy is minimized by moving *most* mass *inwards*, and a small amount *outwards* to large radii. We can then ask: If this occurs then how would the angular momentum be redistributed? Below we sketch out a physical argument that will help us see the answer to this question.

2.9.2 *Kinematic and Turbulent Viscosity*

Consider a rotating disk of material. At any given location there may be a radial velocity gradient. This is certainly true for a Keplerian disk, where gas or particles follow Keplerian orbits, so $\omega = \left(\frac{GM}{r^3}\right)^{1/2}$ and $v = \omega r$. To keep things simple we will also first assume the motion of the disk material to be **laminar**. This means that if we look at any adjacent, imaginary, layers of the disk material (adjacent annuli in this case with a radial geometry) there is no disruption between the layers—they smoothly transition. In such a case the only motions that we would assign to an individual particle, or small "packet" of material are its rotational velocity and a random velocity due to a finite thermal energy.

Consider then two such identical mass packets of material, labeled 1 and 2. Across the imaginary interface of a small segment of disk layers, drawn in Figure 2.17, there can be an equal but opposite flux of packets due to random, thermal, velocities. In this case there is no net change in the density of the disk material due to the exchange of packets. However, if $v_1 \neq v_2$ there *will* be a net change in angular momentum in each layer.

The flux of packets can be written as $\sim n v_T$, where n is the number of packets per unit volume and v_T is the mean random velocity of the packets. Averaged over time this flux will therefore carry an excess or

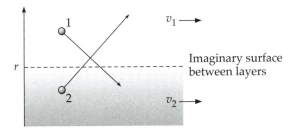

Figure 2.17. A schematic of the exchange of material "packets" and momentum across an imaginary surface within a disk. At any given radius r a small packet may move to greater or lesser radius. If there is a velocity gradient with radius (indicated by the velocities v_1 and v_2) then that packet will alter the net momentum at its new radius. On a small scale it can be argued that it is equally likely for a packet 1 to change position as it is for a packet 2. Thus the density of a disk is maintained, but the angular momentum distribution may be altered.

deficit of momentum (Δp per packet) in and out of the layers. This will vary according to the velocity gradient dv/dr in the disk:

$$\Delta p \sim ml\frac{dv}{dr}, \tag{2.37}$$

where m is a packet mass and l is the mean distance a packet travels before interacting or dissipating (we could treat the packets as particles with the same result). In other words, l is the packet **mean free path**:

$$l \sim \frac{1}{n\sigma}, \tag{2.38}$$

where σ is the cross-sectional area for interaction of a packet. All of this yields an expression for a **coefficient of kinematic (or shear) viscosity**:

$$\kappa_v \sim nv_T\Delta p \sim \frac{v_T m}{\sigma}\frac{dv}{dr}, \tag{2.39}$$

which is *independent* of density! Thus, the strength of this type of mechanism for the exchange of angular momentum due to velocity shear is dictated by the disk composition and temperature (or turbulence, see below), and by the velocity field, *not* by the density distribution of the disk. In a Keplerian disk, therefore, the velocity decreases with radius and so kinematic viscosity will transport angular momentum *outwards*.

Given this physical picture we can then argue that, in order to lower the energy of a disk system, then *most* of the *mass* must move *inwards*, while (because of the above argument) the angular momentum is transported *outwards* and may be net conserved. This is therefore a plausible, zeroth-order, physical argument for why our Sun is not spinning once a second, and why the "disk" mass in the Solar system is concentrated towards the center. The bulk of the angular momentum in the system is carried by the minority of the mass, on orbits far from the central mass. It also fits beautifully with our earlier description of the inside-out formation of stars, since the collapse of a slowly rotating, centrally concentrated cloud of gas will produce the "seed" stellar embryo that will then accumulate further mass as the disk evolves dynamically.

It turns out that while this is pretty accurate as a general rule, the precise mechanisms that dominate the transport of angular momentum may be rather more complex. For example, rather than relying simply on the above kinematic viscosity (which is very small in real astrophysical situations), it is necessary to invoke more efficient momentum transfer due to structure within the disk. One potential source of additional viscosity is due to **turbulence**. Turbulence in a fluid or gas is a flow regime characterized by non-laminar behavior, and chaotic, stochastic variations in the properties of the system (meaning that the properties change with time in a way that is only partially determined by the system parameters, and otherwise appears random). Turbulence usually kicks in when the velocity of a laminar flow is increased beyond some critical point—depending on the fluid or gas characteristics, such as density.[1] In turbulent motion, eddies or vortices tend to form at the boundaries

---

1. The onset of turbulence is generally indicated by the **critical Reynolds number**, where the Reynolds number is the dimensionless ratio of inertial forces to viscous forces.

between moving layers. This can greatly increase the transfer of momentum by effectively increasing the viscosity of the medium (imagine a series of interacting tornadoes!).

Furthermore, in the case of a proto-planetary system, rather than being simple and uniform, a disk may develop spiral density wave patterns (not unlike those of the vastly larger Milky Way spiral galaxy) that work to transport mass and momentum. We will discuss this phenomenon more in the next chapter. There are yet further complications that enter into this description of a collapsing, rotating cloud. While a disk is forming there will still be material infalling from larger radii (c.f. CLASS 0-1 proto-stellar systems). This material will accrete *onto* the disk. Indeed, since the disk material itself will be moving inwards and falling onto the forming proto-star, the entire system is often termed an **accretion disk** (with many analogs in other astrophysical phenomena). The environment around a forming proto-star is such that magnetic fields, ionization processes, and turbulence become very important in the dynamical evolution of such a disk. These are phenomena that are often best dealt with through detailed computer simulation and fluid dynamics, and we will not delve very far into their investigation. Suffice to say that, although the mechanisms of disk evolution sketched out above certainly help in "de-spinning" the central proto-stellar object, the most likely (and potentially observable) mechanism to dominate in this particular task is magnetic coupling between the inner part of the disk and the proto-star. This has the potential to drain off angular momentum, and possibly even dissipate it through radiation mechanisms associated with the magnetic fields themselves and with the observed bipolar outflows or jets.

Finally, an issue that we will touch on briefly below is that some two-thirds of all local stars exist in *binary* systems, some close, some far apart, but all gravitationally bound. Although binary stars may form through gravitational capture after the stars themselves have formed, they may also form through the dynamical evolution of a collapsing, rotating structure. In fact, as we describe in §2.10 below, this latter route may be the most likely.

We have now set at least some of the stage for what follows in a proto-stellar system, and in the next chapter we will pick up part of this story in our study of planet formation.

2.10   Binary and Multiple Star Systems

In the preceding efforts we have conveniently left out the observational fact that the majority of bright stars exist in binary, or multiple systems, some very close together indeed. How do binary systems form in the above picture? At the time of writing, as with so much in the field, this is still a matter of intense debate.

Here we provide a brief discussion of the current thinking on the subject. In later chapters we will occasionally bring up the issue of planets in binary star systems. History may prove this to be a gross underestimate of the importance of binary, or multiple, stellar systems. Indeed, on the basis of observation and theory it has been postulated that binary star formation is actually the *primary* route by which stars are formed, and that single stars are really an offshoot of this process (Mathieu 1994).

There are three broad categories for possible binary star formation mechanisms (e.g., see Tohline 2002): **Capture**, **Prompt fragmentation**, and **Delayed breakup**. In capture, stars form preferentially as single objects. Later, through dynamical interactions, they become gravitationally bound into pairs. While there are certainly mechanisms for this to occur—including energy dissipation through tidal effects between stars, three-body encounters, and disk interactions in proto-stars—it appears to be too inefficient in typical environments to explain the number of observed binary systems. The only possible exception is in small, dense clusters of stars.

Prompt fragmentation builds on some of the collapse phenomena we have discussed above. In brief, this type of fragmentation does *not* occur during the initial free-fall collapse, but rather after a rotating, flattened structure has formed. Prompt fragmentation can occur immediately after this stage has been reached, but only if the final collapse is relatively uniform and takes place essentially in unison—especially if things are still isothermal. The Jeans unstable fragments must collapse rapidly in order to avoid merging back together again. The most favorable conditions for this to occur arise when there are significant departures from spherical symmetry in the initial cloud, but no strong *central* concentration of density. In this latter case the central core tends to dominate and is simply surrounded by an accretion disk of material—effectively erasing other possible centers for collapse. Delayed breakup is a process that may occur when a relatively dense, rotating, central core has collapsed and continues to accumulate high angular momentum material from the cloud. Initially the mass of this core is large compared to that of the surrounding disk, but if this changes as the disk grows from infalling material, then the disk itself may become dynamically unstable to large, non-axisymmetric density structures. At this point the disk itself may break up into major pieces—potentially forming binary or multiple proto-stellar structures. This is fundamentally different from earlier notions of "fission," in which a dense core effectively splits up as it spins—much like a drop of water. While this can happen to water, which is an essentially *incompressible* fluid, a forming stellar core is *compressible* and can lose angular momentum through its interaction with the surrounding disk material.

2.11   Star Formation in a Crowd

A significant fraction (possibly 80–90%) of stars are formed in groups and clusters (with about 100–1000 stellar members) embedded within the denser regions, or cores, of GMCs (see for example Adams et al. 2006 and Megeath et al. 2007). In the most massive cores ($M > 50M_\odot$) it is likely that Jeans fragmentation (see above) is one possible mechanism for the formation of stars—together with those related to magnetic fields. One consequence of this type of formation environment is that the most massive and UV luminous ("hot") stars (generally labeled as **OB stars** in reference to B stars of $3–20M_\odot$ and O stars of $20–100M_\odot$) exert considerable physical influence on proto-stellar and proto-planetary systems (see also Chapter 3). Extreme and far UV photons from young OB stars can photoionize the surface of molecular clouds, creating flows of ionized gas and eroding the clouds. This radiation can also influence the gas that is collapsing into proto-stellar structures—possibly limiting the mass of stars that are formed there. OB stars may also play a role in triggering star formation (see above), as fronts of ionized gas may be preceded by shock fronts. Thus it seems likely that the most massive stars in any dense environment may play a major role in determining the local star formation outcome—and (see Chapter 3) the planet formation outcome.

It can be surprising to learn that current evidence suggests that the Sun probably formed within a denser stellar group or cluster—long since dispersed by dynamical motions in the Galaxy. The most compelling piece of this evidence is the inferred presence of **short-lived radionuclides** (SLRs) during the formation of solids that are now seen in meteorites (see also Chapter 3 and Chapter 8). In particular, it has been established that ^{60}Fe was present in the early solar system.[2] This isotope of iron can have arrived in the forming solar system only from the wind and/or ejecta of at least one nearby supernova (implying massive stars). Other evidence includes the measurement of ^{18}O and ^{17}O isotopic fractionation relative to the more abundant ^{16}O (see also Chapter 8). Intense UV radiation in the proto-stellar/proto-planetary disk (Chapter 3) can result in the preferential incorporation of these heavier oxygen isotopes into the water that eventually participates in the formation of solids. The apparently uniform distribution of radioactive ^{26}Al (half-life approximately 700,000 years) in the early Solar System (inferred from decay products in meteorites) is also consistent with an external source—namely other very nearby stars.

2.12   Brown Dwarfs to Planets

In the next chapter we will tackle planet formation. Before we do so however, it is worth pausing to reconsider the way in which we defined planets in Chapter 1, and the star formation mechanisms described above. In particular there is a very interesting (and currently unresolved) question about what happens at the lowest mass end of star formation— in the realm of brown dwarfs, with masses between 13 times the mass of Jupiter and $0.08M_\odot$. Our initial assumption might be that these form much like their higher mass stellar relatives, and could themselves be parent bodies for planets. There is certainly good observational evidence for brown dwarfs forming at the center of circumstellar disks (e.g., Luhman et al. 2007). This may be the primary route to their formation, however it may also be true that some objects in this mass range share formation characteristics with those of giant planets—which as we will

---

2. ^{60}Fe is an extinct radioactive isotope of iron with a half-life of some 1.5 million years.

see in the next chapter have much to do with the nature of proto-stellar disks and solids. At some level this becomes a semantic discussion—there may be a continuum of processes by which condensed bodies form from interstellar clouds, with certain mechanisms dominating at the high and low mass ends, but becoming more interchangeable at the classical boundaries between stars, brown dwarfs, and planets.

References and Suggested Reading

Adams, F. C., et al. (2006). Early evolution of stellar groups and clusters: environmental effects on forming planetary systems, *Astrophysical Journal*, **641**, 504.

Luhman, K. L., Joergens, V., Lada, C., Muzerolle, J., Pascucci, I., & White, R. (2007). The Formation of Brown Dwarfs: Observations, *Protostars and Planets V*, 443.

Lynden-Bell, D., & Pringle, J. E. (1974). The evolution of viscous discs and the origin of the nebular variables, *Monthly Notices of the Royal Astronomical Society*, **168**, 603 .

Mathieu, R. D. (1994). Pre-Main-Sequence binary stars, *Annual Review of Astronomy and Astrophysics*, **32**, 465.

Megeath, S. T. (2007). Cool stars in hot places, ASP conference series, *Cool Stars* **14**, Ed. G. Van Belle, see also online astro-ph/0704.1045v1.

Shu, F. H., Adams, F. C., & Lizano, S. (1987). Star formation in molecular clouds: Observation and theory, *Annual Review of Astronomy and Astrophysics*, **25**, 23.

Tohline, J. E. (2002). The origin of binary stars, *Annual Review of Astronomy and Astrophysics*, **40**, 349.

Problems

2.1 Describe the observed nature of proto-stellar systems—include a discussion of relevant timescales during their evolution and their classification at different stages. Can you give one, very simple, physically motivated reason for the difference in evolutionary timescale between a high-mass proto-stellar system and a low-mass one? Justify your answer.

2.2 Let us suppose that a shock wave traverses a system of interstellar clouds (the shock wave comes from, say, a nearby supernova) and mechanically compresses the clouds so that their radii change by a factor $1/b$, where $b > 1$. This is the kind of scenario that is sometimes invoked as a trigger for the collapse of clouds and the formation of proto-stellar systems. Is this as straightforward as it sounds?

(a) Assume that before compression the clouds are *below* the Jeans Mass (all clouds have the same temperature T and density ρ) by a factor $1/\alpha$, where $\alpha > 1$. If the compression is rapid and T remains constant then what is the necessary condition/relation between b and α for collapse?

(b) It is probably incorrect to assume that T remains constant, so let us suppose that the shockwave *adiabatically* compresses the clouds. Treating the clouds as an ideal gas with adiabatic index $\gamma = C_p/C_v = 5/3$ work out the relationship between the initial and final temperatures for a given b. Having done that you can work out by what factor the Jeans Mass changes—does it go up or down?

(c) Instead of increasing the thermal energy (increasing T) of the cloud, the compression could do work in ionizing the gas. If the cloud was initially entirely neutral, and of cosmic element abundance, but was completely ionized following the shock wave, determine the minimum value of b required to ensure collapse.

(d) By now you should be wondering how you can get the clouds to collapse at all (ignoring complications such as magnetic fields). It is a good question, so let us think about some of the possibilities. The cloud can cool via radiation, by what factor does its blackbody luminosity change after compression (assuming adiabaticity)? Describe how this could lead to collapse sooner than if the shockwave had not passed through the system.

Other possibilities in this simple picture are that the clouds are themselves within a larger, lower density cloud of gas. If the shock wave preferentially heats up that larger cloud then the gas pressure external to the smaller clouds or clumps goes up, and will act like an extra, "mock gravity" compressive force, thereby lowering the effective Jeans Mass. Then, there are possible chemical routes, for example, if the passage of the shock partially ionizes the gas then in some situations the ions can catalyze molecular reactions that then soak up thermal energy and actually make the clouds cool much more rapidly *and* increase μ by building bigger molecules—again, lowering the Jeans Mass.

2.3   Describe the major differences observed between proto-stars and pre-main-sequence stars. How might this relate to the observation that most of the angular momentum in the solar system lies in the planetary orbits, while most of the mass lies in the Sun?

2.4   (a) The Jeans Mass of a spherical, uniform density (ρ), gas cloud is

$$M_{\text{Jean}} = \left(\frac{5kT}{G\mu m_{\text{H}}} \right)^{3/2} \left(\frac{3}{4\pi\rho} \right)^{1/2}.$$

Derive the expression for Jeans Radius.

(b) Assume uniform, isothermal, non-rotating, Jeans collapse. What radius gas cloud would a brown dwarf of mass $0.07 M_\odot$ and temperature 1000 K form from? (Assume $\mu = 1$.)

(c) The *smallest* observed molecular clouds (Bok globules) are ~ 1000 a.u. in radius, with a wide range in density and T—why is this a problem for (b) above? What additional physics must the model include to allow brown dwarfs to form via gravitational collapse? Describe the chain of events involved.

2.5 (a) A planet of mass m_p orbits a star M_* at distance r. The total energy of the system is

$$E = \frac{1}{2}\mu v^2 - \frac{GM\mu}{r},$$ (2.40)

where M is the *total* system mass, v is the orbital velocity, and μ is the reduced mass: $\mu = \frac{m_1 m_2}{m_1 + m_2}$. Given that $M_* >> m_p$ rewrite this expression in terms of M_* and m_p. Now, by recognizing the right-hand-side terms in this equation, and using the Virial Theorem, derive Kepler's orbital law: $\omega = (GM_*/r^3)^{1/2}$, where ω is the orbital angular velocity.

(b) An object of mass fm_p collides with the planet and sticks to it, where f is a constant. Assume that the collision imparts *no* angular momentum and that the net angular momentum of the planet–star system is *conserved*. Use Kepler to derive a relationship between the old orbital radius (r_1) and the one that the planet will end up in (r_2) (ignore planet spin).

(c) Now, find an expression in terms of f for the *ratio* between the initial total energy of the system (E_1) and the final, E_2. (*Hint:* use the Virial Theorem to simplify E).

2.6 This is a rather artifical question to examine the issues of how a rotating system adjusts when something is altered. Let us suppose we have a thin, narrow *ring* of material (e.g., small particles) rotating at a distance r around a star of mass M_*. The mass of the ring is M_{ring} and $M_{ring} << M_*$. Its initial angular frequency is ω.

From some external region surrounding this system an amount of material is deposited uniformly onto the ring such that it carries no additional angular momentum (e.g., drops onto the ring along the spin-axis or z direction) and does not alter the width of the ring, but it does increase the ring mass to αM_{ring}.

Assume that the ring system adjusts in such a way that the total angular momentum is conserved after the mass deposition.

(a) Derive an expression for the new radius of the ring in the special case when $\omega = 1$.

(b) Suppose that the system adjusts so as to conserve angular momentum, but to *reduce* the total energy, i.e., $E_{final} < E_{initial}$, does the final radius increase or decrease? What might happen if the width of the ring could change?

Hint for Exercise 2.6: reduced mass will be M_{ring}

2.7 Describe the roles that magnetic fields play in both star formation and circumstellar disk evolution.

2.8 Using the review article "The Origin of Binary Stars" by Tohline (2002) (see references), summarize the prevailing ideas on how binary star formation occurs. Discuss in more detail the premise that binary star formation is the primary branch of star formation processes in the universe.

CHAPTER 3

Planet Formation

3.1 Introduction

The origin of planets is central to astrobiology. While we may speculate about life in forms other than those we currently know of, it still holds that of all astrophysical environments, planets have a unique potential to provide habitats for organisms. They can offer a rich chemical environment, with a thermodynamic state suited to complex molecular reactions. They can also offer relative stability in terms of environment, while still providing energy sources to drive biology—starlight and volcanic or tectonic activity being primary examples. Planets however come in many forms, and we need to understand this in more detail. We also need to understand the mechanisms that determine the orbital architecture and chemical composition of planetary systems—since this directly impacts the nature of their surface or interior environments.

Through an investigation of observations and theory in Chapter 2 we have drafted a picture where a proto-stellar system consists of a central object surrounded by a flattened disk of rotating material. We have furthermore rather implicitly assumed that somehow this disk of material (which initially consists of gas and dust) turns into a collection of planets and other bodies as described in Chapter 1.

At the time of writing it would be fair to say that the precise modes and details of planet formation are still highly uncertain. Nonetheless, considerable progress has been made in developing the mathematical and physical foundations for a complete theory of the process. It is also true that, however elegant some of this work is, we will not be able to sift out the truth until we have observed many systems other than our own, caught at different stages in their evolution, and with different initial conditions. Indeed, we may need to adjust some of our preconceptions. Since we inhabit the surface of a "finished" world (one which is changing at a much slower rate than it was four billion years ago) we tend to assume that this is the most interesting stage of a planet's history—or at least where we would like our models to lead us. Taking a contrarian view though, we might equally argue that planets such as our Earth are simply the left-over "fossil" bones of a much more lively past. In this case, the true pinnacle of organization was the gas-rich proto-planetary disk and the complex chemistry and thermodynamic activity taking place therein. Astrophysically this is almost certainly true; the period during which a star is on the Main Sequence until it finishes hydrogen burning is arguably the dullest part of its lifetime, with the least real interaction with either its immediate surroundings or the rest of the galaxy. From the point of view of astrobiology this is of course a bit of a facetious argument. A planet like the Earth has certainly not been inert for the past four billion years, and while life may have emerged quite soon after the planet's formation, its linkage to planet-wide systems of climate and **bio-geomorphism** has changed and evolved dramatically (Chapter 10). Nonetheless, an awareness of our innate bias in these matters is important.

The purpose of this chapter is much more straightforward. Here we will lay out the rough groundwork for understanding some of the mechanisms that produce planets, and discuss these in the context of both our own solar system, and the emerging characteristics of other solar systems. Some of the physical arguments (perhaps all of them) are highly simplistic, but they still serve as a basis from which to start.

3.2   Planet Classes and Formation Scenarios

3.2.1   Gas-Giants and Ice-Giants

Until recently our only guide to the natural classification of planets has come from our own Solar System. As we shall see (in particular in Chapter 4), this undoubtedly creates many biases. Nonetheless, it does seem that our immediate surroundings contain quite a decent spread in basic planet types. Dominating the planetary mass of our system are

Table 3.1. Selective data table for giant planets in the solar system. Primary gas composition is given by volume. Heavy element abundances, denoted as metals, are given relative to the Solar value, which comprises some 2% by mass of stellar material. Data taken from NASA's NSSDC databases

Property	Jupiter	Saturn	Uranus	Neptune
Atmospheric H_2 by volume	89%	96%	83%	80%
Atmospheric He by volume	10%	3%	15%	19%
Metal abundance (relative to Sun)	5	15	300	300
Mean density (g cm^{-3})	1.3	0.7	1.3	1.6
Number of satellites	63	60	27	13

the **giant planets**: Jupiter ($318M_\oplus$), Saturn ($95M_\oplus$), Uranus ($15M_\oplus$), and Neptune ($17M_\oplus$). In Table 3.1 we list some of the basic compositional properties and characteristics of these worlds.

All four giant planets have deep atmospheres dominated by molecular hydrogen and helium. In the case of Jupiter and Saturn the total mass of each planet is also dominated by the mass of the atmosphere. The composition of these atmospheres is very similar to that deduced for the proto-stellar disk gas based on solar values, although somewhat metal enhanced. This appears to support the notion that these giant planets (and in particular Jupiter and Saturn) accumulated vast amounts of "primordial" gas from the proto-planetary disk in the same proportions as accumulated by the proto-Sun. For Jupiter and Saturn it is speculated, but by no means confirmed, that their cores consist of rocky material of mass $\sim 10 - 15 M_\oplus$. Under the immense pressure of the overlying gas (4 million atmospheres for Jupiter) there will also be a region surrounding this core of **liquid metallic hydrogen**, which may play a role in the powerful magnetic fields of these planets. Both Jupiter and Saturn exhibit lower atmospheric helium content than the solar proportion—this is thought to be a consequence of helium "rain-out" at high pressures towards the planetary cores.

Uranus and Neptune are distinct by virtue of their greatly enhanced mean metal abundance (Table 3.1). This is a direct consequence of their much smaller outer atmospheres in comparison to Jupiter or Saturn. Beneath these atmospheres lie solid worlds, which may be very similar—although perhaps less massive—to the cores of Jupiter and Saturn. For both Uranus and Neptune the interior composition likely consists of rocky material and ices amounting to some $10M_\oplus$ in mass. These cores are in turn surrounded by a significant mantle of water, ammonia, and methane of $\sim 1M_\oplus$ in mass.

These four giant objects can therefore be placed into two distinct classes. Jupiter and Saturn are considered **gas giants**, and Uranus and Neptune are **ice giants**. As we will discuss later on, it is possible that Uranus and Neptune represent "seed" objects for gas giants, but for some reason did not accumulate as much hydrogen or helium as Jupiter or Saturn.

3.2.2   Formation Scenarios

Before talking about smaller worlds it is helpful to plunge in and consider the possible formation pathways to giant planets. The fact that the elemental composition of the gaseous atmospheres of giant planets (in our Solar System) appears to match that expected in the earlier proto-planetary disk strongly suggests that these worlds must complete their formation during the period that a *gaseous* proto-stellar disk exists (before being dissipated). From observational constraints (Chapter 2) we know this to be on average $\sim 3 \times 10^6$ years, and no more than 10^7 years for systems producing stars of $\sim 1M_\odot$. At present there are two schools of thought on the route to giant planet formation.

The first is that they can form *directly* from the young, gaseous, proto-stellar disk through **gravitational instabilities** akin to the Jeans instability discussed in Chapter 2. This leads to massive gaseous "fragments" in the Jupiter mass range. The advantage of this mechanism is that it can happen fast in the early disk, well within 3×10^6 years. Once the gas has begun to collapse and contract, solid material can then be accumulated quite efficiently in its gravitational potential well. This then is a "top-down" formation process—whereby the solid core of giant planets is formed *after* the bulk of the mass (in gas) is in place. In order to produce ice giants in this scenario a further mechanism must be invoked to strip away the outer parts of proto-gas giants. **Photoevaporation** by ultraviolet radiation from other nearby, young, high-mass stars is a potential candidate, and might preferentially strip off gas in a system beyond some critical distance from the proto-star (see §3.7.1 below).

The second possible route is a multistage process starting with the co-agulation of microscopic dust grains (Chapter 7), leading to small bodies called **planetesimals** (rather arbitrarily thought of as 1 km and larger in size, see below). The further collision and merger of these objects into solid cores of some $10 - 15M_\oplus$ is then followed by the gravitational *accretion* of gas onto these massive cores. Above some critical mass— including both gas and solids—the accretion of gas can proceed extremely rapidly owing to hydrostatic *instability* (Bodenheimer & Pollack 1986). The object quite literally siphons up as much gas as is available from the surrounding disk. This is termed the **Core Instability Model** or the **Core Accretion—Gas Capture** model and has the appeal that a similar solid material coagulation mechanism must be in part responsible for the growth of terrestrial-sized planets. The biggest disadvantage of this hypothesis is that it requires the solid core to be formed very quickly, before the gaseous disk is dissipated—in order for it to accrete gas. This is a real challenge to the model, although, as we will see below, various factors can mitigate this problem.

At the time of writing it is this latter theory of core accretion that is most generally favored (although it is quite possible that both mechanisms can operate, albeit perhaps under different circumstances). Among the reasons for considering core accretion as the dominant route to forming giant planets is the observation that stars of higher metal content are more likely to harbor giant planets (Chapter 4). If the stellar metallicity reflects that of the proto-planetary disk then higher metal content would suggest more solid material in the proto-planetary disk and hence, for the core-accretion model, an increased likelihood of the rapid formation of giant planet embryos. There are of course potential problems with this scenario, not least of which is the limited set of planetary systems from which this conclusion is drawn. It is then possible that it is a consequence not of the efficiency of planet formation, but of the mechanisms which determine the final orbital architecture, which in turn determines our ability to detect planets (see for example Boss 2006). The gravitational instability model also does not make any prediction about the mass function of planets (i.e. the relative numbers of high and low mass objects), while observations seem to indicate that lower mass (giant) planets are more numerous than high-mass planets.

However, some common mechanisms must exist between the core accretion model and models of the formation of smaller, rocky worlds, and it is on these mechanisms that we will first focus.

3.2.3 Terrestrial/Rocky Planets

As with the giant planets, our own Solar System becomes the de facto guide to classification of smaller worlds. Mercury, Venus, Earth, and Mars are generally labeled as **rocky planets**, since they harbor relatively tenuous atmospheres, amounting to less than $\sim 10^{-7}$ of the total planetary mass (Table 3.2). They are abundant in heavy elements and have well differentiated interior structures (§3.7.3) with iron-rich inner cores. Although in our Solar System the rocky worlds are 10–100 times less massive than the giant planets this should not be assumed to indicate that intermediate planets cannot exist in other systems. Indeed, as we will see (Chapter 4), there is almost certainly a continuum of planet masses between these worlds and the gas and ice giants.

It is also worth pointing out that some of the *satellites* of the giant planets in our solar system are themselves comparable in size to the rocky planets. For example, Ganymede (Jupiter) has a mass of $0.025 M_{\oplus}$ and a radius of $0.41 R_{\oplus}$. It therefore has a larger radius than Mercury ($0.38 R_{\oplus}$) although a much lower density (1.9 g cm^{-3}) due to a dominant water-ice composition. Titan (Saturn) has a comparable radius and mass to Ganymede, and harbors a thick, cold atmosphere with a surface pressure of about 1.4 Earth atmospheres. Although we will not discuss the possible formation pathways to these moons here (see instead Chapter 10), it is clear that they can amount to a population comparable to what we often consider as small rocky worlds in a system. This becomes

Table 3.2. Selective data table for rocky/terrestrial planets in the solar system. Data taken from NASA's NSSDC databases

Property	Mercury	Venus	Earth	Mars
Mass ($M_{\oplus}$)	0.055	0.82	1.0	0.11
Total mass of atmosphere (kg)	$< 10^3$	5×10^{20}	5×10^{18}	3×10^{16}
Mean density (g cm^{-3})	5.4	5.3	5.5	3.9
Number of satellites	0	0	1	2

relevant in the broader astrobiological context of harbors for life (Chapter 10).

Unlike giant planets, there appears to really be only one major pathway to forming rocky planets on the scale of $\sim 1M_{\oplus}$. This involves the accumulation, or coagulation, of solids from dust-sized particles upwards into large bodies over periods of some 10^6–10^8 years. The same basic mechanism must apply to the formation of giant planet cores in the core accretion model. We investigate some simple descriptive models for this process in the following sections.

3.3   Coagulation of Solids

Our assumption in what follows is that the proto-planetary disk consists of gas (99% by mass) and dust/solids (1% by mass). This estimate is based on observations of typical proto-planetary systems. Such disks are initially about a *third* the mass of the final, central, star and extend to at least 100 AU in radius. In our own Solar System the mass of all solid bodies is only some $\sim 0.01M_{\odot}$. Clearly then, there has been very significant redistribution and dispersal of a major part of the initial proto-planetary disk. We will deal with this issue in some more detail in §3.7.1 and in later chapters, since it is of tremendous importance in understanding the chronology of planet formation.

We will initially focus on what may happen to the solids in this structure. In reality, especially in the very earliest stages of a protoplanetary system, the interplay between the gas and solids is extremely important. This is true for both dynamics and the chemical and thermal processing of material (Chapters 7 and 8). It also creates the potential for great complexity, and so we will begin by simplifying the problem and considering just the basic mechanics of getting the solids to stick together and make larger objects.

Perhaps the *most* uncertain part of this entire process is how microscopic dust (Chapter 7) initially coagulates into small solid bodies. This is a tricky problem. We shall, for now, content ourselves by assuming the following picture: Although microscopic dust must be strongly coupled to the gas component of a disk through molecular interactions (c.f. Brownian motion), *if* dust grains are able to stick together—or grow by gathering molecules from the gas (Chapter 7)—they can form larger objects that feel the gas pressure less severely. As the grains grow they are more detached from the vertical, pressure-supported distribution of gas, and can settle, or **sediment** out, towards the central plane of the disk (c.f. Chapter 2). Thus, the initial density of *solids* will tend to be naturally concentrated towards the disk vertical midplane. The agglomeration of grains will continue, and from current theories appears to take place quite rapidly. This leads to a vast population of objects of cm to meter sizes. Once these have begun to merge we reach the stage that is commonly considered as the primary place from which to seriously consider planet formation. This is where small bodies on kilometer scales (which we have termed **planetesimals**) collide and coalesce into larger objects.

A stunning example of what at least one type of planetesimal may be like is the asteroid **Itokawa**. This object was the subject of intense study by the Japanese space probe Hayabusa. Rather than being a dense solid, Itokawa appears to be a **rubble pile** (Fujiwara et al. 2006, Miyamoto 2007) with a density of only ~ 1.9 g cm^{-3}. It consists of material ranging from large boulder-like rocks to millimeter size granules, held together rather loosely by its self-gravity. It also shows signs of having been assembled from some pieces that were in an early, perhaps larger object that was broken up during a collision. In Figure 3.1 a picture taken by Hayabusa clearly indicates this lumpy, loose composition.

3.3.1 Gravitational Focussing and Growth

We begin this discussion by considering the collision, and merger, of spherical objects. We assume that there is a population of such objects—planetesimals—within the disk and that their space density is high enough that to first order we can just consider a small part of the disk in a simple, linear, geometry. In the most basic situation, for a low-mass object, then we define the **geometric** collisional cross section of that object as just the projected area: πR^2 (where R is the mean radius). As the object moves along at some velocity v it will collide with anything that comes into the volume swept out by this area. In this picture we will assume that collision equals coalescence with 100% efficiency. Many things are missing from this model, including the possibility that planetesimals will shatter, or fragment, upon collision, but it is the simplest tractable situation. To further simplify things we consider a primary

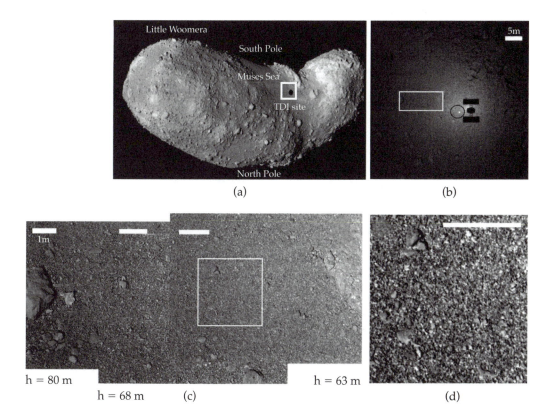

Figure 3.1. A series of images of the asteroid Itokawa taken by the Hayabusa ("falcon") spacecraft at different altitudes. The locations of images B, C, and D are indicated by the white rectangles in the prior images. Image A shows the full size of Itokawa, some 550 m long and 224–298 m across. Its "bent potato" shape is a consequence of its formation from smaller bodies that have not settled into a spherical configuration. Image B was taken on approach—the shadow of Hayabusa is clearly visible near the center. The dark circle indicates the landing site. Image C is a composite of images taken at altitudes of 80, 68, and 63 m. The rubble and pebble-like composition of the asteroid is clearly seen. Finally, Image D is a blow-up showing the granular nature of the surface in detail. Different shades indicate varying surface colors and compositions. (From Yano et al., SCIENCE 312:1350–1353 (2006). Reprinted with permission from AAAS.)

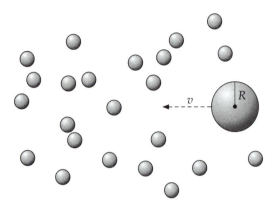

Figure 3.2. Schematic of a planetary embryo moving through part of a proto-planetary disk populated by planetesimals, with a mean relative velocity v. The cross section of interaction when the embryo is small (< 10 km) is just its geometrical projected area: πR^2.

object which we label as a **planetary embryo**, which is typically larger than the general population of planetesimals. The embryo sweeps up and accumulates the smaller bodies within the system (Figure 3.2).

As an object (embryo) grows more massive, its gravitational influence becomes important (for planetary material this occurs roughly when the object is ~ 10 km in size). This gives rise to an *enhanced* cross-sectional area due to **gravitational focussing**. In other words, the gravitational pull of the object allows it to grab planetesimals from beyond its geometric cross section.

Here we give a simple derivation of the approximate form of this gravitational enhancement. Consider the situation illustrated in Figure 3.3. Initially two masses are at large separation. For the sake of this example we treat one object (the planetary embryo of mass M_e, radius R_e) as being at rest in our coordinate reference frame. The other object (a planetesimal, mass m) has an initial, linear velocity v_0 towards M_e, with an offset, or **impact parameter** of b. We will assume that the embryo (M_e) is significantly larger than the planetestimal (m). This is purely for

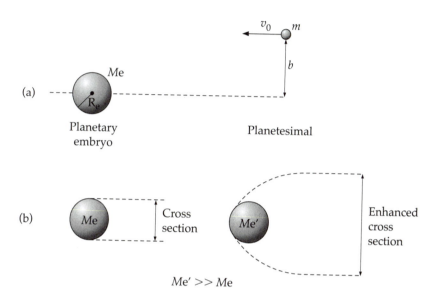

Figure 3.3. An illustration of the model situation used to evaluate the gravitational focussing factor. (a) A planetesimal (mass m) is initially at a large distance from a planetary embryo (mass M_e, radius R_e) and is moving with velocity v_0 towards the embryo, but offset by an impact parameter b. (b, left) At low embryo masses the impact parameter b for collision will lie within the cross-sectional area of the embryo. (b, right) At high embryo masses the cross section is enhanced due to gravity and for collisons b can now lie within a larger region.

convenience in the derivation, we could equally leave things in a more general form.

The initial total energy of the system is just the kinetic energy of mass m, namely $\frac{1}{2}mv_0^2$. Assuming that energy is conserved, then when mass m makes its closest approach to M_e at some distance r_{min}, the total energy will be

$$\frac{1}{2}mv_0^2 = \frac{1}{2}mv_{max}^2 - \frac{GM_em}{r_{min}}, \qquad (3.1)$$

where v_{max} is the velocity of m at closest approach. We also assume that angular momentum will be conserved. Initially the total system angular momentum is $L = mbv_0$ (since $L = m\omega r^2$ and $v = \omega r$ in general). If the angular momentum at closest approach is $mr_{min}v_{max}$ then clearly:

$$v_{max} = \frac{bv_0}{r_{min}}. \tag{3.2}$$

Substituting for v_{max} in Equation 3.1 we can rearrange to obtain

$$b^2 = r_{min}^2 \left(1 + \frac{2GM_e}{v_0^2 r_{min}}\right). \tag{3.3}$$

The factor on the right-hand side multiplying r_{min}^2 is the gravitational enhancement or **gravitational focussing** factor for the collisional cross section. If M_e is small then $b = r_{min}$, and we recover the geometric cross-section for collision. If however M_e is sufficiently large then the *effective* impact parameter b is enhanced.

In the case of an actual collision then it must be true that $r_{min} = R_e$ (assuming the relative size of mass m is small, and ignoring more complex situations such as orbital capture). We then recognize that since $\left(\frac{2GM_e}{R_e}\right)^{1/2} = v_\infty$ (where v_∞ is the surface escape velocity of mass M_e), the final cross-sectional area for collision must be

$$\pi R_e^2 \left(1 + \left(\frac{v_\infty}{v_0}\right)^2\right), \tag{3.4}$$

since, in the absence of gravity, $b = R_e$ must be true for collisions. In applying this to planetesimal capture by a planetary embryo we can replace v_0 with the mean mutual velocity v and v_∞ with the mean mutual escape velocity at the point of contact.

For the computations below we expand our expression back out and define the gravitational focussing factor as

$$F_g = \left(1 + \frac{2GM_e}{R_e v^2}\right), \tag{3.5}$$

where M_e and R_e are now the embryo mass and radius respectively.

As the embryo sweeps through a swarm of smaller planetesimals of *mass space density* ρ_s (g cm^{-3}) then the **feeding rate**, or rate of change of embryo mass, can be defined as

$$\frac{dM_e}{dt} = \pi R_e^2 \left(1 + \frac{2GM_e}{R_e v^2}\right) \rho_s v. \tag{3.6}$$

This immediately raises an interesting question. At first glance it would appear that, according to Equation 3.6, any massive embryo will actually find itself growing at an *ever increasing* rate dM_e/dt. Let us look at this a little closer. First, we again consider the situation of *weak* focussing, that is, $F_g \sim 1$.

In this case $dM_e/dt \propto R_e^2$. However, it is of course true (assuming sphericity) that $M_e \propto \frac{4}{3}\pi R_e^3 \rho_e$, where ρ_e is the mean density of the embryo material. Thus $R_e \propto M_e^{1/3}$ and

$$\frac{dM_e}{dt} \propto M_e^{2/3}, \tag{3.7}$$

and the *fractional rate of change* in mass is then just

$$\frac{1}{M_e}\frac{dM_e}{dt} \propto M_e^{-1/3}. \tag{3.8}$$

Thus, for weak focussing the *relative* growth of the embryo does in fact *slow* as it grows larger.

However, let us now consider full-blown *strong* focussing, $F_g \sim (2GM_e)/(R_e v^2)$. In this case $dM_e/dt \propto M_e^{1/3}M_e$ and

$$\frac{1}{M_e}\frac{dM_e}{dt} \propto M_e^{1/3}. \tag{3.9}$$

In other words, for strong focussing the relative growth of the embryo *increases* with time—it experiences **runaway growth**! In fact, in this case a single embryo will rapidly outgrow all other bodies by gobbling up material—if there is no restriction on the planetesimals available to it, which we will discuss later.

3.3.2 Size Distributions

Before dealing further with this progression of planetary embryo growth we need to ensure that our mental picture of the situation at hand is correct. It is not realistic to assume that the system is as simple as a single embryo growing by sweeping up identical planetesimals. Planetesimals, we assume, have a distribution of masses (we will further consider this below); furthermore, the planetesimals themselves will be growing by sweeping up other planetesimals. In fact this leads to an interesting observation. Let us suppose that the density ρ_e is in fact the material density of *all* objects, planetesimals and embryos, in the system. We can then further rewrite the growth rate expressions in terms of the rate of change of size, or physical *radius* of an object. Since $dM_e/dt = dR_e/dt \cdot dM/dR_e$ and $dM/dR_e = 4\pi R_e^2 \rho_e$ then

$$\frac{dM_e}{dt} = \frac{dR_e}{dt} \cdot 4\pi R_e^2 \rho_e \qquad (3.10)$$

Equating this expression with the weak focussing rate from Equation 3.6 above we obtain

$$\frac{dR_e}{dt} = \frac{\rho_s}{4\rho_e} v \qquad (3.11)$$

which implies that the rate of change of radius of *all* objects, *regardless* of size, is the same! This tells us something about the way in which the distribution of object sizes must change with time (at least in the weak focussing regime). For example, during the period that a 1-km radius planetesimal grows to 101 km, a 10-km embryo will grow to 110 km. Initially these objects differed in size by a factor of ten, after growth they differ by only ten percent. Thus, smaller objects effectively *catch up* with larger bodies and the distribution of object sizes skews towards larger objects. In fact the same is true for the mass distribution of objects, however, the reason for this is a little more complicated, and requires an explicit consideration of the *population* of objects. Below we illustrate this in more detail.

3.3.3 Coagulation Equations

The full solution to determining how many objects there are of a given mass as a function of time in our simple model of planet growth is clearly complex. Ultimately it requires the full computer simulation of a system. However, an intermediate step can be made that helps further illustrate the situation via use of a "semi-analytic" description of the system of bodies. We outline such an approach here. The idea is that, as small things merge, then the population statistics of planetesimals change— fewer small things, more large things. This can be made explicit in terms of a **Coagulation Equation**. It is assumed that there is a minimum mass object, or building block, in the system: m_1. All object masses are then integral multiples of this mass. Then, if there are n_k bodies of mass $m_k = km_1$, the rate of change of n_k can be written as

$$\frac{dn_k}{dt} = \frac{1}{2} \sum_{i+j=k} A_{ij} n_i n_j - n_k \sum_{i=1} A_{ki} n_i, \qquad (3.12)$$

where A_{ij} is the collisional probability, or rate coefficient, of merger between planetesimals of mass m_i and m_j within some fixed volume V. So, the first right-hand-side term describes the *increase* in n_k due to the collision and merger of smaller bodies and the second term describes the *decrease* due to incorporation into larger bodies. As before, we do not include the possibility of disruption or fragmentation during collisions; however, it should be clear that this can be incorporated by adding terms to Equation 3.12.

It turns out that there is at least one analytic solution to this equation. The simplest is when $A_{ij} = \alpha$ (a constant), which will certainly hold in early stages where all objects are of similar size. In this case, if n_0 is the initial number of bodies of mass m_1, then the number of bodies with mass $m_k = km_1$ at time t is

$$n_k = n_0 f^2 (1 - f)^{k-1}$$

where

$$f = \frac{1}{1 + \eta/2}$$

with $\eta = \alpha n_0 t$, and f is the fraction of bodies of mass m_1 which have not yet had a collision at time t (e.g. Inaba et al. 2001., and the excellent Armitage 2007). We can use this solution to simply investigate what happens to the *distribution* of the number of planetesimals of a given mass, n_k, as a function of time. In Figure 3.4 the number of objects n_k is plotted versus time for an arbitrary range of k ($k = 1, 10, 100, 1000$). All units are relative and can be ignored. Clearly there are initially no objects

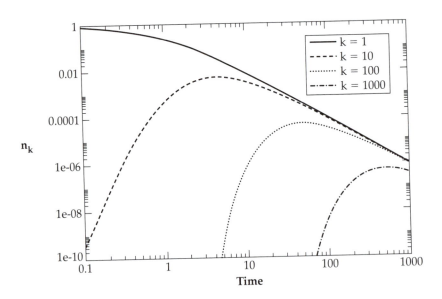

Figure 3.4. The evolution of the distribution of planetesimal masses as a function of time is plotted according to the simplified coagulation equation. The number of objects n_k (in arbitrary units) is plotted against time (in arbitrary units) for a range of relative masses—denoted by k where mass is actually km_1, and m_1 is a unit mass. Initially there are no objects with mass greater than m_1. As time goes by the number of smaller masses decreases, while the number of larger masses increases—although all mass scales peak in number at certain periods.

with $k > 1$ (i.e. $m > m_1$). As time goes by the number of the smallest $k = 1$ objects declines while the number of larger objects grows. In fact, in a similar way to the phenomena discussed in §3.3.2 above, we can also see that over time the total number of objects of different masses equalizes. In Figure 3.5 we show this more explicitly by plotting the *ratio* of different n_k's (i.e. n_k'/n_k) as a function of time. All curves tend asymptotically to unity. Thus, while the number of *high* mass (high k) objects continues to grow with time, the actual "mass function" tends to flatten out.

Obviously this is a highly oversimplified model. In general it will not be true that $A_{ij} = \alpha$, rather α will be a function of k, and fragmentation will further modify things. Nonetheless, the model does give an indica-

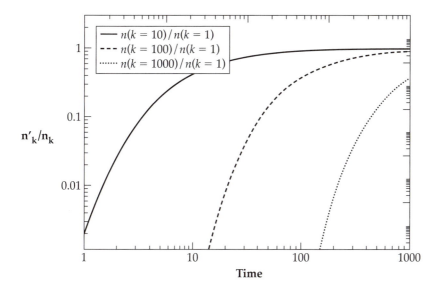

Figure 3.5. The ratio of the number of planetesimals at different mass scales (Figure 3.4) is plotted as a function of time, according to the coagulation equation. All curves asymptote to unity with increasing time, hence the population of planetesimals tends to even out both in number of a given mass, and in physical radius (Equation 3.11). We note that this ignores effects such as fragmentation.

tion of the general way in which a population will skew towards larger masses with time. This is important for seeing how the entire system converges on fewer, but larger, objects as time goes by.

3.4 Stages of Growth

The ideas sketched out in the previous section lead to a basic model for embryonic planet growth that is central to the core accretion model for giant planets, as well as the formation of terrestrial type worlds. In this model there are three distinct stages that an embryo may (or may not) pass through. We have already considered the first two.

(1) **Orderly growth**: Weak gravitational focussing. The relative growth of an embryo slows with time, and all objects tend to even out in size as smaller ones catch up with larger ones.

(2) **Runaway growth**: Strong gravitational focussing. The growth rate accelerates for massive embryos for which $F_g > 1$—typically leading to a single embryo dominating a particular region of the protoplanetary disk.

The third possible stage includes more complex mechanisms:

(3) **Oligarchic growth**: An embryo is massive enough to actually begin to influence the relative velocities of the planetesimal population, well beyond even its gravitationally enhanced cross section of interaction (see below). Embryo growth actually *slows* compared to (2), but still exceeds that of (1) (since v is enhanced).

Oligarchic growth really comes in two flavors: (a) **Shear-dominated**, and (b) **Dispersion dominated**. In both cases, large bodies (embryos) are dynamically "heated" by interactions with each other, and "cooled" by dynamical friction from small bodies. In case (a) the random velocity dispersion of the small bodies is *less* than some factor times the escape velocity from the surface of the large bodies. In case (b) the random velocity dispersion of the small bodies is *greater* than this escape velocity. One can then see that in case (a) the large body may create "bulk" velocity fields in the small bodies, while in case (b) the small bodies tend to be scattered by the large bodies. A system such as this can, in principle, eventually attain dynamical equilibrium by equipartition of energy between all bodies. We discuss this in §3.4.1 below.

The full solution to a given case of embryo growth will clearly depend on the initial conditions—specifically the mass distribution of planetesimals, their physical properties (such as density) and the greater environment (for example whether or not gas is present). Computer simulation is the only complete way to examine possible growth scenarios. In Figure 3.6 we sketch out an example what the outcome may be. In this case we again see that the distribution of object masses in a system shifts increasingly towards high masses as objects merge. Occasionally a single body, or embryo, experiences runaway growth and can rapidly increase in mass by factors of 100 above the rest of the population. This offers the hope that giant planet cores might form rapidly after all, but we need to consider further issues before considering this in detail.

3.4.1 Velocities and Equipartition

As large and small bodies jostle and collide, their individual velocities will be altered. Although we will not go into any great depth on this

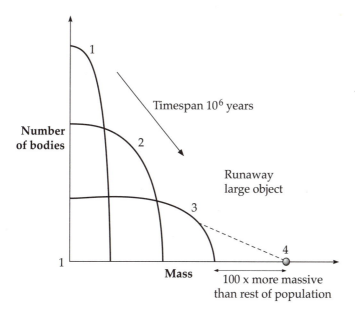

Figure 3.6. A sketch of the results of a typical numerical simulation of planetesimal merger and embryo growth. The number of bodies is plotted as a function of mass for three timesteps spanning approximately 10^6 years. By the end of the simulation one object has experienced runaway growth, and has become a factor 100 more massive than the next most massive objects.

issue it is nonetheless helpful to investigate it a little in order to develop our mental picture of solid body coagulation and dynamical evolution. We appeal here to a simple description by Goldreich, Lithwick and Sari (2004), also sketched out in Figure 3.7. We assume that the mean random velocity of a planetesimal (mass m) is u, for planetary embryos (mass M_e) the mean random velocity is v. For a head-on *elastic* (i.e. energy is conserved) collision the net relative velocity between the planetesimal and embryo is therefore just $u + v$. To first order the larger embryo will *lose* momentum by an amount $\sim m(u + v)$ due to the collision. The alternative is that the planetesimal velocity u is greater than v and so it can catch up to the embryo and collide "tail-on" (Figure 3.7). In this case the embryo will *gain* momentum by $\sim m(u - v)$.

Head-on collision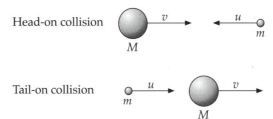

Tail-on collision

Figure 3.7. A schematic of the two collision situations considered between a planetary embryo and a planetesimal. First, in a head-on (elastic) collision, the embryo will to first order lose momentum $u \sim m(u + v)$. Second, in a catch-up, or tail-on (elastic) collision the embryo will to first order gain momentum $\sim m(u - v)$.

Thus, in a weak focussing regime, the rates of head-on and tail-on collisions will simply be $\sim n_s R^2(u + v)$ and $\sim n_s R^2(u - v)$ respectively, where n_s is the number density of planetesimals (number per unit volume) and R is the embryo radius (c.f. §3.3.1). Therefore the rate of momentum loss or gain is simply

$$M_e \left(\frac{dv}{dt} \right)_{\text{head-on}} = -n_s R^2 m (u + v)^2, \tag{3.13}$$

and

$$M_e \left(\frac{dv}{dt} \right)_{\text{tail-on}} = +n_s R^2 m (u - v)^2, \tag{3.14}$$

in either case. Because we are treating this as a lowest order approximation we can further say that the *net* rate of momentum change is

$$M_e \left(\frac{dv}{dt} \right)_{\text{net}} = M_e \left(\frac{dv}{dt} \right)_{\text{head-on}} + M_e \left(\frac{dv}{dt} \right)_{\text{tail-on}}. \tag{3.15}$$

If we perform this summation using Equations 3.13 and 3.14 and drop higher powers of u or v as well as ignore numerical factors in the approximations, we arrive at the *fractional* rate of change of embryo velocity:

$$\frac{1}{v}\left(\frac{dv}{dt}\right)_{net} \sim -n_s R^2 u \frac{m}{M_e}. \tag{3.16}$$

This is very interesting. If the mean velocities of small and large bodies are comparable then Equation 3.16 is telling us that the large bodies, or planetary embryos, will tend to be slowed, or **dynamically cooled**, by collisions with planetesimals. At the same time, if energy is conserved in the system, the planetesimals must be **dynamically heated**. Equation 3.16 also tells us that if n_s, u, and m are constant then, to first order, the embryo will be brought to a standstill if it collides with approximately its own mass (M_e) in planetesimals. If however the embryo is initially moving slowly, then the planetesimals will tend to impart momentum and speed it up. Clearly then there must be some intermediate dynamic equilibrium for the system, between that of stationary large bodies and significant velocity large bodies. It can be shown (using random walks, Chapter 9) that this occurs when the kinetic energy of the large bodies is equal to that of the small bodies—in other words when **energy equipartition** is achieved between the populations.

In general, for proto-planetary systems, larger bodies have greater energies than the planetesimals, and so dynamical *cooling* is most likely to be occurring for planetary embryos. For planetesimals the situation is a little more complex, and in fact it appears that while dynamical heating or cooling by embryos can occur, the dominant effect on mean velocities is actually from **viscous stirring**. Stirring occurs because in a Keplerian system large bodies can increase the random velocities of smaller bodies (Goldreich, Lithwick, and Sari 2004). It is here that we can see the connection with the oligarchic growth regime described in §3.4 above. In essence, a large body resulting from runaway growth will begin to dynamically heat the planetesimals within its orbital zone via viscous stirring. From Equation 3.6 we can see that in the strong focussing regime this is going to actually act to slow growth to a more orderly rate as the mean mutual velocity (v) increases.

Of course, there may also be several large embryos, or oligarchs, competing with each other. It seems likely however that, given limited feeding grounds, one or more may emerge as dominant over time in any given region of the disk.

3.4.2   The Endpoints of Growth

Clearly the growth of a planetary embryo must end at some point. Precisely when, and at what mass this occurs, is critical for understanding the success of a given model. The most obvious endpoint will be when the surrounding raw material—the planetesimals—runs out for a given object. The **feeding zone** of an embryo is initially just defined by the region swept out by its effective cross section. This is illustrated in Figure 3.8 for a situation where the feeding zone has grown to exceed the simple geometric cross section. At any stage the maximum possible "zone of influence" of an embryo is related to the distance from the embryo outside of which orbits about it are unstable due to the tidal effects of the parent star—which we have thus far ignored. This is known as the **Hill Sphere Radius** and is really just a special case of the restricted gravitational 3-body problem, representing the gravitational tug-of-war between (for example) a planet and its parent star in a rotating frame. The Hill Sphere Radius is given by (see also Chapter 10):

$$R_H = a_e \left(\frac{M_e}{3M_*} \right)^{1/3} , \tag{3.17}$$

where a_e is the orbital radius of the embryo and M_* is the stellar mass (the denominator is strictly speaking $3(M_e + M_*)$, but since $M_* \gg M_e$ we simplify to $3M_*$).

Armed with this definition we can make a crude, first-order argument about why runaway, or oligarchic, growth slows or stops. As described in §3.3.1, the effective *requirement* of runaway growth is that

$$\frac{dM_e}{dt} \propto M_e^{4/3}. \tag{3.18}$$

Now, the Hill Sphere Radius obeys $R_H \propto M_e^{1/3}$ and thus $dM_e/dt \propto R_H^4$ is *required* in order to maintain runaway growth. But clearly the *available* planetesimals can increase only as fast as the total possible cross-section: πR_H^2 (assuming that the disk vertical thickness is larger than R_H).

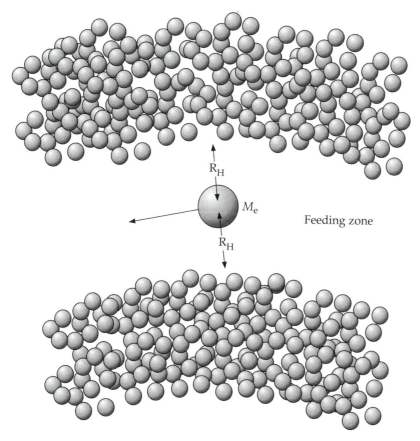

Figure 3.8. An illustration of what is meant by the "feeding zone" of a planetary embryo. To first order an object may be able to capture and merge with material within a Hill radius (R_H). Thus, it will tend to sweep out an annulus in the disk of planetesimals as shown here. Material must enter, or be perturbed, into this zone to stand a chance of being subsumed into the growing proto-planet. More detailed simulation suggests that the actual, effective, size of the feeding zone is more like $\sim 8R_H$ in total width—due to stirring of the planetesimal population by the proto-planet.

Thus the requirement for runaway growth is not met, even by using the maximum possible cross section of the Hill Radius—so in effect, runaway growth occurs in the regime where the gravitationally enhanced cross-section is *between* πR_e^2 and πR_H^2. Oligarchic growth begins to kick in as the velocity field of the planetesimals begins to become appreciably disturbed beyond R_H (recall that R_H defines the sphere of potentially *stable* orbits, the embryo can certainly exert influence beyond this). One can then see that planetesimals will be perturbed so that their orbits cross into the region defined by R_H and growth will continue, albeit at a reduced rate. In fact, as we will describe below, more thorough numerical simulations indicate that the effective feeding zone for a planetary embryo is indeed somewhat larger than $2R_H$.

Further complications to the question of when embryo growth through coagulation ends (and there are many) include the fact that the planetesimal population is itself presumably changing (as described in §3.3.3). The consequence of ongoing planetesimal clumping or growth is that the *space density* of the entire system will decrease with time. Thus accretion onto a given embryo will become more sporadic and random (i.e., **stochastic**).

As we begin to consider these later stages of planet formation, there are some additional important facts that we can now recognize. The analytic form of R_H is actually telling us a little more about the nature of planet formation. Since $R_H \propto a_e$ we immediately see that (assuming embryos stay in their initial orbits, which is something we will explore a little later) the final mass of an embryo must *increase* with distance from the parent star—*if* the planetesimal population is more or less constant with orbital semi-major axis a_e.

We can also make a crude estimate of this final embryo mass—often called the **isolation mass** since it is when the embryo is truly isolated from all other objects. If we assume that the disk is relatively thin, then the distribution of mass within it is conveniently given by the **surface mass density**: $\sigma(r)$ (simply the vertically integrated mass: g cm^{-2}). Numerical simulations lead to a feeding zone of size $B R_H$ from the embryo—where $B \sim 4$. Thus, if $B R_H$ is still small compared to the orbital distance a_e we can see that the *total* available mass, or **isolation mass**, is

$$M_e^{iso} = 2\pi a_e \times 2B R_H \times \sigma. \qquad (3.19)$$

Thus, after a little re-arrangement:

$$M_e^{iso} = \frac{(4\pi B a_e^2 \sigma)^{3/2}}{(3M_*)^{1/2}}. \tag{3.20}$$

This yields, for a forming embryo at 1 AU, around a 1 $M_\odot$ star, with a mean planetesimal $\sigma = 3$ g cm^{-2} an isolation mass $M_e^{iso} \simeq 0.013 M_\oplus$. At 4 AU this increases to approximately 1 $M_\oplus$ if σ is constant.

Within a range of reasonable σ, this suggests that for Earth-type planets the final processes of mass accretion are *not* part of the regimes described above, but rather must involve the stochastic merger of ~ 100 larger (lunar mass) objects. Furthermore, at late stages (generally considered to be 10–100 Myr), terrestrial sized embryos/planets can also be *scattered* within a system due to impacts and close encounters with other large bodies. Scattering can therefore mix up any elemental/chemical segregation that was present in the original proto-planetary disk (§3.7.2).

3.5 The Rate of Formation as a Function of Position

In order to fully address the questions that we brought up earlier concerning the rates at which giant planets form—and competing models for their formation—we need to delve a little deeper into the factors that govern the availability of planetesimal material to a growing embryo. This will also have implications for the processing of material in the proto-planetary disk, as well as the composition of objects.

3.5.1 The Minimum Mass Solar Nebula

There is a fundamental uncertainty in our discussion of planet formation thus far. We have not considered the true mass distribution, or composition, of a generic proto-planetary disk. Observations of proto-planetary systems are beginning to yield some insight (Chapters 2, 4 and 8), but for our own Solar System we have no direct record of the proto-planetary structure. In the absence of this information we need to appeal to a simplified, but justifiable model. This is an empirically motivated model for the proto-planetary disk known as the **Minimum Mass Solar Nebula (MMSN)**. The basic idea is straightforward. We take the known distribution of *mass* in our Solar System (i.e. the planets and the asteroid belt) and "smear" it out, or smooth it mathematically, to construct a continuous, azimuthally symmetric, surface density function ($\sigma(r)$). We then *add in* enough hydrogen and helium to obtain a **solar element abundance** in this disk. The logic here is simply that the planets represent the material retained in the inner solar system following the heating up and eventual ignition of the proto-sun. This material is therefore biased towards metals, since lighter elements will be lost. The Sun, however, in at least its outer atmosphere, must still exhibit the protostellar mix of elements (§3.2.1). Adding in the primordial hydrogen and helium is the obvious thing to do to arrive at an estimate of the initial mass of the proto-planetary disk. Thus, smearing the planets out yields an empirical fit to the *form* of σ_{MMSN}, and adding in the additional mass gets the *normalization* of σ_{MMSN} (e.g. Hayashi 1981).

Specifically, for the Solar System, we arrive at:

$$\sigma_{\mathrm{MMSN}}(r) = \sigma_0 \left(\frac{r}{r_0}\right)^{-n}, \tag{3.21}$$

where σ_0 and r_0 are empirically fit, and $n \simeq 3/2$. Typical fits yield

$$\sigma_{\mathrm{MMSN}}(r) = 1700 \left(\frac{r}{1\,\mathrm{AU}}\right)^{-3/2} \mathrm{g\ cm^{-2}}. \tag{3.22}$$

This form of the MMSN corresponds to a total metal content of approximately $60 M_\oplus$ and a 100-to-1 gas to solids ratio, in other words approximately $6000 M_\oplus$ in gas (corresponding to $\sim 0.018 M_\odot$).

Recalling the growth rate $dM_e/dt \propto \rho_s \pi R_e^2 F_g v$: the mass density ρ_s can be related to the *surface density* of material in a protoplanetary disk—where we now include both gas and solid material. In other words, since the surface density (σ) is just the vertically integrated mass per unit area;

$$\rho_s \propto \frac{\sigma \omega}{v}, \tag{3.23}$$

where ω is the disk angular velocity at a given radius. We will assume circular Keplerian motion for the disk, namely: $\omega = \left(\frac{GM}{r^3}\right)^{1/2}$. Thus $dM_e/dt \propto r^{-3}$. Not surprisingly then, the rate of growth is a decreasing function of radius in the MMSN—although the final isolation mass may still increase with radius.

On the face of it this exacerbates the problem for the core accretion model of giant planet formation, since the cores must form quickly ($< 10^7$ yrs), before the disk gas is gone. In order to address this we need to consider the proto-planetary disk in even more detail.

3.5.2   The Temperature Structure of the Proto-planetary Disk

A proto-planetary disk is a complex structure (see below and Chapter 8), nonetheless, some major characteristics are set by relatively simple physics. The temperature structure will be determined largely by composition (gas versus solids), hydrostatic equilibrium of the gas, and the stellar irradiation.

Here we sketch out a first-order approach to finding the radial temperature structure of gas in a proto-planetary disk—this is very basic, and ignores a host of things, but it does give an idea of how to begin to tackle the problem. In Chapter 8 we will revisit disk temperature structure in much more detail.

The disk at this stage is a mix of gas and dust—dominated by the gas component. It also has a finite thickness—something that we have not emphasized until now. If σ is not too large (e.g. if the disk mass is less than some $0.1 M_*$) then we can neglect the self-gravity of the disk compared with the gravitational influence of the central proto-star. Thus, the gravitational force *perpendicular* to the plane of the disk is approximately *just* the vertical component of the proto-star's gravitational field (Figure 3.9).

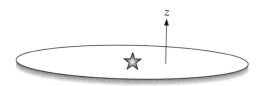

Figure 3.9. The simple geometry used to evaluate the vertical (z axis) hydrostatic equilibrium of gas in the proto-planetary disk. If the total disk mass is less than about $0.1 M_*$ then the vertical gravitational acceleration is dominated by the central object (proto-star):

This can be written in terms of the vertical gravitational acceleration g_z:

$$g_z = \frac{GM_*}{r^2}\left(\frac{z}{r}\right) = \omega^2 z, \tag{3.24}$$

where z is the vertical distance from the plane of the disk. If we further ignore any *vertical* variation in the gas temperature at any given radius then we can apply hydrostatic equilibrium (Chapter 2) to determine the gas pressure at the disk midplane at any radius. Consider the vertical hydrostatic equilibrium:

$$\frac{dP}{dz} = -\rho g_z. \tag{3.25}$$

Now, to first order the pressure gradient is just

$$\frac{dP}{dz} \sim \frac{P_{surface} - P_{center}}{z_{surface} - 0} \sim \frac{-P_{center}}{z_{surface}}. \tag{3.26}$$

Thus, $P_{center} \sim z_{surface}\rho g_z$. For an ideal gas $P = \rho kT$ and so $kT_{center} = z_{surface}g_z$, and since $g_z = \omega^2 z$ we see that

$$T_{center} \propto r^{-3/2}. \tag{3.27}$$

In other words, the vertical hydrostatic equilibrium of the gas demands that the midplane gas temperature *decreases* with increasing radius. It is worth noting that the precise temperature structure of a gaseous proto-planetary disk is still a matter of active investigation. Nonetheless, more sophisticated modeling than the simple example above implies that (for our own MMSN), at a distance somewhat greater than that of Mars (~ 3 AU), the disk temperature drops below 170 K. This is the laboratory measured **water-ice sublimation point** in a vacuum (in reality, for real bodies, such as small grains this temperature can be lower). In other words, beyond this distance in the disk, water can condense out of the gaseous phase into a solid. Not surprisingly, this distance has come to be known as the proto-planetary **snow-line**.

The precise location of the snow-line is uncertain for our own system. Indeed, models of proto-planetary disks that include the effects of larger amounts of dust—that can modify the thermal characteristics of the disk—suggest that the snow-line could have been at much lower radii, even at 1 AU. The ultimate test of all such models will be the comparison with actual temperature and water-phase observations of real proto-planetary disks. We discuss this in more detail in Chapter 8.

3.5.3 Implications of the Snow-Line for Giant Planets

In considering solid body coagulation we have assumed that the surface mass density of the disk in solids is either a constant or decreasing function of radius. We have also ignored the contemporaneous presence of gas, since we have assumed that until a very massive (e.g. $10M_\oplus$) embryo forms, the gas simply does not accumulate much around the solid bodies. However, we now see that beyond a certain radius in the disk, water will start to condense out as ice (other volatiles are also present, for now we will focus on water). This means that the surface density of solid material will *increase*. With more thorough modeling we find that this can provide as much as 4–5 times *more* solid material for an embryo to sweep up than would otherwise exist. On the face of it this offers a possible, or at least partial, solution to the core-accretion model's dilemma in making large cores quickly enough—this also places giant planet formation at the larger radii where we find such worlds in our own Solar System today.

3.5.4 Further Implications—Water Delivery

If giant planets form at or beyond the snow-line they may scatter water-bearing material to the inner system. This provides one possible mechanism for delivering water to the surface of terrestrial-type planets at an early epoch. Indeed, it appears that objects that have likely formed in the more distant proto-planetary system at much lower temperatures (e.g. comets) have high D/H ratios compared to those on Earth, due to their low-temperature chemistry (which preferentially sequesters the heavier D atoms). By comparison, objects thought to have formed closer to the giant planets, such as Jupiter, have D/H ratios in closer agreement with those on Earth. Thus, if such objects were indeed scattered inwards to Earth and deposited their water on its surface, this would be one possible explanation for the matching D/H ratios. This is still a controversial subject, limited in part by our still incomplete information on the composition of small Solar System bodies. We will discuss this issue more in Chapters 8 and 9.

3.6   Planetesimal and Proto-planet Migration

A full treatment of the origins and evolution of planetary bodies must include the full dynamics (and more realistically, magnetic fields, ionization structure, and chemical evolution) of the system of the disk and planets. At minimum, and especially if we are to understand giant planet formation, we must consider the dynamics related to the gaseous component of the proto-planetary disk. In the following sections we present a highly simplified overview.

3.6.1   Disk Drag

We have already shown (§3.5.2) that a pressure gradient ($P(r)$) must exist in the disk gas. Recalling hydrostatic equilibrium in a *non-rotating* system, then we have

$$\frac{d^2 r}{dt^2} = \frac{1}{\rho} \frac{dP}{dr} + \frac{GM}{r^2} = 0 \qquad (3.28)$$

In a *rotating* system however there is an additional **centrifugal acceleration**: $\omega^2 r$. Thus

$$\frac{1}{\rho} \frac{dP}{dr} + \frac{GM}{r^2} = \omega^2 r \qquad (3.29)$$

The consequence, if one solves this equation for the orbital motion of the gas, is that the gas rotates *slightly slower* (in general) than pure Keplerian orbital motion. Typically the rotation is about 0.5% slower. This is akin to the gas experiencing a "reduced gravitational force" at any given radius in the disk, and thus $\omega_{gas} < \omega_{solid}$.

A solid body in orbit will therefore encounter a **headwind** as it moves through the gas. It will experience aerodynamic *drag* and will begin to lose energy and spiral inwards. The rate of this depends strongly on object size and velocity (c.f. embryo growth) and leads to a **differential drift** between planetesimals and larger planetary embryos. This effect can actually enhance collision rates by bringing more planetesimals into the embryo's feeding zone. Massive bodies are less perturbed, as are small particles—which are strongly coupled to the gas through molecular forces. However, in the regime of 1-meter-sized bodies the drag can reach a maximum. For example, a 1-meter body at ~ 1 AU embedded in a MMSN proto-planetary disk will drift at $\sim 10^6$ km per year, implying a very short 100-year infall timescale.

At first glance this presents a potential difficulty, since it suggests that planetesimals, and in particular their precursors, will be dragged into the parent proto-star very efficiently and the disk will be cleaned out of raw material. In order for this material to survive, objects need to grow beyond the high-drag 1 meter scale quickly. In fact, it may be the very act of migrating inwards that solves this problem, since this will tend to increase the likelihood of collision and merger, thereby enabling objects to quickly grow to a safe size. It appears that this can happen, but it is by no means clear cut at this time.

3.6.2 Planet Migration

The idea of planet migration had surfaced well before measurements of exoplanet systems existed that supported the idea (e.g. Ward 1986, and see Chapter 4). There are two principal types of planet migration resulting from the interaction of planets with the proto-planetary disk, known as **Type I** and **Type II**, and there are some variations or special cases of both.[1] These forms of migration almost exclusively result in the **inward** motion of the planet. The details are complex and so here we present a very watered down version of affairs (see also Armitage 2007 for useful information).

As a starting point it is useful to again consider the most basic facts about a gas-disk/planet system. As we have already seen, the gas disk will rotate some 0.5% slower than Keplerian at any given radius. For large planets this may not matter too much; however, their gravitational field will exert an influence on disk material and vice versa—making the phenomenon of dynamical **resonances** important.

To give the simplest example. A **mean-motion resonance** occurs in an orbital system when the orbital angular velocity of one object is an integer factor times that of another. For example, in a 2:1 resonance for two bodies orbiting a central mass, one body will complete 2 orbital

---

1. For example, runaway, or Type III migration.

revolutions in precisely the time it takes the other body to complete 1. As a consequence the two bodies will continue to find themselves at precisely the same relative location again and again. Much as with resonance in simple systems like the classic harmonic oscillator or forced pendulum—small perturbations, such as those due to the mutual attraction of the two bodies to each other, can grow. A great example of a mean-motion resonance is that of the innermost Galilean moons. Io, Europa, and Ganymede are in a 1:2:4 resonance (known as a Laplacian resonance), which is responsible for maintaining their non-circular orbits (Chapter 10).

In the case of a planet-disk system we can see that a similar resonance can exist when $\omega_{gas} = n\omega_{planet}$, or where $\omega_{planet} = n\omega_{gas}$, where n is an integer. In the former case the gas will be interior to the orbit of the planet in a Keplerian system, and in the latter it will be exterior. In fact, for a simple 2:1 resonance ($n = 2$), assuming Keplerian motion for both planet and gas, the inner resonance will occur for a radius $r_{gas} = 0.63\, r_{planet}$ and the outer at $r_{gas} = 1.59\, r_{planet}$. Although this is not a very interesting example it does reveal something of relevance for the following discussion. The inner resonance is much closer radially than the outer. Thus, if the resonances result in amplification of any perturbations to the disk gas or the planet then there are intrinsic asymmetries to the likely influence of any resulting structures.

Type I migration

In Type I migration the planet, or embryo, is generally no more massive than about $10 M_{\oplus}$ and remains fully embedded in the gaseous disk, with no gap between it and the disk gas.

The type of resonance that becomes important in this case is generally a **Lindblad resonance**. In essence this is a resonance where the *difference* in orbital velocity (ω_p) of the planet and the **pattern speed** of the disk (where the pattern speed ω_b is the angular velocity of any density perturbation, e.g. a spiral wave, in the disk) is related in a simple way to the **epicyclic frequency** κ of the system. The epicyclic frequency relates to *radial* perturbations of a mass or parcel of gas. In other words it describes the natural oscillation rate of material in the radial direction. Specifically

$$m(\omega_p - \omega_b) = \pm\kappa, \qquad (3.30)$$

where m is an integer, and

$$\kappa^2 = 4\omega_0^2 + r_0 \frac{\partial \omega^2(r)}{\partial r} \bigg|_{r=r_0}, \qquad (3.31)$$

and r_0 (ω_0) is the radius about which a particle/parcel is perturbed, where angular momentum is conserved in the system. For Keplerian orbital motion $\kappa = \omega_0$. To phrase this differently, there is a natural frequency for radial oscillations in the disk gas. If the orbital frequency of a massive proto-planet is in resonance with this natural frequency then small perturbations in the disk can be strongly amplified. The actual consequence of this becomes much more apparent if one looks at the results of numerical simulations, illustrated in Figures 3.10 and 3.11.

Put simply, the planet creates a spiral density wave, or wake, in the disk due to gravitational torques and the Lindblad resonances. Once set in motion, the spiral wake in turn exerts a torque on the planet. The leading (inner) wake acts to push the planet to larger radii by boosting angular momentum, while the trailing (outer) wake acts to pull the planet inwards by reducing angular momentum. In general, for proto-planetary disks, the torque exerted by the outer Lindblad resonance wake is larger, and thus the planet migrates inwards.

With more work, a migration *timescale* (τ_m) can be estimated for this Type-I migration phenomenon. Specifically,

$$\tau_m \approx \frac{2}{c_1} \left(\frac{M_*}{M_p}\right)^2 \left(\frac{M_p}{\sigma R^2}\right) \left(\frac{H}{R}\right)^3 \omega^{-1}. \qquad (3.32)$$

Here the parameter c_1 is the torque asymmetry between the disk inwards of the proto-planet and outwards, σ is the disk surface mass density, R is an initial, or current, orbital radius of the proto-planet (assuming circular orbits), and H is the scale height (Chapter 9), or characteristic thickness of the disk. Typical values for a $M_p \sim 1 M_\oplus$ proto-planet at 5 AU yield $\tau_m \sim 10^5$ years; for a $10 M_\oplus$ proto-planet also at 5 AU then $\tau_m \sim 10^4$ years. Thus, Type I migration appears to be quite rapid relative to the disk lifetime.

If such migration were highly efficient then it might suggest that many reasonably large proto-planets simply don't survive. They will migrate in through the disk and wind up being accreted by the central

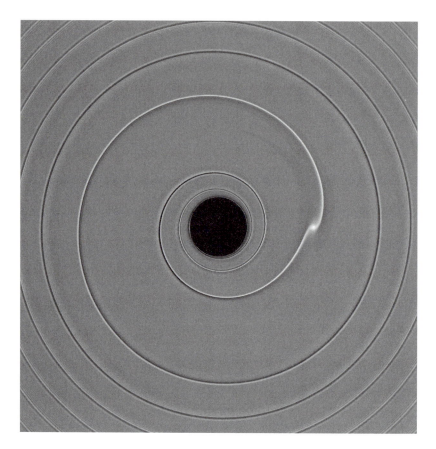

Figure 3.10. An illustration of the results of a two-dimensional computer simulation of a massive (< 1 M$_\oplus$) proto-planet in a gas disk, seen from above. Brighter regions correspond to higher disk surface density (σ). The central stellar object is not shown, and the central disk is shown truncated. The planet at center-right is orbiting clockwise and is exciting a Lindblad resonance with the disk which creates the two bright density enhancements, or spiral wakes. Now that the disk is non-uniform it will exert a non-zero torque on the proto-planet. As we describe, and further illustrate in Figure 3.11, this will result in the orbital decay, or inward migration, of the proto-planet. In this case, where the disk gas is continuous across the orbit of the proto-planet, the migration is known as Type I (figure courtesy of F. S. Masset (CEA), private communication).

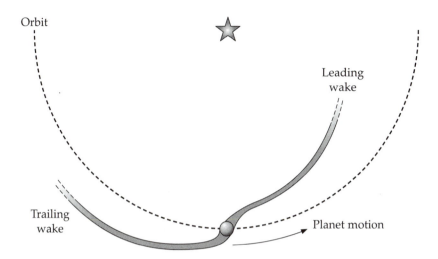

Figure 3.11. In this schematic diagram we further illustrate the phenomenon shown in Figure 3.10. The orbiting proto-planet excites an inner and outer density wakes in the disk gas. The leading inner wake exerts a positive torque on the proto-planet to boost angular momentum. However, the trailing outer wake exerts a negative torque on the proto-planet to reduce angular momentum. In general, for proto-planetary disks, the outer wake exerts a greater torque and hence the planet suffers orbital decay on quite short timescales (Equation 3.32).

proto-star. This may indeed happen, but it is also likely (if not necessary to match observations of exoplanets, Chapter 4) that the migration does not always continue up to that point. Many possible halting mechanisms exist. For example, the disk may simply thin out close to the star (σ becomes small) and migration slows. In very close orbits then tidal bulges (Chapter 9) may be raised on the parent proto-star itself and angular momentum can then be transferred *outwards* to the planet— slowing its migration. Furthermore, such planet–disk resonances may be dampened or disrupted due to the presence of other proto-planets in the system.

The proto-planet is also not going to remain at a fixed mass during all of this. Indeed, if it gains sufficient mass during its migration (which moves the proto-planet through fresh disk fodder for accretion, thereby altering the isolation criterion) then the sheer mass of the proto-planet may alter its interaction with the disk, as we will see below.

Type II migration

For proto-planets, or the forming cores of gas giants, that have masses greater than about $10M_\oplus$, the gravitational torque on the disk due to the proto-planet is sufficiently large to push open a **disk gap** around the proto-planet.

In essence the planet speeds up disk material exterior to its orbit and slows material interior to its orbit, and so an empty zone appears (see Figure 3.12). In this case the nearby resonance structures that occur in Type I migration don't form in the same way, since there is no disk material that close to the proto-planet. Instead, the massive proto-planet acts like a **momentum bridge** in the disk. The situation is illustrated in Figures 3.12 and 3.13. Viscosity in the disk (Chapter 2) tends to *close* the gap in the disk via angular momentum transport, whereas the gravitational torque from the planet tends to open up the gap. A dynamical equilibrium is reached when the effective viscous torque *equals* the planetary torque. Thus, viscous torques from the inner disk are transferred to planetary torques, which are then transferred back to viscous torques in the outer disk.

Consequently, the giant proto-planet is effectively *locked into* the natural viscous evolution of the proto-planetary disk! We already know that as such a disk evolves then mass is transferred *inwards* and angular momentum is transferred *outwards*. Thus, the proto-planet will migrate inwards on the dynamical timescale of the disk evolution—which is *greater* than 10^5 years. Type II migration is therefore somewhat less of a challenge to the survival of proto-planets than Type I migration.

3.6.3 Consequences and Complications of Planetary Migration

One clear consequence of planet migration is that massive, gas giant or ice giant planets can arrive at a final orbit much closer to their parent star than would otherwise be expected from our previous discussions. As we will see in detail in Chapter 4, this indeed appears to be the situation in many exoplanet systems. Giant planets are seen in orbits with periods of only a few days, placing them closer to the parent star than Mercury is to the Sun in our own system. These objects have been labeled **hot**

99 orbits

Figure 3.12. An illustration of the results of a two-dimensional computer simulation of a massive ($> 1M_J$) proto-planet in a gas disk. Brighter regions correspond to higher disk surface density (σ). The central stellar object is not shown, and the central disk is shown truncated. The planet at center-right is orbiting clockwise. In contrast to the situation of Type I migration, the more massive proto-planet here has a profound effect on the distribution of disk gas, and effectively creates a gap. Spiral density waves are again excited in the disk, but owing to the clearance of material around the planet's orbit they do not exert the same torques on the planet. Instead the planet acts as a momentum bridge, connecting the outer and inner disk, and therefore migrates in step with the overall disk evolution. In this case the migration is known as Type II. Figure courtesy of F. S. Masset (CEA), private communication.

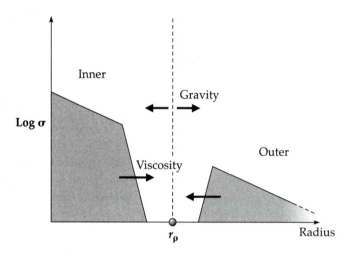

Figure 3.13. In this schematic view we illustrate the nature of Type II migration. The disk surface mass density is plotted (logarithmically) versus radius. The proto-planet is in a circular orbit at r_p, denoted by the vertical dashed line. Viscosity in the disk acts to close the gap via angular momentum transport. The gravitational torque of the proto-planet on the inner and outer disk acts to open the gap. A balance is reached where the viscous torque equals the proto-planet torque and the proto-planet becomes locked into the overall evolution of the disk (Chapter 2).

Jupiters—although they actually represent a range of planetary masses and compositions. Since this phenomenon was observed, and quite naturally linked to planet migration, it has been speculated that such systems might be barren for the presence of rocky planets at an Earth-like distance from the parent star. The argument is that a migrating giant planet, or embryo, would "sweep" clear forming objects between its initial and final orbital radii (see also Chapter 9). In fact, it appears that while the migration of giant worlds may reduce the likelihood of terrestrial-type planets in a system, it by no means eliminates them. Detailed numerical simulations employed to investigate this suggest that at least a *third* of systems with massive planet migration may still harbor Earth-type worlds (e.g. Raymond et al. 2006). The reason for this is that a significant amount of material in the proto-planetary disk can survive the passage of a migrating planet, and subsequently form rocky worlds—even replete with water. It would however be fair so say that, as with so much in this field, this is not the only answer. It may still be true that giant migrating planets are better at sweeping systems clean—we will not know for sure until we detect Earth mass worlds elsewhere.

Among the complications to planet migration are the situations in which there are *multiple* giant planets or embryos in a system—several of which may be undergoing migration. The interaction, and orbital resonances, beween the giant planets themselves can dramatically alter the evolution of a system—potentially even inhibiting disk-induced migration.

3.7 Final Stages of Planet Formation

3.7.1 Photoevaporation of the Disk Gas

A central part of the physics involved in planet formation is the dispersal of a major fraction of the initial proto-planetary disk material—in particular the gas. We have briefly mentioned the role of photoevaporation but have not discussed this in any detail (see also Chapter 8).

The dispersal of the disk directly influences the formation of planets. As we have shown, for giant worlds, the timescale of disk dispersal is critical in determining whether a significant amount of gas can be accreted. Even more fundamentally than this—when the solids in a disk are in microscopic dust form, their strong coupling to the gas component implies that if the gas is removed, then the dust will go too. In such a case, *no* planets will get a chance to form. We have also seen that the final orbital architecture of a system is profoundly effected by the presence of a substantial disk—producing the orbital migration described above.

Potential mechanisms for dispersing the gas in a proto-planetary disk include accretion onto the central proto-star, outflows created by protostellar winds, tidal stripping due to close passage of other stars or proto-stars, and photoevaporation. Although all these phenomena may occur, it seems that the most efficient is likely to be photoevaporation.

Both proto-stars in their later stages and massive young stars produce a significant flux of ultraviolet (UV) photons. With a range of energies, UV photons can ionize hydrogen, photo-dissociate molecules, and ionize heavier atoms such as carbon. These same UV photons can also heat the gas to temperatures as high as $\sim 10^4$ K, both directly and through the photo-electric effect on dust particles. If we consider the vertical hydrostatic equilibrium of the disk gas given by Equation 3.25 then we can see that if the gas is heated by an external source it will no longer be in equilibrium and will quite literally **evaporate** from the surface of the disk. However, from the form of the vertical gravitational acceleration (Equation 3.24), it is clear that the evaporation will be easier *farther* from the central stellar object. Thus, if the UV photons originate from the proto-star then disk evaporation will begin only at some distance out in the system. If on the other hand the proto-planetary system is in close proximity to other UV bright stars then the disk may also be eroded from the outside-in. More detailed calculations predict mass loss rates due to the central proto-star between $\sim 10^{-10}$ and $\sim 10^{-5} M_\odot$ per year, for low and high mass protostars respectively. These rates are in good agreement with the lifetimes of disks inferred from observations (e.g., Hollenbach et al. 1994).

3.7.2 Late-Stage Dynamics

For solid planetary embryos, the nature of runaway growth and the limits of feeding zones described above necessarily lead to massive objects with fairly regular orbital spacing. However, as the planetesimal population (and gas component) diminishes there are fewer "damping" forces (e.g. dynamic cooling, §3.4.1) to help maintain these large embryos in stable, close-to-circular orbits. In essence, the embryos can begin to perturb each other and enter into **crossing orbits**. Their subsequent evolution is based upon close encounters—with violent and inelastic collisions. Detailed numerical simulations indicate that many planetary embryos are essentially *randomly scattered* in a system. Not only does this imply that some embryos are ejected from the inner system, but that there is also significant *chemical mixing* of any earlier chemical gradients in the disk (Chapter 8). The implications for the final orbital architecture of a system may also be that the observed range of eccentric orbits are generated by the gravitational interactions of proto-planets (e.g. Juric & Tremaine 2007). This is directly relevant to the observations of exoplanets discussed in the next chapter. The relative circularity of our own Solar System may point towards either a late gas dispersal, a dynamically "inactive" system, or a large amount of mass in the late-stage population of planetesimals, that would dampen any induced eccentricities (e.g. Ford 2005).

In Chapter 9 we will discuss in more detail the origin of the Earth's moon. It appears likely that the Earth–Moon system is a direct consequence of late-stage planetary embryo collision. The inclination or obliquity of the giant planets in our system (e.g. Uranus) may also be a consequence of late-stage collisions.

If gas or ice giant planets form in a system (which they must have done by this stage) then they will tend to exert considerable influence over the dynamical evolution of the smaller, rocky, worlds. As we discuss below, and in later chapters, this is certainly true for our own Solar System. Of particular interest is the role that Neptune seems to have played in helping set the characteristics of the Kuiper belt (Chapter 1). It seems likely that Neptune was originally formed at a larger orbital radius (~ 50 AU, e.g. Brown 2004). In the presence of a disk composed of smaller embryos and planetesimals interior to Neptune's orbit there would have been resonance interactions leading to the pushing of small bodies *outwards* onto eccentric orbits, while Neptune would have migrated inwards to conserve angular momentum. Thus, both the final orbital configuration of Neptune and the origin and distribution of Kuiper belt objects are linked to late stage dynamics.

Giant planets may also contribute to the final buildup of smaller worlds, by acting through **secular resonances** (i.e., a resonance in which the *precession* of orbits is synchronized). These will tend to concentrate planetesimals or embryos in regions of the disk, thereby enhancing their final collision rates. Such a mechanism may have allowed Jupiter to encourage the final assembly of Mars in our solar system.

3.7.3 Radionuclides, Internal Differentiation and Late-Time Bombardment

Rocky worlds such as the Earth appear to have highly structured interiors (Chapter 1). Denser material forms the cores of such planets, and lighter material tends to form the outer layers. Forming a large solid body through the mechanisms we have discussed involves the release of a lot of energy, from collisions, natural radioactivity, and even chemical reactions. A growing world can be heated to the point where substantial parts of its interior can melt.

Short-lived radionuclides (SLRs; see Chapter 2) are thought to have played a central role in setting the internal temperature of planetesimals and young planets in our solar system, and by extrapolation should be important in other systems too. The radioactive decay of ^{26}Al (to ^{26}Mg), and to a lesser degree ^{60}Fe (to ^{60}Ni), in the forming solar system should have provided the largest source of energy for heating planetesimal interiors, and would have contributed a major heating component to the interior of larger proto-planets. As an example, heating from radioactive decay in the young Earth prior to 4.5 Gyr ago is estimated to have been at least five times the present level.

Models that incorporate SLR heating (see for example Bizzarro et al. 2005 and Hevey and Sanders 2006) suggest that rocky bodies larger than about 20–80 km in radius that formed within 3 million years of the initial proto-stellar collapse should have been extensively melted in their interiors (at a temperature of $\sim$1850 K all potential rock compositions will melt). Smaller objects would not provide enough thermal insulation to keep their interiors hot. Indeed, the heating and melting could have been so efficient that these large planetesimals could have spent a few hundred thousand years as molten spheres, with a thin outer crust of solids. As the radionuclides decay with half-lives of 0.73 to 1.5 Myr (Chapter 2) the heating declines and these planetesimals would have cooled from the outside in. Planetestimals that formed later (after about 3 Myr) would have been unlikely to ever attain such high internal temperatures as the SLRs would already be greatly reduced in abundance. This is consistent with the evidence in our own Solar System that some planetesimals must have been internally differentiated (see below), which others (presumably later forming) were not (Chapter 8).

With a molten interior, heavier elements (metals) will sediment out and sink towards the center of an object. In particular, for rocky objects, elements that can chemically bond with iron are known as **siderophiles** (e.g., nickel, Ni) and tend to sink to the center. Lighter elements that bond with silicates are known as **lithophiles** and will tend to rise to the outer regions. This process of **internal differentiation** creates a layered object, with density increasing from the surface inwards, and with a clear segregation of elements, unlike the undifferentiated, homogeneous nature of unmelted objects. In Table 3.3 the estimated internal layering of the Earth is listed as an example (see also Chapter 1).

Table 3.3. The approximate compositional layering of the Earth as a consequence of internal differentiation.

Layer	% mass of planet	Dominant composition
Inner core	∼2%	Fe–Ni alloy (solid due to pressure)
Outer core	∼31%	Fe–Ni + O and S (liquid)
Lower mantle	∼52%	Si, Mg, O and small amounts of Fe, Ca, Al
Transition region	∼8%	"
Upper mantle	∼10%	" (partially molten)
Crust	∼1%	" plus lighter elements

In gas or ice giants there is also likely to be significant internal structure following similar physical principles.

Energy is constantly being radiated away from an object not in thermal equilibrium with its surroundings (to first order as a blackbody). For terrestrial planets the large insulating effect of outer layers of regolith (Chapter 8), and eventually a possible atmosphere due to outgassing of volatiles (e.g. H_2O, CO_2), act to slow cooling to timescales well beyond a few million years (as experienced by smaller planetesimals).

In the case of the early Earth, it was likely repeatedly melted again through impact events (including that which formed the Moon, Chapter 9) and formed its major internal structures, such as a core, and a cooled solid outer crust, by about 4.5 Gyr ago. It also appears that a significant water ocean and outgassed atmosphere could have been in place by then. Following this, an intense, so-called "late heavy bombardment" of smaller but significant bodies continued up to about 3.9 Gyr ago (Chapter 8).

3.7.4 The Mass-Radius Relationship for Planets

At the simplest level it is easy to see that planets of different masses and compositions should be of different sizes. For example, terrestrial mass planets of different composition should have different mean densities. Computing this very basic quantity does however require a certain amount of sophistication. Fortunately we have already seen the basis for such calculations in the form of the hydrostatic equilibrium embodied by Equation 2.7 in Chapter 2. For both rocky and gas-dominated planets we simply need to choose appropriate equations of state (i.e., $P(T)$ relationships) to capture the necessary physics (see e.g. Fortney, Marley, and Barnes 2007, and Chapter 6). For gas-dominated planets we also need to include a knowledge of the *age* of the planet, since the cooling of the gas envelope with time (via blackbody radiation to first order) plays a major role in determining the overall size of the object (strictly speaking the cooling of rocky/icy worlds should also be modeled, however this has a more secondary effect on overall size). In Figure 3.14 some examples of the mass–radius relationships for different planet types of age 4.5 Gyr are shown, together with the location of solar system objects in this parameter space. For the gas-dominated planets in this figure, sophisticated atmospheric models are employed to determine the radiative cooling properties. For the curves corresponding to terrestrial-type worlds (iron, rock, or water dominated) all radii increase with mass approximately as $M^{1/3}$ below a few 10s of Earth masses. Beyond this mass then non-classical pressure[2] becomes important in the planetary interiors, as atoms are increasingly pressure ionized. Thus, the mass–radius relationship flattens off at around $1000M_\oplus$, and radii could even *shrink* as $M^{-1/3}$ beyond this.

What is clear from Figure 3.14 is that the mass and radius of a planet act as powerful diagnostics of planetary composition (and indeed age for gas giants). We will consider this again in Chapter 4 when applied to real exoplanet data.

3.7.5   Debris Disks

In some nomenclatures, the remains of the proto-planetary disk, including planets and asteroids, are termed a **debris disk**. One interesting feature of such a disk is that the continuing collision of small rocky bodies will generate new dust. The sublimation or evaporation of cometary material will also release dust back into the system. We know this to be the case for our own system due to the observation of **zodiacal dust** (Chapter 7), and increasingly we find similar situations in exoplanet systems—even if we cannot see the planets, we can see the infrared glow

---

2. Fermi or degeneracy pressure, due to the exclusion principle which forbids fermions, such as electrons in this case, from occupying the same quantum states.

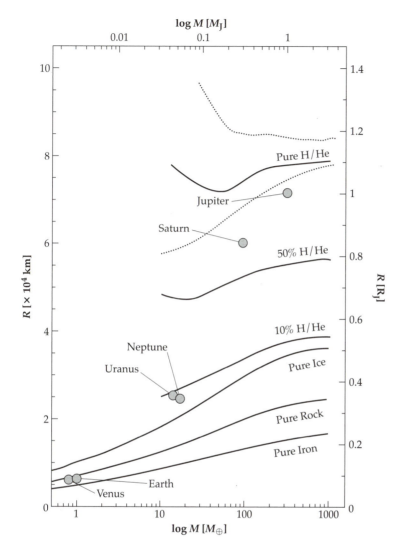

Figure 3.14. An example of the mass–radius relationships for planets of varying composition (based on Fortney, Marley and Barnes 2007, and Gillon et al. 2007). All curves correspond to theoretical estimates. Lowermost three curves (from bottom upwards) correspond to compositions of pure iron, pure "rock" (i.e. silicate), and pure water ice. The next two solid curves correspond to objects with increasingly significant H/He atmospheres (10% and 50% by mass) and cores composed of heavy elements. The upper most solid curve corresponds to a *pure* H/He planet with no heavy element core. All three of these giant planet models assume that the planets are orbiting 0.1 AU from a solar-type star and therefore correspond to moderately hot Jupiters. The dotted curves illustrate the effect of stellar heating, the lowermost is for a pure H/He planet at 10 AU from a solar-type star and the uppermost curve is for the same composition but at 0.02 AU from the star—a very hot Jupiter. For the purposes of illustration the locations of Venus, Earth, Jupiter, Saturn, Uranus, and Neptune are shown in this plot. It is apparent that the basic composition of these worlds can be quickly deduced from their location on the mass–radius plot.

of the debris dust. Since stellar radiation pressure and wind will tend to remove such new dust over time, its observation leads us to conclude that these systems indeed contain rocky bodies—which are colliding and grinding down to replenish the dust.

3.8   Planet Formation Summary

Planet formation is a complex business, and will result in a great diversity of planetary systems. Nonetheless, we can summarize some of the principal features of the models and observations that we have at present.

Following the collapse from a molecular cloud core, the formation of a solar-mass proto-stellar object is fast, occuring in well under a million years. Subsequently, the clock is ticking for planet formation, since the major fraction of the circumstellar or proto-planetary disk will be dispersed (primarily via photoevaporation) in less than 10 Myr (for a solar mass system). Through the coagulation of dust into small (approximately centimeter) sized bodies and its settling to the disk midplane, the process of solid body merger and growth can produce km size planetesimals and even $\sim 0.1 M_\oplus$ planetary embryos in less than 1 Myr. The nature of growth through mergers naturally produces a population of objects of similar size. If the core accretion–gas capture model is correct then gravitational focussing enables runaway growth, followed by oligarchic growth, and can produce giant planet cores of $\sim 10 M_\oplus$ in only 1–3 million or so years more. These can accrete the massive gaseous atmospheres that we see in our own gas giants, Jupiter and Saturn. Further out in the system, where the disk density is lower, embryo formation is slower and the chances of accreting gas is diminished. Again, in our own system, the ice giants Uranus and Neptune were likely gas starved for this reason. By 10 Myr the gas, and much of the dust, in the disk will have been dispersed and late-stage, violent, collision-driven planet formation will take place.

In Figure 3.15 we present an example of the full range of planet masses and orbital radii resulting from a series of core-accretion-based gravitational simulations that incorporate much of the physics we have described in this chapter. It is clear that a remarkable diversity of planets can result—even without considering characteristics such as detailed

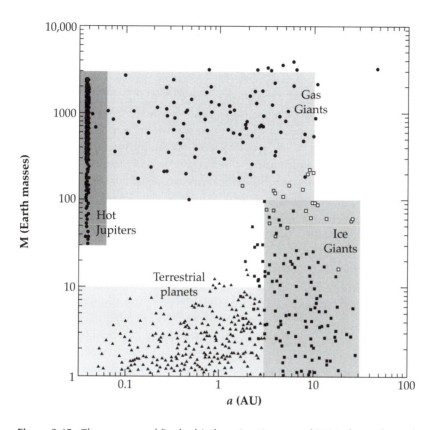

Figure 3.15. The masses and final orbital semi-major axes of 3000 planets formed from a series of computer simulations are plotted. The simulations model the core-accretion gas-capture process for planet formation, and include proto-planet/disk interaction, as well as disk temperature structure. Four principal classes of planet are indicated: Terrestrial planets include rocky and ice-laden worlds up to $10M_{\oplus}$ in mass. Ice giants have gas mass less than their solid mass. Gas giants have gas mass greater than their solid mass, and span a wide range of orbits due primarily to migration. Hot Jupiters (see Chapter 4) are gas, or ice, giants that have migrated to with ~0.04 AU of the parent star. Planets represented by open symbols around $100M_{\oplus}$ have a gas mass between 1 and 10 times their solid mass—by comparison all gas giants have gas masses greater than 10 times their solid mass. (Adapted from Figure 10 of the "Precursor Science for the Terrestrial Planet Finder" NASA document (2004), JPL 04-014, edited by Lawson, Unwin, and Beichman).

compositional variation or complex disk physics such as magnetohy-drodynamics. Planet masses can range from $1M_\oplus$ (and below) to a few thousand $M_\oplus$—corresponding to the upper end of what we consider to be planets (below $13M_J$, c.f. Chapter 1). For rocky worlds, or lower mass ice giants, orbits can range from less than 0.1 AU to some 30 AU. With migration the orbital range of gas giants can range from less than 0.04 AU to some 10 AU. Worlds unlike those in our on Solar System, such as "super-earths," or "dwarf" ice worlds, and "hot Jupiters" (Chapter 4), can clearly also be abundant.

Timescales appear to be critical in successful planet formation through the core accretion model. Several things can act to enhance the rate at which planetary embryos can grow, and in the case of giant plan-ets, may help ensure that gas rich worlds are quickly enough. First, the temperature structure of the protoplanetary disk implies that beyond the snow-line in a system there is a factor 4–5 enhancement of solid material. Second, the nature of planet migration is such that an embryo can, in some circumstances, quite quickly move inwards to new feeding grounds, as well as be supplied by the gas drag infall of planetesimals, thereby sustaining a higher rate of growth (although it can also lead to the *loss* of food). Third, as a giant planet core accretes a large gaseous atmosphere its effective cross section will be increased. It is not clear which of these mechanisms really helps the most, but it seems likely that at least one, possibly in combination with others, must play a role if the core accretion picture is correct.

Finally, we have (again) not considered the prospects for planet for-mation around stars in binary or multiple systems. It is unclear how efficient this might be, or exactly how it depends on the wide range of possible stellar configurations (e.g. close binaries, wide binaries etc.), but it is certainly possible. Current observations contain an increasing number of systems where a planet hosting star is part of a moderately close to wide binary or triple stellar system, with typical separations ranging from approximately 50 to 7000 AU.

References

Armitage, P. J. (2007). Lecture notes on the formation and early evolution of planetary systems, ArXiv *Astrophysics* e-prints, arXiv:astro-ph/0701485.

Bizzarro, M. et al (2005). Rapid timescales for accretion and melting of differentiated planetesimals inferred from ^{26}Al-^{26}Mg chronometry, *Astrophysical Journal Letters*, **632**, 44.

Bodenheimer, P., & Pollack, J. B. (1986). Calculations of the accretion and evolution of giant planets: The effects of solid cores, *Icarus*, **67**, 391 .

Boss, A. (2006). "Giant planet formation: Theories meet observations." In *Planet Formation, Theory, Observations, and Experiments,* Eds. Klahr, H., & Brandner, W., Cambridge University Press.

Brown, M. E. (2004). The Kuiper Belt, *Physics Today*, **57**, 49.

Ford, E. B. (2005). What do multiple planet systems teach us about planet formation?, *New Horizons in Astronomy*, ASP conference series, see astro-ph/0512635.

Fortney, J. J., Marley, M. S., & Barnes, J. W. (2007). Planetary radii across five orders of magnitude in mass and stellar insolation: Application to transits, *The Astrophysical Journal*, **659**, 1661.

Fujiwara, A., et al. (2006). The rubble-pile asteroid Itokawa as observed by Hayabusa, *Science*, **312**, 1330–1334.

Gillon, M., et al. (2007). Detection of transits of the nearby hot Neptune GJ 436b, *Astronomy & Astrophysics*, **472**, L13.

Goldreich, P., Lithwick, Y., & Sari, R. (2004). Planet formation by coagulation: A focus on Uranus and Neptune, *Annual Review Astronomy and Astrophysics*, **42**, 549.

Hevey, P. J., & Sanders, I. S. (2006). A model for planetesimal meltdown by ^{26}Al and its implications for meteorite parent bodies, *Meteoritics & Planetary Science*, **41**, 95.

Hollenbach, D., Johnstone, D., Lizano, S., & Shu, F. (1994). Photoevaporation of disks around massive stars and application to ultracompact H II regions, *The Astrophysical Journal*, **428**, 654.

Hyashi, C. (1981). Formation of the planets: Fundamental problems in the theory of stellar evolution; *Proceedings of the Symposium*, Kyoto, Japan (A82-34012 16-90) Dordrecht, D. Reidel Publishing Co., 113–128.

Inaba, S., Tanaka, H., Nakazawa, K., Wetherill, G. W., & Kokubo, E. (2001). "High-accuracy statistical simulation of planetary accretion: II. Comparison with n-body simulation," *Icarus*, **149**, 235.

Miyamoto, H. (2007). Regolith migration and sorting on asteroid Itokawa, *Science*, **316**, 1011–1014.

Raymond, S. N., Mandell, A. M., & Sigurdsson, S. (2006). Exotic Earths: Forming habitable worlds with giant planet migration, *Science*, **313**, 1413.

Ward, W. R. (1986). Density waves in the solar nebula—Differential Lindblad torque, *Icarus*, **67**, 64.

Yano, H. et al. (2006). Touchdown of the Hayabusa spacecraft at the Muses Sea on Itokawa, *Science*, 312, 1350.

Problems

3.1 Using a combination of source material describe the currently popular theories of terrestrial and gas giant planet formation, paying attention to the possible differences between these regimes and how planetary migration may help explain current extra-solar planet observational results (see also Chapter 4).

3.2 (a) Briefly describe the Core Accretion model for gas giant planet formation. What is the "snow line" in a proto-planetary disk, and how might this critically assist gas-giant formation? (include discussion of timescales).

(b) The feeding rate for a solid planetary embryo (M_e, R_e) sweeping through a volume of planetesimals is given by

$$\frac{dM_e}{dt} = \pi R_e^2 (1 + \frac{2GM_e}{R_e v^2})\rho_s v \tag{3.33}$$

where ρ_s is the mass density of planetesimals and v is the relative velocity of the embryo and the planetesimals. By referring to the components of this equation, *demonstrate* the orderly and runaway growth regimes for a planet embryo. Give two physical arguments for why runaway growth must eventually slow down.

(c) Describe the physical mechanism which can make $\sim$1 meter planetesimals migrate inwards in a gas disk. Describe the (different) physical mechanisms by which a *large* proto-planet can migrate inwards.

As a planetary embryo accretes planetesimals its mean density is likely to increase as its self-gravity compresses the "rubble pile" of material. In the case of water ice the compression can also change the ice phase and increase the mean density of the embryo. How would such a change in mean density alter the subsequent growth rate of the embryo?

3.3 A full derivation of the Hill Sphere Radius involves solving the restricted 3-body gravitational problem. A simpler, approximate, derivation involves recognizing that R_H is the distance from an object at which the orbital period of a small test particle *equals* the orbital period of the object itself around the star. Assume a circular orbit and derive the expression for R_H.

3.4 Consider the possible effects of *fragmentation* during planet coagulation. Specifically-discuss the potential effect of an increasing fragmentation likelihood during runaway growth. You can tackle this analytically or numerically, or by careful discussion. Numerical simulations suggest that ultimately the final number of large bodies formed in a system by runaway and oligarchic growth is not strongly dependent on whether or not fragmentation occurs. Why is this? [*Hint:* consider space densities].

3.5 Beyond the snow line in a proto-planetary disk the relative surface density of solid material increases by approximately a factor of 4. Using the MMSN model of Equation 3.22 and assuming that the gas-to-solid ratio *within* the snow line (i.e., in the hotter region of the disk) is 100–1 by mass, and that the snow line occurs at 4 AU, estimate what the isolation mass would be for a planetary embryo at 5 AU around a $1M_\odot$ star.

3.6 The sublimation point of methane (CH_4) is approximately 90 K. If a proto-planetary disk has a temperature of 170 K at a radius 4 AU, compute the radius at which methane will freeze-out in this system (assume a simple midplane temperature function).

By referring to data on our own solar system discuss the implication of methane freeze out for the composition of objects. What other compounds will have frozen out at larger radii? Again, make use of solar system data to discuss evidence for this zoning of sublimation points. [*Note:* for this exercise you should consider moons as forming in the ambient temperature of the proto-planetary disk].

3.7 The idea of the minimum mass solar nebula (MMSN) is a useful one and could be extended to other systems. By referring to Chapter 4, and the astronomical literature, consider what the MMSN might be in the exoplanet systems Upsilon Andromedae and 55 Cancri. In both cases it is believed that the majority of the disk mass that was locked up in planets has been observed. Compare these minimum disk mass models to that of our solar system. [*Note:* this may be expanded into a longer term project].

3.8 Use Figure 3.13, the paper by Fortney, Marley & Barnes (2007), and the online catalog of extrasolar planets (http://exoplanet.eu/) to deduce the composition of the ten most recently detected *transiting* exoplanets (see also Chapter 4). To do this correctly you will also need to take into account the stellar insolation of each planet.

3.9 Describe the mechanisms thought to operate in Type I and II planet migration. If substantial numbers of proto-planets do somehow migrate and fall into the central proto-star what might be an observational consequence for the mature stars?

A gaseous proto-planetary disk is, in reality, unlikely to be perfectly smooth and in laminar flow (Chapter 2). Indeed, there may be turbulent radial zones in a disk as well as "dead zones." Forming planets may migrate but then pile-up in dead zones at certain radii. Discuss what might happen in such a situation, consider the pure gravitational interaction of these objects—what will they do to each other?

CHAPTER 4

Extrasolar Planets

To consider the Earth as the only populated world in infinite space is as absurd as to assert that in an entire field of millet, only one grain will grow.

—*Metrodorus of Chios (Fourth Century B.C.)*

4.1 Introduction

For at least the last two thousand years, and likely for much longer, the idea that other worlds, and in particular other "Earths," exist has seemed such a natural extension of the order of things—as demonstrated by our own situation—that it has largely escaped true scientific scrutiny. Is it correct to think that other planets are a naturally occuring phenomenon around other stars? The previous chapters include, it is hoped, a fairly convincing theoretical case for this being so—although the question of whether other "Earths" truly exist remains open (Chapters 9 and 10). A theory is, however, nothing if it cannot be verified. Although we have from the outset discussed the fact that worlds beyond our solar system are real, we have not yet examined the evidence.

We are fortunate, at the present time, to be witnessing the nascent steps towards the routine detection and observation of planets around other stars. Extrasolar planets, or **exoplanets**, are now a well established fact. Nonetheless, their detection is a great challenge and requires a full armory of astronomical techniques, applied with modern technology. In this chapter we will discuss some of the approaches that are employed to detect and characterize exoplanets. We will place an emphasis on the physics behind these techniques rather than the technical details—since techniques can come and go, but (we hope) physics remains the same. We will also inspect selected results from efforts to detect exoplanets. As we also discuss elsewhere, planets are not as readily classified by simple physical characteristics as, say, stars are. Thus, much of what current exoplanet detections tell us, and much of the motivation to further such observations, comes from the study of **comparative planetology**. In fact, we have already succumbed to this approach by (trivially) choosing mass units such as M_J or $M_\oplus$ to describe other worlds. Our natural bias is to compare our own system of planets to others. Yet, as mentioned in Chapter 1, our first peek into the population of other worlds indicates that our own solar system may not represent the norm. It is critical that we remain aware of this fact in designing and implementing searches for exoplanets.

4.2   Indirect Planet Detection

Detecting planets directly is tough, as we demonstrate in §4.2.1 below. For this reason, an array of "indirect" approaches are also utilized, which circumvent many of the problems associated with direct detection. We will effectively consider only three general types of indirect planet detection methods; two of which are based upon gravitational influence, and one of which is based upon the geometrical blocking of light from other objects.

4.2.1   The Basic Problem

Why should planets be hard to detect around other stars? There are two reasons: brightness, and the wave nature of light. Consider a Jupiter-like planet orbiting a star like our Sun at about 1 AU distance. The planet will both *reflect* light and *emit* light (due to its finite temperature—to first order as a blackbody emitter). We will cover this in much more detail in Chapter 6. Irrespective of the details of this planet, its net luminosity will be approximately 10^{-9} times the luminosity of the parent star. This contrast is somewhat improved in the infrared (where most of the *emitted* radiation of the planet will be), where it is some 10^{-6} of the parent, but it is still enormous. In addition, the apparent separation on the sky of a star and its planets is small. This can be seen readily by recalling the definition of a parsec (the distance at which 1 AU presents a 1 arcsec angle on the sky). A planet such as the one above will therefore present a source some million to a billion times fainter than its parent star, and separated by only a small angle, such as 0.05 arcsec for a star merely 20 parsecs distant.

Why should this be problematic? After all, we routinely study things in astronomy that differ in apparent brightness by factors of more than 10^{10}. The principal, intrinsic difficulty lies with the wave nature of light, and the resulting "imperfect" nature of our astronomical instruments.

Any optical instrument suffers from what is known as the **diffraction limit**. Put simply, an incident wavefront of light will undergo diffraction due to the finite *size* of the optic—be it an aperture, mirror, or lens. This diffraction results in an interference pattern which *fundamentally* limits the resolving power of the optic. In other words, there is a finite, inescapable "blurring" of the final image (often known as the "**Airy disk**" or spot; see Figure 4.1). This is given by

$$\theta_{(radians)} = \frac{1.22\lambda_{(cm)}}{D_{(cm)}}, \tag{4.1}$$

where θ is the diffraction limited resolution (in radians), λ is the wavelength of the light passing through the optic, and D is the diameter of the telescope, or optic aperture.

Applying this formula to a modestly sized astronomical telescope with an aperture of, say, 2.5 meters diameter, observing at a typical optical wavelength of 5000 Å, we find $\theta \simeq 2.5 \times 10^{-7}$ radians, or some 0.05 arcseconds. This is what we said above would be the separation of a 1 AU planet 20 parsecs away—so this would be *just* resolvable by such a telescope.

At first this doesn't sound so bad. The problem is that this assumes a *perfect* telescope. Sadly no telescope is perfect; every additional lens, filter, even the image detector itself further blurs the image. Furthermore, any telescope on Earth must peer through the terrestrial atmosphere—which drastically distorts any wavefront of light passing through it. To put this in perspective, the best typical resolving power of a basic ground-based telescope, when the atmosphere is fairly stable, is usually some 0.5 arcseconds—largely due to the atmospheric distortions. This is a solid factor of ten away from the number above. Furthermore,

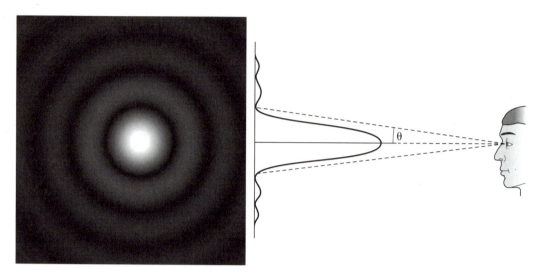

Figure 4.1. Illustration of the Airy diffraction pattern. As photons pass through the aperture of an instrument (e.g., telescope) they are diffracted and the resulting interference pattern results in a point source being blurred to a "sinc" function in the image plane (left panel). The angular radius to the first minimum of the azimuthally averaged intensity is generally used to define the effective resolution θ of the instrument (right panel).

as we previously noted, a companion planet to a solar type star will be a factor 10^6–10^9 times fainter—and both it and the star will be blurred by the finite resolving power of a telescope. Smearing together the light from two such objects makes the direct detection (imaging) of the fainter planet essentially impossible.

There are solutions to some of these problems, including putting telescopes in space, or using **adaptive optics** on Earth which attempt to correct for the distorting effects of the atmosphere (see §4.3.2 below), in concert with the construction and use of larger telescopes. In both cases, very significant improvements can be made—albeit at great technological cost. Ultimately though, the diffraction limit of Equation 4.1 is an insurmountable hurdle. Improving it requires either moving to shorter wavelengths (such as the ultraviolet, which is not optimal since the contrast between star and planet is typically poorer than it is in the red or infrared part of the spectrum) or building bigger aperture telescopes. This can include those which produce an *effective* large aperture by combining the signal from many smaller, physically separated telescopes, exploiting interferometric techniques. We will discuss some of these issues later on below.

For all of these reasons, the use of alternate—less direct—techniques for finding and characterizing exoplanets is a major topic. The relative technical cheapness of such methods also enables them to be applied to very large sets of potential host stars for planets.

4.2.2 Dynamical Perturbations

Consider two objects in orbit about each other; one is a star, and one is a planet—of masses M_*, and M_p respectively. The **center of mass** (or **barycenter**) of the system can be considered as the stationary point about which either mass is orbiting (see also Chapter 2). The center-of-mass condition (Figure 4.2) is given by $M_* r_* = M_p r_p$, where the total separation of the masses is always $r = r_* + r_p$. Since stellar masses are significantly larger than planetary masses we often make the approximation that the center of mass is at the stellar center. This is reasonable to do for the discussion of planetary orbits, but it does neglect the very real perturbation felt by the star. For example, considering just the Sun–Jupiter 2-body system we find that $r_* \sim 1.07 R_\odot$. Thus the Sun actually orbits a point a little outside of its surface!

In this special case of two bodies, the orbital period of each around the center of mass is the same (P) and is given by the Newtonian version of Kepler's third law (see Chapter 2). We can use this fact to perform a quick derivation of the orbital velocity of the *star* about the center of mass due to the presence of a planet. To simplify the problem we assume purely circular orbits, but all of this is readily generalized to eccentric orbits. In this case, the orbital velocity of the star is

$$v_* = \omega r_* = \frac{2\pi}{P} r_*. \tag{4.2}$$

Since $P^2 = 4\pi^2 r_p^3 / G(M_* + M_p)$ by definition (Chapter 2) we can rewrite this as

$$v_*^2 = \frac{G(M_* + M_p) r_*^2}{r_p^3}. \tag{4.3}$$

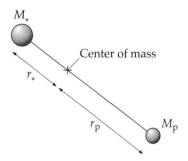

Figure 4.2. An illustration of the definition of center of mass (see also Figure 2.16). For the discussion of dynamical perturbations we assume that mass M_* is greater than mass M_p.

The center-of-mass coordinate system is defined by the condition that $M_* r_* = M_p r_p$ and so substituting for r_* and assuming that $M_p \ll M_*$ we obtain

$$v_*^2 = \frac{G M_p^2}{r_p M_*}.$$
(4.4)

We can immediately see that v_* is indeed going to be small in general. In fact it is really more of a "wobble" induced on the star, since r_* is also going to be small. Hence the star can be considered to be just wobbling about the center of mass of the system. Despite the small size of this dynamic perturbation to the star, detecting the light of the star is easy (in relative terms), and so we might hope to deduce the presence of smaller mass bodies through their gravitational influence by careful monitoring of the stellar photons.

4.2.3   Doppler Shift and Radial Velocity Measurements

The observed frequency of radiation emitted by a source (v_{obs}) moving at a constant velocity is different from that of a source stationary with respect to the observer (v_{source}). The full relativistic expression for the relative frequencies is given by

$$\frac{\nu_{obs}}{\nu_{source}} = \frac{\sqrt{1 - v^2/c^2}}{1 + (v/c)\cos\theta}, \tag{4.5}$$

where v is the source velocity relative to the observer and θ is the angle between the velocity vector and the line of sight to the observer (see inset Figure 4.3). There are really two pieces to this formula—one (with $v\cos\theta$) would arise in the classical physics derivation of the Doppler effect. The other ($\sqrt{1 - v^2/c^2}$) arises from the relativistic **time dilation**. Thus, even if the source were moving entirely *transverse* to the line of sight of the observer (i.e., $\theta = \pi/2$) there would still be a frequency shift: $\nu_{obs}/\nu_{source} = \sqrt{1 - v^2/c^2}$. Obviously in the case of stellar motions due to planets $v \ll c$, and the time dilation is small. In general then (and this is true most astrophysical situations) what we can detect is the line of sight, or **radial velocity** Doppler shift.

Thus for stellar wobble, the *maximum* effect is at the two points in the orbit where the star is moving directly towards or away from the observer ($\theta = \pi$ or $\theta = 0$, assuming that the observer is in the plane of the orbit) in which case Equation 4.5 can be simplified to

$$\frac{\nu_{obs}}{\nu_{source}} = \sqrt{\frac{1 - v/c}{1 + v/c}}, \tag{4.6}$$

and when $v \ll c$ the frequency *shift* $\Delta\nu = \nu_{obs} - \nu_{source}$ can be expressed as

$$\frac{\Delta\nu}{\nu_{source}} \simeq -\frac{v}{c} \tag{4.7}$$

More generally, for motion in an arbitrary direction the non-relativistic expression becomes

$$\frac{\Delta\nu}{\nu_{source}} \simeq -\frac{v\cos\theta}{c} \tag{4.8}$$

and for a circular orbit coplanar with the observer's line of sight then the maximum relative velocity is $v = \pm v_*$ depending on whether the star is moving towards (+ve) or away (−ve) from the observer. The frequency change in the former case is termed a **blueshift** (frequency increases) and in the latter a **redshift** (frequency decreases).

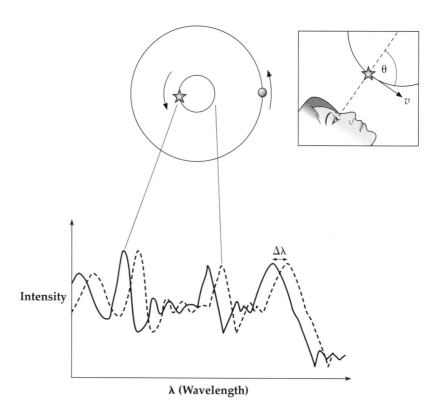

Figure 4.3. A schematic illustration of the spectroscopic measurement of stellar "wobble" due to the presence of a planet [inset: definition of angle θ between line of sight and velocity vector]. Both star and planet orbit about the common center of mass of the system. This is greatly exaggerated here; typically the center of mass will reside close to or within the stellar atmosphere itself. While the star is moving towards the observer its spectrum is blueshifted (to shorter wavelengths), while it is moving away from the observer its spectrum is redshifted (to longer wavelengths). In practice, sophisticated techniques are required in order to achieve the necessary precision to measure $\Delta\lambda$ or Δv.

For stars, we could in principle measure the detailed *spectra* of light and look for red- or blueshifts in the frequencies or wavelengths of known spectral features or lines, whose rest-frame wavelengths can be determined in terrestrial laboratories. In practice, rest-frame wavelengths are known to about 0.001 Å, corresponding to many tens of meters per second uncertainty in velocity. As we will see below, this is insufficient for detecting many planets. Real planet detection using the Doppler effect is non-trivial, and requires exquisite precision and sophisticated techniques (e.g., Butler et al. 1996). These include in-situ reference spectra and the measurement of literally hundreds to thousands of spectral lines in order to lower the experimental uncertainty. One technique involves placing an absorption cell (a transparent cell containing a gas such as iodine) in the path of the light captured by the telescope. This imprints the wavelength information of the absorbing gas on the stellar light, and information about the telescope point-spread-function (see §4.2.1). Other techniques involve combining the stellar light with light from an emission lamp (producing photons at discrete wavelengths) or via cross-correlation methods.

To find a planet-induced wobble we must perform such measurements as a function of time, in order to see the back-and-forth stellar motion (Figure 4.3). It remains to be seen what the ultimate limits are for this technique. As we know from our own Sun, the stellar photosphere can be in a constant state of motion as convective processes cause hot parcels of material to rise, and cold parcels to fall - creating a "granular" surface. Combined with other natural activity - for example, sunspots, flares, pulsations, and differential stellar rotation—this creates a significant amount of "noise" in velocity measurements. The first generation of exoplanet observations appeared to be limited to sensitivities of some $\sim 3 \, \text{m s}^{-1}$ due to stellar activity. However, this is no longer the case, and the sensitivity of Doppler measurements continues to improve.

An Earth mass planet orbiting at 1 AU from a $1 M_\odot$ parent star would induce a wobble of only $\sim 9 \, \text{cm s}^{-1}$, and this is clearly a desirable sensitivity. Terrestrial mass planets orbiting closer in to lower mass stellar parents (e.g., M-dwarfs of masses $\sim 0.1 - 0.5 M_\odot$, see also Chapter 9) could induce a significantly larger wobble (see Exercise 4.2), although the faintness and activity of the stars poses other challenges.

4.2.4 Inclination

Since we measure a line-of-sight radial velocity, the actual geometry of a planetary system is important. In particular, we need to account for the **inclination** of the planet–star orbital plane with respect to our line of sight. Figure 4.4 shows this schematically. We define an **inclination**

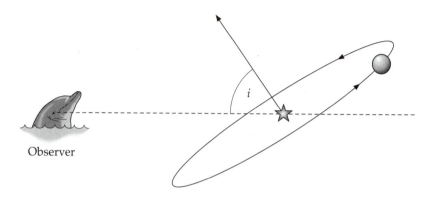

Figure 4.4. Illustration of the definition of orbital inclination—the angle between the normal to the plane of orbit for a planet and the direction of the observer.

angle i, which is the angle between the normal to the system plane and our line of sight. Then the maximum (or peak) radial velocity of the stellar wobble *for a circular orbit* is

$$v_r^{max} = M_p \sin i \sqrt{\frac{G}{r_p M_*}},\qquad (4.9)$$

following Equation 4.4.

If r_p is given in AU this can be written as

$$v_r^{max} = \frac{28.4 \, M_p \sin i}{r_p^{1/2} M_*^{1/2}} \quad (\text{m s}^{-1}), \qquad (4.10)$$

where M_p is in units of Jupiter mass (M_J) and M_* is in units of solar mass ($M_\odot$). Furthermore, applying Kepler's third law, and recognizing that $M_* \gg M_p$, we can write this in terms of the planetary orbital *period* P—which is directly observable through the time variation of the stellar wobble (see below):

$$v_r^{max} = 28.4 \left(\frac{P}{1\text{yr}}\right)^{-1/3} \left(\frac{M_p \sin i}{M_J}\right) \left(\frac{M_*}{M_\odot}\right)^{-2/3} \quad (\text{m s}^{-1}), (4.11)$$

since

$$P = \left(\frac{r_p}{1AU}\right)^{3/2} \left(\frac{M_*}{M_\odot}\right)^{-1/2} \text{ yrs.} \qquad (4.12)$$

Thus, P and v_r^{max} may be estimated by constructing a computer model of an orbiting planet (generalized to deal with eccentric orbits) and varying the free parameters until the predicted Doppler shifts match those observed, within the errors. The stellar mass M_* is generally estimated using detailed stellar models that are calibrated by the enormous body of observational work estimating stellar masses using binary stars (eclipsing binaries in particular). Since the system inclination cannot be determined by Doppler shifts alone this technique provides an estimate of $M_p \sin i$. More details of the practical issues (not least of which are the relaxation of the assumptions of a circular orbit and a single planet) are discussed below (§4.2.6).

4.2.5 Position Wobble

Above we have treated the effect of the stellar wobble on the observed velocity of a star—the same wobble will of course cause the *position* of the star on the sky to vary with time. Clearly the angular movement of a star on the sky will be given by (for small angles) $\theta = r_*/d$ where d is the distance from the observer to the system. In the nomenclature used above,

$$\theta = \frac{r_p}{d} \frac{M_p}{M_*}. \qquad (4.13)$$

If r_p is in units of AU and d is in units of parsecs we can write

$$\theta = 9.551 \times 10^{-4} \frac{M_p}{M_*} \frac{r_p}{d} \text{ arcsec.} \qquad (4.14)$$

Again, applying Kepler's third law we can rewrite this in terms of the orbital period P (in years), which can be measured:

$$\theta = 9.551 \times 10^{-4} \frac{M_p}{M_*^{2/3}} \frac{P^{2/3}}{d} \text{ arcsec.} \qquad (4.15)$$

Thus, if we consider the case of a Jupiter mass planet at 5 AU from its parent $1M_\odot$ star (implying an orbital period of 12 years) at a distance of only 10 pc from Earth, then the position wobble of the star would be $\theta \sim 0.0005$ arcseconds. Attaining such astrometric precision is extremely challenging from ground-based observations. Space-based instruments should be able to attain micro-arcsecond position measurements, and offer a more viable approach. Astrometric monitoring also allows direct constraints to be made on orbital inclination.

4.2.6 Application of Dynamical Measurements

Typically, the radial velocity method for measuring planet-induced wobble is the most sensitive of the dynamical approaches, since it is independent of the distance to the exoplanet system. This is readily understood by considering the velocity and position perturbations in a given case. For example, a Jupiter mass ($1M_J$) planet orbiting at 5 AU from its parent star with a period of 12 years produces a peak $v_r \simeq 13$ ms^{-1}, which is well above any theoretical limiting sensitivity. By comparison, as we saw in the previous section, the corresponding position, or astrometric, wobble is some $\theta \sim 0.0005$ arcseconds, which would be a very challenging measurement to make. If however *both* a radial velocity *and* a position wobble could be measured then the *inclination* of the system could be fully determined and thus M_p could be determined exactly (see Equations 4.11 and 4.15). Ultimately, and especially for terrestrial mass planets, the combination of radial velocity data and space-based astrometric data (§4.2.5 above) can provide the most robust proof of planet detection and characterization.

In reality, when using the radial velocity method, the full orbital configuration must be taken into consideration. Specifically, the assumption of circular orbits must be relaxed. In this case the full expression for v_r^{max} (now in c.g.s. units) is given by

$$v_r^{max} = \left(\frac{2\pi G}{P}\right)^{1/3} \frac{M_p \sin i}{(M_p + M_*)^{2/3}} \frac{1}{(1 - e^2)^{1/2}}, \qquad (4.16)$$

where e is the orbital eccentricity of the planet. In Figure 4.5 we show three examples of the actual observed radial velocity wobble (v_r) as a function of time, and for three different planet eccentricities: $e = 0.0$, 0.34, and 0.67.

The directly measurable orbital parameters are period, time of periastron passage (closest approach to star), the amplitude of the radial velocity variation, orbital eccentricity, the orientation of periastron, and the center-of-mass velocity. In all real cases, the actual estimation of fundamental parameters (e.g., $M_p \sin i$, a, e) requires some quite sophisticated modeling. For example, to determine M_* we must often rely on *stellar models*, which fully determine the internal structure of a main-sequence star of a given luminosity, temperature, and elemental composition. This presents a limiting step to the precision of planetary masses determined by Doppler measurements. Stellar masses determined this way are uncertain to a level of approximately 10%. Factors such as stellar metal content, rotation, helium abundance, age, and magnetic fields all contribute to this uncertainty.

To extract parameters from the curves shown in Figure 4.5 one then typically constructs a computer model which adjusts the unknown variables (including the number of planets) in order to best fit these curves. The parameters are then "measured" as being those which provide the best fit, and their random errors can then also be estimated. In order to successfully fit radial velocity curves there are several additional parameters that must be incorporated. For example, the *mean anomaly* of the orbits and their *longitude of periastron* must also be accounted for. These correspond to the position of a planet in its orbit at the start of the data acquisition and the orientation of the periastron of the orbit with respect to the observer's line of sight, respectively.

Further complications include the presence of *multiple*, massive worlds in a system. An example of the resultant radial velocity curve, as well as the deduced orbital configuration, is shown in Figure 4.6 for the Upsilon Andromedae system. In this case, *three* planets are thought to provide a good fit to the observed radial velocity curve. The innermost has an orbital period of only 4.6 days, but a mass of $0.72 M_J$ (modulo an unknown $\sin i$ factor). The intermediate planet has a period of 242 days and mass $1.98 M_J$, and the outermost has a period of 1269 days and mass $4.11 M_J$.

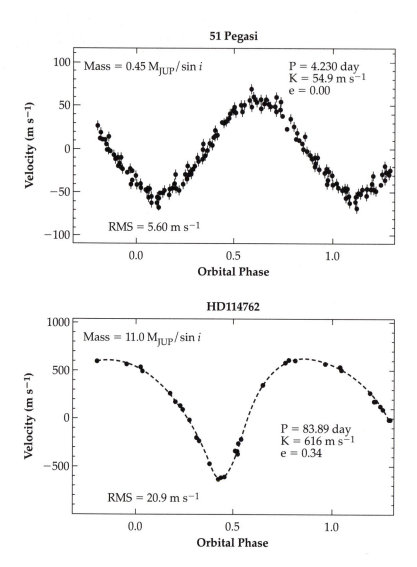

Figure 4.5. Three examples of real radial velocity measurements and the best-fit models. The stellar velocity relative to a mean velocity is plotted as a function of time. In the top panel the planet (51 Pegasi b, mass $0.45 M_J / \sin i$) has a period of only 4.23 days and zero orbital eccentricity. In the middle plot a much more massive planet (HD114762b, $11 M_J$) is shown (note the greatly increased radial velocity variation) with an orbital eccentricity of $e \sim 0.34$. The velocity curve is no longer sinusoidal since the planet spends less time at periastron than at apastron. In the final panel an intermediate mass planet is shown (HD2039b, with a longer period of ~ 3 years) with an eccentricity of $e \sim 0.67$, clearly showing the rapid swing-by at periastron and the slow approach to apastron (plots based upon Butler et al. 1996 published data).

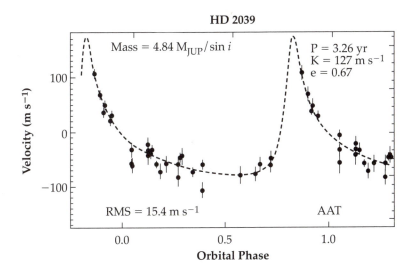

Figure 4.5 *(continued)*.

4.2.7 *Observed Exoplanet Systems*

The first exo-planet discovered using the radial velocity technique (and indeed the first "normal" exo-planet, since previous detections were of "planets" orbiting pulsars[1]) was 51 Pegasi b (see Figure 4.5). It was detected in 1995 by Mayor and Queloz of Geneva, at the Haute-Provence observatory (which must be counted as one of the all time great places for situating an observatory, if human pleasure is a criterion). At the time of writing this technique has several (meaning more than one) hundreds of detections to its credit, including many multiple planet systems (all systems may of course contain multiple planets, but they have simply not been detected yet). One of the most profound, but in retrospect not surprising, discoveries has been just how different most of these systems are when compared to our own, although some share characteristics. An excellent example is the 55 Cancri system, some 13.4 pc away (as well as

---

1. These "planets" may in fact correspond to objects formed from *post* supernova rocky debris in the pulsar systems (Wang, Chakrabarty, & Kaplan, 2006)

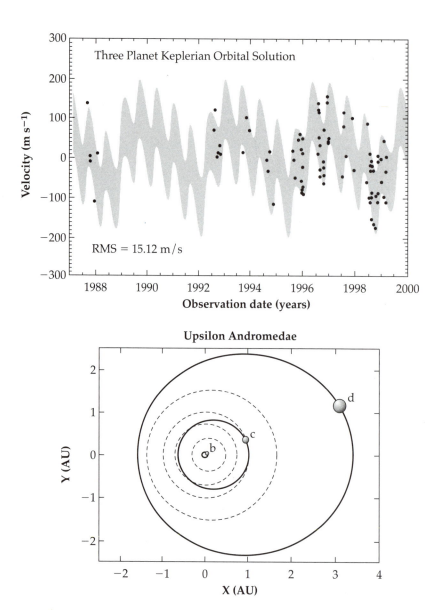

Figure 4.6. The system of Upsilon Andromedae. Upper panel shows the radial velocity curve data and model fit. Three periods are superimposed so that the star exhibits an elaborate "jiggling" with time. Despite this complexity, a multiparameter model can be successfully matched to the data. The lower plot illustrates the estimated orbits of the three planets (solid curves). Plotted for comparison (dashed curves) are the orbits of Mercury, Venus, Earth, and Mars. Since all three planets detected in the Upsilon Andormedae system are gas giants, it clearly represents a very different type of solar system than our own (adapted from plots and data in Butler et al. (1999)).

Table 4.1. The mass and semi-major axes of the
planets of 55 Cancri, deduced from radial velocity measurements. The orbital
eccentricity, e, is also tabulated (from
Fischer et al. 2007).

Name	$M_p \sin i$ (M$_J$)	a (AU)	e
55Cnc b	0.82	0.12	0.02
55Cnc c	0.17	0.24	0.09
55Cnc d	3.84	5.77	0.03
55Cnc e	0.034	0.038	0.07
55Cn3 f	0.14	0.78	0.2

Upsilon Andromedae from the previous section). At the present count
this is thought to contain 5 planets, which we list here in Table 4.1.

The usual notation of letter assignments for planets (b, c, d, e, f) is
used, indicating the order of discovery. The name 55 Cancri corresponds
to the parent star itself. Planets b and c in the 55 Cancri system may in
fact be in a 3:1 mean motion orbital resonance (see Chapter 3)—both are
giant worlds within 0.24 AU. Planet e is some $11M_\oplus$, but orbiting only
0.038 AU from 55 Cancri itself, which is a G-dwarf star of mass $1.03M_\odot$.
Such a world is a real conundrum—a "super-Earth," very close to the
parent star. One possible explanation for the origin of 55Cnc e is that it
formed further out in the system, and migrated inwards, either losing
any accreted gaseous envelope due to evaporation, or never managing
to acquire one (Chapter 3). An interesting, although possibly irrelevant,
fact is that 55 Cancri is actually a binary star. While the above planets
have been detected around the Sun-like (G dwarf) primary, there is a
secondary companion which is a lower mass M dwarf, some 1100 AU
away.

So here is a system that in some ways bears a passing resemblance to
our own (a giant planet at about 5 AU, inner worlds that are less massive,
relatively circular orbits), but in others is extraordinarily different. It has
massive worlds close to the parent star, and a probable mean motion
resonance between major planets. The Upsilon Andromedae system is

even less like our own. As we will see below, however, these systems are *not* unusual in comparison to other exoplanetary systems.

Radial velocity measurements have provided several fundamental insights to the properties of planets in general, and we list these here (e.g., see Marcy et al. 2006):

- The population distribution of all known planetary masses follows a functional form of approximately $dN/dM \propto M^{-1.1}$, and may in fact be bi-modal, reflecting gaseous and rocky planets.

- Beyond about 0.5 AU from the parent star (for at least F, G, and K stars) the number of planets is statistically uniform (flat) with increasing semi-major axis. For periods of 2–3 days there is however a "pile-up," or enhancement of the number of objects.

- Hot Jupiters—or migrated giant planets within 0.1 AU of the parent star (see also Chapters 1 and 3)—exist around some 3% of high metallicity F, G, and K stars (see also transits, below), and about 1% of low metallicity stars.

- Eccentric orbits are common, although there are many fewer high eccentricity planets than low eccentricity planets. The median eccentricity of known exoplanets is approximately 0.2, the mean is 0.24. Excluding hot Jupiters, the mean increases to 0.3.

- The overall likelihood of *giant* planets around a star increases with increasing *stellar* metallicity.

- Multiple planets must be common, and are often in resonant orbits - most likely due to differential planet migration.

While many of these empirical facts have yet to be demonstrated as a natural consequence of our ever-evolving models of planet formation, they nonetheless point towards the beginning of a genuine science of comparative planetology. In Figure 4.7 we plot the estimated mass and orbital semi-major axis for the present sample of radial-velocity detected planets in the same format as our model predictions shown in Figure 3.15 (Chapter 3). Some of the "families" of expected planet types are well sampled by this technique. In Figures 4.8 and 4.9 we plot two further sets of radial velocity planet properties. In Figure 4.8 the estimated orbital eccentricity is plotted against semi-major axis a. As

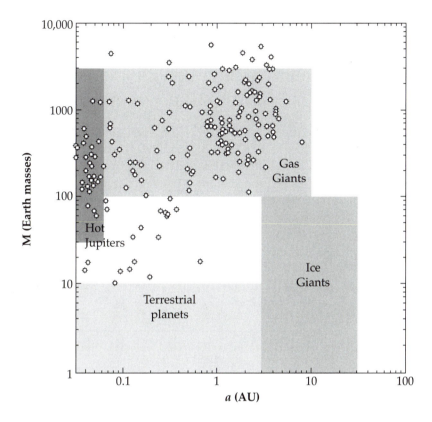

Figure 4.7. The estimated mass (modulo sin i) is plotted versus the estimated orbital semi-major axis for exoplanets detected through radial velocity measurements. This figure can be compared with the example theoretical expectations show in Figure 3.15. The limitations of the radial velocity technique (and the finite amount of time that stars have been monitored for) results in some of the apparent clustering of data points to high mass, small a systems. Nonetheless, we can see that two distinct populations of giant planets are observed—gas giants, and the so-called hot Jupiters.

described in the above list, the mean eccentricity excluding hot Jupiters corresponds to $e \sim 0.3$. It is interesting to note in this plot the moderate clustering of points towards $a < 0.2$ AU, and the general trend for the upper range of eccentricity to *decrease* with decreasing a. Assuming that there are no observational biases causing this distribution, then the ex-

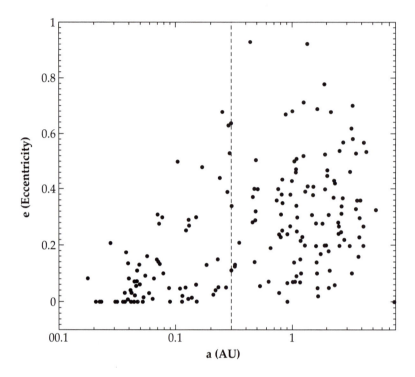

Figure 4.8. The estimated orbital eccentricity is plotted versus the estimated orbital semi-major axis for the sample of exoplanets shown in Figure 4.7. As described in the main text, there is a systematic trend towards lower eccentricity at small orbital radii, and a "pile-up" of objects with 2–3 day periods.

planation may originate in the orbital evolution of systems. As a planet approaches the parent star (i.e., during migration, Chapter 3) it will become increasingly subject to tidal forces (see Chapter 9, 10, and §4.3.1 below), which dissipate energy. Two consequences are that the planet will tend to become tidally locked (i.e., in a synchronous orbit where the planet spin period equals its orbital period) and its orbit will tend to be **circularized** over time. Circularization will drive e to zero, and reduce the semi-major axis. Thus, in Figure 4.8 the general trend for decreasing e is potentially a consequence of tidal dissipation. The apparent "pile-up" of objects within $a \sim 0.2$ AU is greatest at 2–3 day orbital periods,

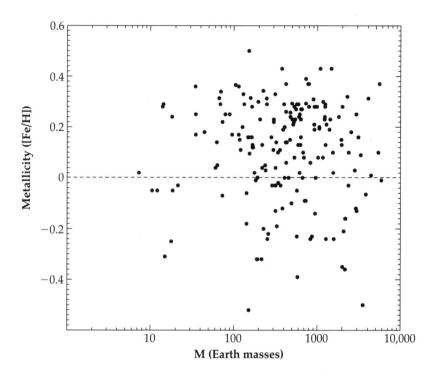

Figure 4.9. The stellar metal abundance (in terms of Fe abundance relative to solar) is plotted versus exoplanet mass (modulo sin i) for the sample used in Figures 4.7 and 4.8. There is a trend towards a greater likelihood of the detection of planets in systems with higher stellar metallicity (and presumed higher circumstellar disk abundance and hence enhanced solids in the core accretion model of planet formation). Planets nonetheless still occur around stars with sub-solar metal abundance.

and may relate to the mechanisms that halt planet migration (at least on short timescales) as discussed in Chapter 3.

In Figure 4.9 we plot the measured *stellar* metallicity relative to solar abundance [2] versus the estimated planet mass (modulo sin i). The origin of the fifth statement in the empirical list of properties given above

---

2. In the usual astronomical units of $[Fe/H] = \log_{10}[(NFe_*/H_*)/(NFe_\odot/H_\odot)]$, where N indicates the number of atoms (i.e., NFe_*/H_* indicates the ratio of the number of

should be immediately apparent. The majority of massive exoplanets are seen in systems with greater metal abundance (i.e., above the dotted horizontal line). It is interesting to note however that planets are nonetheless still found around stars with metallicities a third of that of the Sun (i.e., [Fe/H] = −0.5).

4.2.8   Gravitational Microlensing

In the previous section we discussed radial velocity and position wobble measurements. These are techniques very much suited to the one-by-one study of stars, owing to the technical requirements for very high precision time-resolved stellar spectroscopy, or very high precision astrometric measurements. By contrast, other techniques may be more suited to asking *statistical* questions about the net population of exoplanets in our Galaxy. **Gravitational microlensing** is such a technique; it also provides an excellent excuse to learn something about the nature of general relativity and the lensing of light in general by astrophysical bodies.

Before dealing with *micro*lensing let us consider simple lensing. Gravitational lensing is the **achromatic** deflection of light due to gravity (Einstein 1915). A basic schematic of what happens is shown in Figure 4.10.

We refer the reader elsewhere for a full description of the physical concepts behind general relativity. For our purposes here it should be enough to consider that the path followed by freely falling objects is in general found by solving Einstein's field equation for a given distribution of mass (or energy). For the "simple" case of a spherically symmetric mass one solution is that due to Karl Schwarzschild. This yields the **Schwarzschild metric**, which describes the geometry of space–time surrounding such a mass, and demonstrates that space–time is distorted, or curved, by the presence of the mass. If the path of a photon passing the mass is calculated using this metric it is found that the photon is deflected in physical coordinates, as illustrated in Figure 4.10. The

---

iron to hydrogen atoms in a star) and [Fe/H] = 0 therefore corresponds to a solar iron abundance.

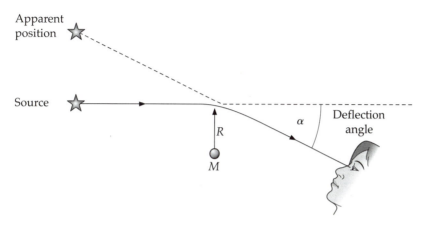

Figure 4.10. The achromatic deflection of light by a mass M between a source (e.g., a distant star) and the observer. In a fashion analogous to classical optics, the mass M acts as a lens. In this example the resultant image of the source is deflected from its unlensed position by an angle α.

deflection angle is given by (assuming weak gravitational fields)

$$\alpha = \frac{4GM}{c^2R},$$
(4.17)

where R is the impact parameter, and we note that the Schwarzschild radius is $2GM/c^2$. As an example of the size of the deflection, consider that the α of a distant star seen close to the edge of the Sun (for example during a solar eclipse) is some 1.74 arcseconds. In fact, just such a deflection was the first direct confirmation of general relativity, and was made in 1919 by Sir Arthur Eddington during solar eclipses seen in Brazil and the Gulf of Guinea on the west coast of Africa.

4.2.9 Magnification

How does this phenomenon help us search for exoplanets? In order to understand this we need to consider the consequences of lensing. Much of this discussion is really just the same as in classical optics. Convex lenses *magnify* sources from the point of view of the observer. Specifically, more photons arrive at the image (observer) than they would otherwise, although the **surface brightness** (the amount of energy per

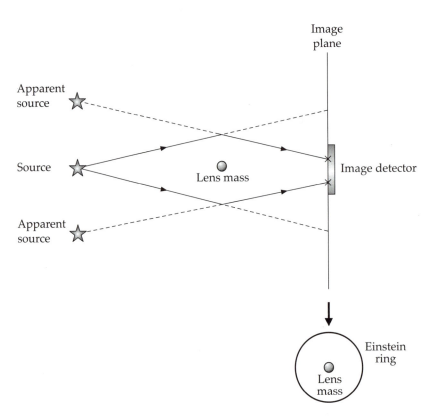

Figure 4.11. An illustration of the nature of image magnification with gravitational lensing. The photons from a distant source that would otherwise pass outside the range of an image detector are deflected onto it, and the apparent angular position of their origin is much greater than it would be otherwise (locations labeled as "apparent source"). The net result is the collection of more photons from an apparently extended image: magnification. In the case where a symmetric source is *directly* behind a point-like lens mass then the image appears as a circle, or Einstein ring.

unit image area, per unit time, per unit solid angle) is conserved. This is illustrated schematically in Figure 4.11, where a lens precisely along the line of sight to a distant source can produce an extended image containing more photons than would otherwise have arrived at the detector.

The specific image geometry in Figure 4.11 arises if both source and lens are symmetric and point-like. The observer will then see a so-called **Einstein ring**, where the source appears as a ring of light around the lens.[3] In this situation the **Einstein ring radius** (R_E) is given by

$$R_E = \left[\frac{4GM_L}{c^2} \frac{(D_S - D_L)D_L}{D_S} \right]^{1/2},$$
(4.18)

where M_L is the point-lens mass, and D_S and D_L are the distances from the observer to the *source* and *lens* respectively. It should again be noted that $2GM_L/c^2$ is just the Schwarzschild radius of the lens mass. If we then write the ratio of lens and source distances as $\beta = D_L/D_s$ we find the projected ring radius at the distance of the lens is

$$R_E = 8.1 \left(\frac{M_L}{M_\odot} \right)^{1/2} \left(\frac{D_S}{8 \, \text{kpc}} \right)^{1/2} [(1-\beta)\beta]^{1/2} \quad \text{AU}.$$
(4.19)

Of more direct astronomical use is the apparent angular size of the Einstein ring, which is given by

$$\theta_E = \frac{R_E}{D_L} = 1.0 \left(\frac{M_L}{M_\odot} \right)^{1/2} \left(\frac{D_L}{8 \, \text{kpc}} \right)^{-1/2} (1-\beta)^{1/2} \quad \text{milliarcsec.}$$
(4.20)

Here we immediately see why, for stellar and planetary masses, we are dealing with **micro-lensing**. For a $1M_\odot$ lens mass 8 kpc distant, and $\beta \sim 1/2$, the Einstein ring radius is only 10^{-3} arcseconds! Thus, for these mass scales the actual lensed image will (with all typical optical telescopes) remain unresolved, and the only effect seen will be the *magnification* of the source brightness as more photons are diverted into the observer's line of sight (see below).

The precision of the alignment required for strong lensing to occur and the constant relative motion of objects in the Galaxy implies that microlensing is, in general, a transient phenomenon. The characteristic timescale for lensing events can be evaluated in terms of the source

---

3. If, in this instance, the source and lens are not perfectly aligned to the observer, then multiple, asymmetric images will be seen rather than a single ring.

travel time (projected velocity v) in the sky across the Einstein ring:

$$t_E = \frac{69.9 \left(\frac{M_L}{M_\odot}\right)^{1/2} \left(\frac{D_S}{8\text{kpc}}\right) [(1-\beta)\beta]^{1/2}}{(v/200 \text{ km s}^{-1})} \quad \text{days.} \quad (4.21)$$

Thus, for events such as the lensing of a star in the Galactic bulge (which provides the best backdrop of sources in terms of density on the sky) by a foreground object, the timescale is a few tens of days in total. Given the need for precise alignment between source and lens it is found that the chance of background Galactic bulge stars being lensed by foreground stellar systems at any given time is $\sim 10^{-6}$. Thus, millions of stars must be monitored in order to stand a chance of catching lensing events as they occur.

4.2.10 Interpreting Microlensing Events

Figure 4.12 illustrates a simplified model for describing lensing by the introduction of the concepts of a **source plane**, a **lens plane**, and an **image plane**. Such a model works quite well *if* the distance between the planes is large. A full mathematical treatment of lensing leads to locations in the lens plane that formally produce *infinite* magnification (e.g., Levine, Petters, & Wambsganss 2001).

These locations of formally infinite magnification can be described as "**critical**" curves, or lines, in the *image* plane (i.e., the location of the observer). They also correspond mathematically to locations in the *source* plane known as "**caustics**." This is all a rather artificial construct—there are not "real" planes in space, but it is a useful simplification for computing the effects of lensing. To illustrate this, in Figure 4.13 we sketch out an example for the case of a point-like lens, and a disk-like (i.e., finite extent) source. As the source approaches the (point-like) caustic, two images are seen, each close to the critical curve (Einstein ring). If the source moves along the track shown in Figure 4.13 then the *total* intensity of light measured in the image plane will vary as illustrated in Figure 4.14. While infinite magnification (μ) would formally occur if the central caustic was crossed, in reality lenses and sources are of finite size and so the intensity curve would be similar to that shown here.

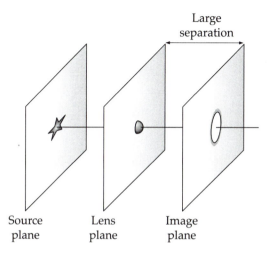

Figure 4.12. An illustration of the basis for one of the mathematical treatments of gravitational lensing. If the distances between source and lens and lens and image/observer, are large then we can treat the deflection of light from objects in the source plane as being due to passage through the lens plane onto the image plane. In this scheme then locations are found in the lens plane known as critical curves, which correspond to formally infinite magnification. Corresponding locations in the source plane are known as caustics. Thus, for a fixed lens distribution, if a source resides on a caustic in the source plane its light will pass through the corresponding critical curves on the lens plane—which describe the image location and shape as seen in the image plane.

This is the basis of the idea of gravitational microlensing—which we will expand upon below—a background object passes through or near the caustic region of a closer object (the lens) and is consequently lensed, producing a time-dependent light curve with a strong, *achromatic*, amplification of the observed brightness. The achromatic nature of lensing is critical for verifying that an event is unlikely to be due to intrinsic stellar variability in either the source or lensing system.

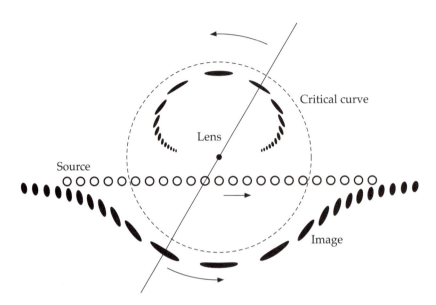

Figure 4.13. The critical curve (dashed circle) is shown for a point-like lens. The caustic is point-like and aligned with the lens. As a source passes (left to right) close to the caustic, two images (dark ellipses) are seen. At any time the source, lens, and images lie on a single line, as indicated (adapted from Paczyński 1996).

Before we deal with the particular characteristics that apply to the detection of exoplanets it is useful to also consider what is perhaps a rather more intuitive approach to understanding the formation of the critical and caustic curves shown in Figure 4.13. Rather than conceptualize the lensing in terms of light rays, or photons, it can be useful to instead consider the **time-delay** effect on **wavefronts**. In Figure 4.15 the basic idea is shown schematically. In this case we consider the lens as introducing a **lag** in the wavefront. The lensing mass introduces a time delay to the passage of the wavefront, since signal propagation is slowed in the presence of a gravitational field (gravitational time dilation). The photons then effectively pile up due to the distortion of the wavefronts and the observer will see a magnification. In Figure 4.15 we also show a close-up of the wavefront distortion.

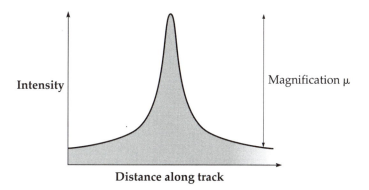

Figure 4.14. The integrated intensity of the lensed source shown in Figure 4.13 as a function of position along its track. The magnification μ is defined as the ratio of the intensity at a given location to that of the unlensed source. As the source approaches the caustic μ increases. If the source crossed the caustic there would be formally ∞ magnification. In practice neither source or lens is actually point-like (i.e., of zero size) and so the intensity curve is smoothed out.

The wavefront passing close to the lens will actually **self-intersect**, and it is this which gives rise to the very characteristic shape of locations where the magnification is formally infinite (which we previously referred to in terms of caustics and critical curves). In fact, in Figure 4.16 we can use this picture to understand both time delays and multiple imaging in lensing.

Observer A experiences only one passage of the wavefront and therefore sees only one image. Observer B, inside the caustic envelope, will be crossed 3 times by different sections of wavefront, each of which appears to come from a different direction due to the lensing distortion. Thus, this observer will see three different images, appearing at different *times* if the source is either intrinsically varying in brightness or is moving through the source plane. Observer C will see multiple images simultaneously—or if the source/lens system is perfectly circularly symmetric along the axis with C, this observer will witness an Einstein ring.

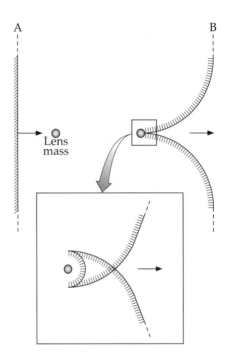

Figure 4.15. An illustration of the time delay, or lag, introduced into a propagating wavefront of light by a point-like lens mass. Before passing the lens mass the wavefront from a distant source (A) can be in phase (a straight front). After passage by the lens mass it becomes distorted (B). In the expanded detail box we see that very close to the lens mass the wavefront is actually folded back on itself so that it self-intersects. Viewed from the direction of propagation the self-intersection forms the characteristic critical curve (c.f. Figure 4.13).

4.2.11 Planet–Star Microlensing

In the case of exoplanet systems there are two possible lensing scenarios: First, the planet and star can be the source, and can be lensed by another object. Second, the planet and star can *themselves* act as the lens.

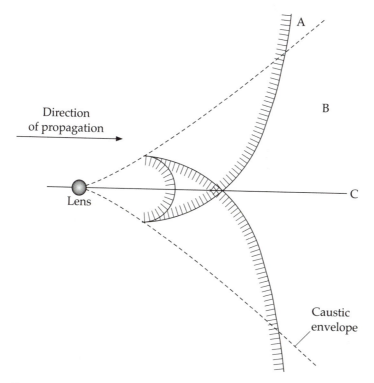

Figure 4.16. Using this wavefront time-delay model we can quickly understand several lensing consequences. As described in the main text, an observer located in the region A (really a volume of space formed by rotating this figure about the lens/obsever axis) experiences only one passage of the wavefront, and therefore sees one image at one time. An observer located in region B is inside the so-called caustic envelope (dashed curves) and will witness three different images at three times as separate parts of the wavefront pass them. Finally, at location C—on the lens/observer axis—an observer will witness multiple images—some simultaneously (forming an Einstein ring).

In either case, the lensing event will occur when the source and lens line up on the plane of the sky close enough for lensing to be significant. This can happen because all objects in our Galaxy are in motion, both around the center of mass of the Galaxy, and with essentially random velocities from the non-zero dynamical temperature of the stellar population (c.f.

the Virial Theorem, Chapter 2). Thus, when such an alignment occurs, which is rare, it typically lasts a short time (by astronomical standards) and will essentially *never* occur again. Microlensing events are therefore unique, but well suited to building up *statistics*. This can yield estimates of the global numbers of planets, and their distribution within the Galaxy, or their relationship to particular stellar properties, such as metallicity. One might ask why we don't just identify the planet–star system subsequently, and make detailed measurements rather than statistical inferences. This is an enormously difficult task, and is only rarely possible, since even a $1M_\odot$ star is relatively faint at distances of many kiloparsecs. Determining the identity of the lensed, or lensing star, is therefore not always possible.

The key in microlensing planet searches is therefore to monitor enormous swathes of the sky, containing many, many stars and to look for the achromatic time-varying brightness of objects. How do we expect to distinguish a planet–star system from just a single star? To answer that, let us consider the situation of a planet–star *lens* which is lensing a background star. In Figure 4.17 we sketch out the critical and caustic curves for a planet–star system where the planet and star are of equal mass—the reason for this is that for a real system the distortion to the critical and caustic curves due to the planet would simply not be easily visible on a diagram this size! Nonetheless, Figure 4.17 illustrates the potential complexity of a light curve when the lens is no longer simple.

The expected light curve shows a characteristic "batman" shape (for want of a more elegant description), where points (a) and (b) correspond to the caustic crossings and formally infinite magnification. If we allow $M_p \ll M_*$, and assume mass A is the star, then the caustic shifts back over to the left and becomes asymmetric. In reality then, when searching for evidence of planets in this scenario one is looking for a small "blip" (Figure 4.18) in the light curve, on top of the magnification due to the parent star. Thus, if the start of the lensing event is caught by a microlensing survey (where the magnification is due almost entirely to the lens star) it can then be monitored closely to look for the "blip" due to any planets associated with that lens star.

In fact, the only way that the planet(s) can be detected in this situation is if they themselves lie close to the critical curves of the lens star, so that the source star is already greatly magnified. For source stars in the

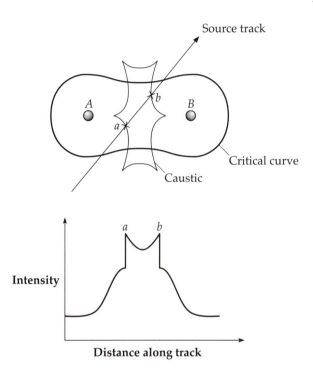

Figure 4.17. An illustration of a two-lens system. The lens masses A and B are equal in this case for purposes of clarity. In the upper panel the resulting caustic and critical curves are shown, together with a sample source track which cuts through the caustic at two locations a and b. In the lower figure the integrated source intensity is plotted versus distance along the track. The formally infinite magnification at caustic crossings is seen at locations a and b. The characteristic shape of this intensity curve allows it to be distinguished from simpler—single lens—events.

Galactic bulge, and typical lens stars, this corresponds to a distance between 0.6 and 1.6 Einstein ring radii from the lens star for giant planets—or a few AU. The precise size and form of the planet "blip" depends primarily on the ratio of planet to lens star masses: $q = \frac{M_P}{M_*}$ and the projected geometry and planet–star separation.

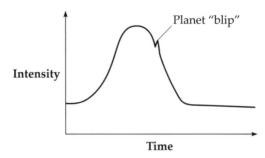

Figure 4.18. A very basic schematic to illustrate the more likely scenario for a planet–star lens system magnifying a background star. The lens star dominates the shape of the intensity curve, producing the primary peak. Depending in detail on the system geometry, the planet lens may (if its caustic curve intersects, or lies close to that of the lens star) produce an additional peak or "blip" during the passage of the source.

In the second lensing scenario, the planet–star system is itself lensed and magnified. Despite the potentially great magnification of the planet–star system, this will be substantially harder to use to detect a planet. This scenario clearly relies on the actual reflected and emitted light of the planet—which we know to be less than a millionth of the parent stars luminosity. Much also depends on the precise timing of the planet–star passage across the caustics and the actual orbital period of the planet (which can cause the planet–star configuration to alter during t_E).

Finally, we note that in either scenario, the specific shape of a microlensing magnification curve is highly dependent on the precise geometry of the lensing event. For this reason, the interpretation of microlensing events is filled with potential pitfalls and requires extensive and careful modeling to ensure that it is done correctly.

4.2.12 Practical Microlensing

The actual detection of planets using microlensing is an ongoing investigation. Current results suggest that *fewer* than 21% of lens stars have

Jupiter mass planets orbiting between 1 and 4 AU. However, the microlensing detection of a $5.5M_\oplus$ planet with an orbital radius of about 2.6 AU (Beaulieu et al. 2006), appears to support the core accretion model for planet formation (Chapter 3) since cool, sub-Neptune mass worlds are therefore (statistically) more common than gas giants.

It is the *ratio* of M_p to M_* that dictates the size and form of the lensing feature in the case of a planet–star lens. Since the parent (lens) star is typically *not* known, the estimate of M_* requires quite extensive, probabilistic modeling based on the overall size of the total microlensing light curve and the expected distribution of stars between us and the lensed objects (e.g., Galactic bulge stars). Similarly, to evaluate more system parameters, such as orbital period, extensive likelihood analyses are required. Despite these complications, there is the potential to exploit the enormous *resolution* gain during caustic crossing (due to the image being spread out). For example, with a star–star lens and a planet–star source it is theoretically possible to even resolve structures like planetary rings.

4.2.13   Transits and Eclipses

The use of transits or eclipses is as ancient as astronomy itself. Solar and lunar eclipses (both partial and total) have been witnessed throughout the Earth's history (and not always by humans). Modern astronomy arguably began with Galileo Galilei realizing that the disappearance and reappearance of the Jovian moons Io, Europa, Ganymede, and Callisto corresponded to their transit of, or eclipse by, Jupiter—thereby indicating that they are in orbit about the planet.

4.2.14   Planet Transits

If a planet passes between the observer and its parent star then the light from the star will be attenuated as the planet passes across the stellar disk (initially assumed to be uniform, Figure 4.19). Typically the attenuation will be significantly less than some 1–2% of the total starlight and, as we will see below, occurs rarely owing to the need for a narrow range of system–observer geometries. The attenuation is, to first order, just the ratio of the planet and star visual disk areas

$$\frac{\Delta L}{L_*} \simeq \left(\frac{R_p}{R_*}\right)^2 . \tag{4.22}$$

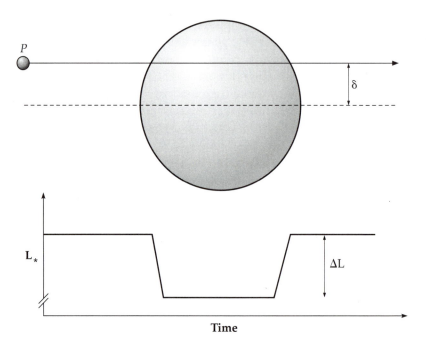

Figure 4.19. The basic model for a planet transiting in front of its parent star is shown here. In the upper plot the planet (P) can, to first order, be considered to be moving along a linear path at a stellar latitude δ. As it transits, a fraction of the stellar light is blocked and the net luminosity of the system appears to drop. This is shown in the lower plot. The shape of the transit curve in this case is just due to the geometric superposition of two uniform, circular disks. The luminosity L_* is reduced by ΔL during full transit.

For a Jupiter radius planet and solar mass star, $\Delta L/L_* \simeq 1\%$ - a change measurable with ground-based observatories. For an Earth-sized planet and solar mass star, $\Delta L/L_* \simeq 0.008\%$. This is small, but certainly not impossible to measure with a dedicated space-based observatory, as long as the eclipse can be repeatedly measured (which can be an issue for long period objects, spanning years). For smaller stars (e.g., M-dwarfs) the transit depth is greater and the potential for detecting terrestrial sized planets is enhanced - however, such stars must be close enough to us to provide sufficient photons to make such measurements, and the finite space density of these systems poses a difficulty in this regard. Nonetheless, although it has many challenges, this technique still offers the possibility of detecting Earth-sized planets, and potentially even smaller objects or features, such as ring systems or even moons (particularly by using the eclipse timing information).

For the linear geometry depicted in Figure 4.19 above (which we will improve upon below)—we ignore the actual orbital figure of the planet and just assume a straight line path in front of the star. The **transit duration** is then easily obtained as

$$\tau = \frac{P}{\pi} \left(\frac{R_* \cos \delta + R_p}{a} \right) \qquad (4.23)$$

where P is the orbital period, a is the planet–star separation at transit, and δ is the mean stellar latitude of the transit (e.g. Sackett 1998). Thus, the duration τ can be directly measured, and the interval between transits also yields P to high precision.[4] The stellar size and mass, R_*, M_*, can be estimated using our detailed knowledge of stellar structure and spectroscopic measurements of the star. Subsequently the planet–star separation a can be determined from Kepler's 3rd law, and R_p can be estimated using the observed attenuation $\Delta L / L_*$, and then the *latitude* δ of the transit can be estimated.

This is therefore very promising, but we should now include some more details of the actual transit geometry. Figure 4.20 illustrates the situation for an arbitrary orbital plane inclination—although we still assume a circular orbit so that there is no issue of the orientation of the semi-major axis of the orbit.

From Figure 4.20 we can see that $\cos i = h/a$ and $\sin \delta = h/R_*$, thus we can eliminate h to obtain

$$\cos i = \frac{R_* \sin \delta}{a}. \qquad (4.24)$$

Hence the latitude of transit and the system inclination are interchangeable in this context.

---

4. Indeed, *variations* in the sequentially measured P can reveal the presence and configurations of *other* planetary bodies in the system, which perturb the primary transiting object (e.g., Agol et al. 2005).

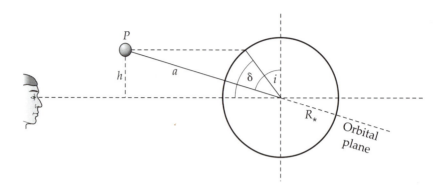

Figure 4.20. The geometry of a transit related to the inclination i of the system. Transit latitutde and inclination are interchangeable since $\cos i = (R_* \sin \delta)/a$.

4.2.15   Transit Probability

We can evaluate a general probability for a given observer to see exoplanet transits based entirely on geometry. Consider the system shown from the point of view of the observer in Figure 4.21. We will further assume that (for a circular orbit) $a \gg R_*$ and that $M_p \ll M_*$ and so $R_p \ll R_*$. Consequently, in order for the observer to witness a transit the orbital inclination i must satisfy

$$a \cos i \leq R_* + R_p. \tag{4.25}$$

If we assume that for *all* stars observed the inclination of planetary orbits is *random* with respect to the observer's location, then $\cos i$ will take all (random) values between 0 and 1. Thus, over some large sample of stars we can calculate a **geometric transit probability**:

$$\frac{\int_0^{(R_*+R_p)/a} d(\cos i)}{\int_0^1 d(\cos i)} = \frac{R_* + R_p}{a} \approx \frac{R_*}{a}, \tag{4.26}$$

since $\int_0^1 d(\cos i) = 1$ over all random inclinations. Therefore we immediately see that on the basis of geometry, *small a* and *large R_** are favored for transit detections. For example, the transit probability for the Earth–Sun system is ~ 0.005, so we would have to monitor *at least* 200 stars in

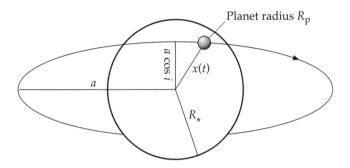

Figure 4.21. A second view of the geometry of a transit at arbitrary system inclination—this time from the point of view of the observer. Clearly, for a transit to occur the condition $a \cos i \leq R_* + R_p$ must hold.

order to stand a chance of detecting just one such transit—assuming all 200 systems indeed harbored Earth-type worlds at 1 AU. This is the basis of most "blind" searches for planet transits—simply monitor as many stars (in the thousands to millions) as possible. The result in Equation 4.26 is however slightly misleading. While it is geometrically correct it does not factor in Equation 4.22 - the larger the star the *smaller* the relative transit depth for planets. Smaller transit depth (dimming) is much harder to detect, even if it is more likely to occur. Additionally, while a $\simeq 1\%$ dimming of a dwarf star may indicate a planetary transit, a similar dimming of a giant star may actually indicate the transit of a stellar companion, not a planet at all.

In principle there could be ways to "pre-select" systems. If the rotational vector of a star can be measured by detailed time-series spectroscopy (which is sensitive to persistent features such as sunspots), and we assume that this likely matches the orientation of the plane of any associated planetary system, then we can estimate the inclination and monitor those systems close to $i = 90°$. The inclination is calculated using the measured line-of-sight stellar rotational velocity, $v_* \sin i$, as

$$\sin i = \frac{v_* P_*}{2\pi R_*}.$$
(4.27)

For solar-type stars however this turns out to be too difficult to measure with any certainty. Another approach might be to use infrared and sub-mm wavelength imaging observations of planetary *debris disks* (Chapter 3) and to again preferentially monitor those systems with inclinations close to 90°.

4.2.16   Transit Duration (Revisited)

The transit (or eclipse) duration is just equal to the fraction of the planet's orbital period during which (Figure 4.21) $x < R_* + R_p$. By further considering the geometry we can obtain an estimate of this, which now includes the inclination information (unlike our crude estimate in §4.2.14). From Figure 4.22 we see that

$$y = \sqrt{(R_* + R_p)^2 - a^2 \cos^2 i},$$
(4.28)

and

$$\theta = \sin^{-1}\left(\frac{\sqrt{(R_* + R_p)^2 - a^2 \cos^2 i}}{a}\right).$$
(4.29)

Therefore, the transit duration is $\tau = \frac{P}{2\pi} \times 2\theta$ and if $a >> R_* >> R_p$ then

$$\tau = \frac{P}{\pi}\left[\left(\frac{R_*}{a}\right)^2 - \cos^2 i\right]^{1/2} \leq \frac{P R_*}{\pi a},$$
(4.30)

where $a \cos i \leq (R_* + R_p)$ for transit and therefore we do not encounter imaginary roots.

This can also be written in terms of the *latitude* of transit across the star (Figure 4.20 above). Recalling that $\cos i = (R_* \sin \delta)/a$ we can then write

$$\tau = \frac{P}{\pi}\left[\left(\frac{R_*}{a}\right)^2 (1 - \sin^2 \delta)\right]^{1/2},$$
(4.31)

$$\tau = \frac{P R_* \cos \delta}{\pi \quad a},$$
(4.32)

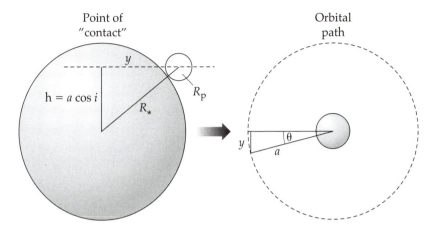

Figure 4.22. An illustration of the geometry used to determine the transit duration as a function of system inclination. The projected distance y represents the time taken from point of contact (left plot) to the midpoint of the transit. This corresponds to an angle θ (right- hand plot, plan view) which to first order can be converted to a fraction of the orbital period and hence transit duration.

which differs from our earlier expression for τ only through the loss of the R_p term. Thus, δ and i are effectively interchangeable in terms of evaluating τ.

Among the approximations made in our above expressions for τ has been the assumption that the path of the planet during transit across the stellar disk is a straight line. It would be better to account for the fact that the transit path will have curvature due to the true orbital path. We simply quote here the full expression obtained in this case:

$$\tau = \frac{P}{\pi} \sin^{-1} \left[\frac{R_*}{a} \left[\frac{(1 + \frac{R_p}{R_*})^2 - (\frac{a}{R_*} \cos i)^2}{1 - \cos^2 i} \right]^{1/2} \right]. \qquad (4.33)$$

Applying this expression we can generate the curves shown in Figure 4.23. What is striking about this plot is that the actual transit duration (τ) is *extremely* sensitive to the geometry as dictated by the inclination i—only a few degrees away from $i = 90°$ and the transit vanishes for all but the smallest orbital radius planet–star systems. We also note that, for

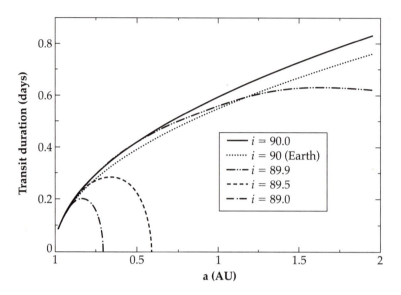

Figure 4.23. Transit duration (τ) is plotted versus orbital radius according to Equation 4.31. Curves correspond to varying system inclinations. The dotted curve is for an Earth-sized planet in orbit around a solar mass (and radius) ($1M_\odot$) star. All other curves correspond to a Jupiter-sized planet in orbit around a solar mass (and radius) star. The dependence of $\tau(a)$ on i is acute. Thus, in addition to the geometric probability of transit according to Equation 4.26, the actual duration of a transit rapidly decreases with changes in inclination. Blind searches for planet transits must therefore monitor many stars with good time resolution.

outer planets (such as Jupiter) with orbital periods measured in years (rather than the days for close-in systems), the probability of observing a system at exactly the time of transit is of course greatly decreased. Surveys for transiting planets (by looking for the characteristic transit light curves) are therefore strongly biased towards planets that are close to their parent stars.

How then does one actually apply this to real measurements of exoplanet characteristics? The basic scheme is straightforward:

1. Measure τ, and with repeats measure P.

2. Measure L_* and take spectroscopic data, then from $L_* = 4\pi R_* \sigma T^4$ and stellar evolution models estimate R_* and M_*.

3. From measured $\Delta L / L_*$ estimate R_p.
4. From M_* and P and Kepler's laws estimate the semi-major axis a (typically assunimg a circular orbit).
5. Calculate δ or i from the above.

Further complications include non-zero orbital eccentricity (which can be allowed for) and the unknown orbit orientation in such cases. Depending on the system parameters it is often possible to follow up transit observations with detailed radial velocity measurements. Since the inclination of the system can be determined from the transit data it is then possible to *directly* estimate M_p.

4.2.17 Transit Light Curves and Other Effects

The detailed *shape* of a transit light curve is in reality a complex function of many variables. However, far from being a problem this actually presents a wealth of information about both the parent star and the transiting planets—although extracting this information may be tricky.

Let us consider some of the principal factors going into the shape of the transit light curve beyond the simple geometric attributes of two overlapping (presumed) circular shapes:

1. Limb darkening of parent star. From the study of basic stellar atmospheres we learn that the outer visible atmosphere of a star— the **photosphere**—has a finite optical opacity, or optical *depth* (see Chapter 9). One consequence of this is that towards the edges of a stellar disk we detect light emitted from *shallower* depths in the atmosphere compared to the center of the disk. The shallower material is cooler, and so for blackbody radiation (where $L \propto T^4$) the outer edges of the disk will appear less luminous. This creates the effect of limb darkening in our own Sun and in other stars. Thus—a typical stellar disk does *not* appear uniform in either temperature or brightness, and so a transit light curve is not simply the geometric obscuration of a uniform disk (Figure 4.24).

2. The wavelength dependence of limb darkening results in the transit changing the net color of the measured starlight as a function of time.

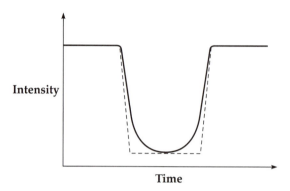

Figure 4.24. An illustration of the general effect of stellar limb darkening on the transit light curve profile. Dashed lines indicate the purely geometric light curve expected for a uniformly emitting stellar disk. The solid curve indicates the actual form typically seen, which includes stellar limb darkening. The precise shape of this curve, and its dependency on wavelength, can be used to fit a model for the stellar disk.

3. Planetary atmospheres. Any finite planetary atmosphere will effect the shape of the transit curve during ingress and egress and the overall depth of the curve—*as a function of wavelength* - due to the detailed properties of light transmitted through an atmosphere.

Of these factors, both 1 and 2 can be largely understood from our models of stellar atmospheres. We can therefore exploit the detailed transit light curves at different wavelengths to help determine the characteristics of planetary atmospheres (see §4.2.19).

More subtle effects due to transits also exist, which may offer additional probes of systems. One of the most interesting actually effects the *radial velocity* measurements of a system. The **Rossiter–McLaughlin effect** is due to the rotation-induced Doppler shift in emission from a stellar surface and its partial obscuration during transit. This is illustrated in Figure 4.25.

In unobscured radial velocity measurements, the spectrum observed is that due to the *integrated* light across the stellar disk, and therefore

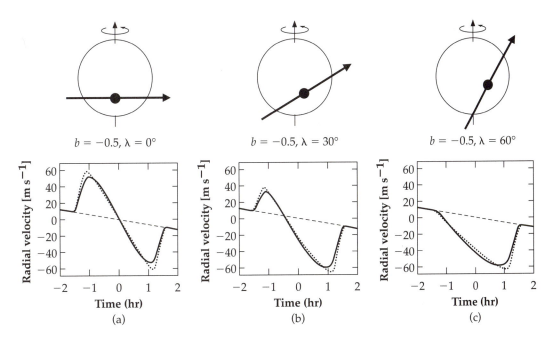

Figure 4.25. In panels (a), (b), and (c) the nature of the Rossiter–McLaughlin effect is shown for increasing values of λ—the angle between the transit path of a planet (or chord) and lines of constant latitude on the star (i.e., a normal to the stellar spin axis). In all three cases a planet transits the star at a fixed distance from the edge of the stellar disk (an impact parameter $b = -0.5$). Considering panel (a): as the planet begins its transit it blocks some of the blueshifted light of the star, hence the spectroscopically measured net velocity of the star appears to be slightly more positive (redshifted) in the lower plot of radial velocity relative to the unobscured value (straight dashed line). Solid and dotted curves in the lower plots correspond to models with and without limb darkening respectively. At the transit center the planet blocks emission that is moving perpendicular to the line of sight and the radial velocity offset goes to zero. As the planet begins its egress from transit it now blocks redshifted emission and the net radial velocity appears negative. The effect of increasing λ is shown in panels (b) and (c) - demonstrating how constraints can be made on its value from detailed spectroscopic measurements. (This figure is based on Figure 2 of Gaudi & Winn 2007.)

includes the Doppler broadening of spectral lines due to the intrinsic rotation of the star. During a transit the obscuration will selectively re-move some part of that spread in wavelength, and as a consequence the apparent shape of a given spectral feature (e.g., atomic line absorp-tion) will change, and its apparent peak may shift. This effect can be exploited - through observing the spectrum when a transit is known to occur—with careful modeling, to reconstruct the *relative orientation* of the stellar spin axis and the planetary orbital axis. This is a very useful piece of information for comparison with our models of planet formation (Chapter 3), since any misalignment between these two axes reflects details of the processes setting the final orbital architecture of a system. Current measurements of the effect (e.g., see Winn et al. 2006) indicate typical spin–orbit alignments to within a degree or so.

4.2.18   The Prototype: Transiting Planet of HD 209458

The easiest way to discuss the practical aspects of transit observations is by examining the prototype of planetary transit measurements. In 2000 it was discovered that the star HD 209458 underwent periodic decreases in its apparent brightness every 3.5 days (Charbonneau et al. 2000, Henry et al. 2000). The transit curve is shown in Figure 4.26.

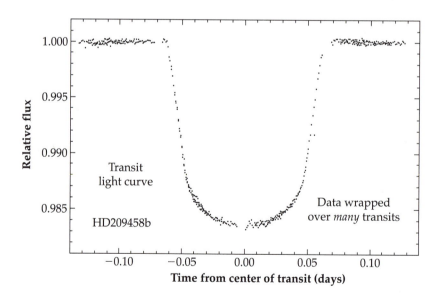

Figure 4.26. The transit curve for the prototype transiting planet HD 209458b (data from NASA's Hubble Space Telescope). Data from multiple transits have been combined here to show the shape of the transit curve in detail. It clearly indicates the stellar limb darkening effects expected (Figure 4.24).

With complementary radial velocity measurements the basic parameters of this system can be determined as $R_* = 1.146 \pm 0.05 R_\odot$, $R_p = 1.347 \pm 0.06 R_J$, $i = 87°$, $a \simeq 0.045$ AU, $M_p \simeq 0.69 M_J$ and $\tau \sim 3$ hours and $P \sim 3.5$ days (Henry et al. 2000). This places the planet, HD 209458b, in the category of a hot Jupiter (see Chapter 3).

In order to search for evidence of an atmosphere, the expected characteristics of a gas giant planet were examined and, within the expected range of temperatures and pressures, it was determined that atmospheric sodium (Na) should produce a very strong absorption feature at a wavelength of 589.3 nm. The presence of such absorption could then be searched for by comparing the transit curve at different wavelengths, including that of sodium. If an atmosphere were present then the transit curve should be systematically *deeper* at the sodium absorption wavelength than at other, arbitrary wavelengths. The test is illustrated and described in Figure 4.27. A preferential sodium absorption was indeed observed, indicating the likely presence of an extended atmosphere around HD 209458b. Subsequent measurements claim the detection of other elements, such as oxygen, hydrogen, and carbon.

The potential diagnostic power of transit curve measurements is great, and this was immediately illustrated by HD 209458b. We can perform a quick estimate of the *density* of this planet, and compare it to that of Jupiter. For Jupiter the mean density is 1.33 g cm^{-3}, for HD 209458b the mean density is only 0.37 g cm^{-3}—and yet these are planets of quite similar mass!

The answer perhaps lies (although see below) in the extraordinarily close proximity of this planet to its parent star, only 0.045 AU. At this distance the stellar irradiation alone produces an equilibrium temperature for the planet's surface of between 900 and 2000 K. By comparison, Jupiter has a mean effective temperature of about 160 K. This might act to both inflate the planetary atmosphere (and hence the inferred planet radius from transits), and alter its atmospheric chemistry. The later measurements of a much more extended envelope of hydrogen around the planet seem to provide some support for this scenario. In the case of HD 209458b it is further expected that the planet is likely to be *tidally locked* with its parent star: i.e., tidal forces will have created a synchronous orbit, where the same face of the planet always points towards the star (like the Moon–Earth system). This may create atmospheric circulation

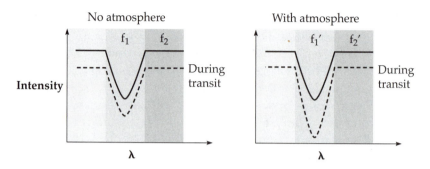

Figure 4.27. A schematic of the method for detecting the presence of a planetary atmosphere during a transit. A strong spectral absorption feature likely to exist in an exoplanet's atmosphere (e.g., Sodium (Na) for a gas giant) is chosen and the net flux is measured from this region and an adjacent region in wavelength before and during transit. The left panel illustrates this in the case of no atmosphere; flux f_1 is measured from wavelengths centered on the spectral absorption feature and flux f_2 is measured from an adjacent band (in practice f_1 and f_2 are normalized to the continuum flux at this wavelength). In this case the ratios of flux before and during transit are equal, $f_1/f_1^{\text{transit}} = f_2/f_2^{\text{transit}}$. In the right panel the situation is shown when an absorbing atmosphere is present. The absorption feature is *stronger* during transit due to the atmosphere. In this case then $f_1'/f_1'^{\text{transit}} > f_2'/f_2'^{\text{transit}}$ and the atmospheric signature can be detected.

patterns and produce energy transport in the planet unlike any seen in our own solar system (see also Chapter 9).

However, other known transiting hot Jupiters exhibit quite a wide range in apparent density, from some 0.3 g cm^{-3} to almost 1.5 g cm^{-3}. This may indicate that stellar heating alone is not responsible. Other possibilities include tidal dissipation of energy driven by the effects of other planets (e.g., forcing an eccentric orbit), planet spin-axes that are not aligned with the orbital axes (motivating the use of the Rossiter-McLaughlin effect, §4.2.17), and of course varying intrinsic composition of the planetary atmosphere and interior.

The importance of planet transit data is further illustrated by considering what is gained with a knowledge of planet size. As we discussed in Chapter 3 (§3.7.4, and Figure 3.14), the mass–radius relationship of planets is a powerful diagnostic of their composition and internal structure. In Figure 4.28 we repeat the theoretical model curves of Figure 3.14, but now include the positions of a few selected exoplanets that transit. It

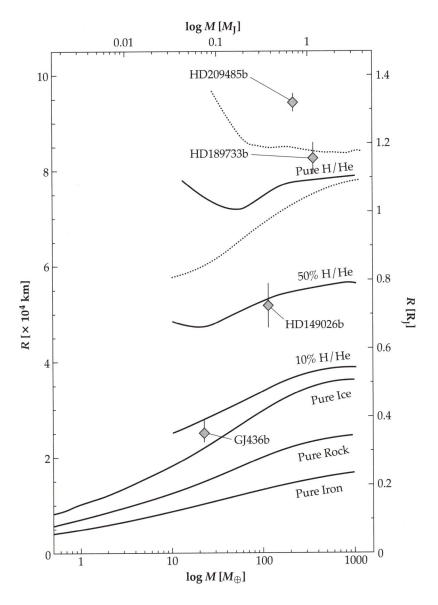

Figure 4.28. The mass–radius relationships for a range of planetary compositions. See Figure 3.14 for a full description of the plotted curves. The planets HD209458b (see text), HD189733b, HD149026b, and GJ436b are also shown to illustrate some of the range of planetary types in known exoplanet systems (based on Gillon et al. 2007).

is immediately apparent that these exoplanets encompass a wide range of planetary types (something not discernible from mass measurements alone).

4.2.19 Planet "Weather"

Once one has a repeatable transit measurement that depends on the structure and composition of a planetary atmosphere, many scientific possibilities open up. Among these is the ability to study time-varying phenomena, such as "weather"—meaning variations in the circulation and structural characteristics of a planetary atmosphere. For hot Jupiters such as HD 209458b the most likely cloud, or condensate, structures are actually condensations of methane or carbon monoxide at lower temperatures (and therefore lower altitudes) and enstatite ($MgSiO_3$) at higher and hotter altitudes. Such clouds are highly opaque, and so their distribution in the atmosphere can drastically alter the transmission of light, and consequently the transit curve—especially around a given spectral feature. This is illustrated in Figure 4.29. High clouds reduce the effect of the preferential atmospheric absorption since they block more light altogether. Lower clouds have less effect.

Because of this it is, in principle, possible for a persistent "weather" pattern—such as a circular system like a hurricane, or Jupiter's Great Red Spot—to be detected through transit observations, as well as other less persistent cloud features. Indeed, highly persistent features seen through transit data could be used to determine the planet spin rate and axis in the same way that sunspots can be exploited on stars.

Other techniques that exploit the transit and photometric properties of a system can also yield information on the surface structures and temperatures of planets, as we describe in the following sections.

4.2.20 Secondary Eclipse

A planet that is seen to transit its parent star will also (typically) be eclipsed by the star during its orbit.[5] The net brightness of the system will therefore be reduced during an eclipse. As discussed above, for re-flected light in the optical band, this will generally correspond to a tiny

---

5. The only time that this may not occur is for eccentric orbits in very particular geometries.

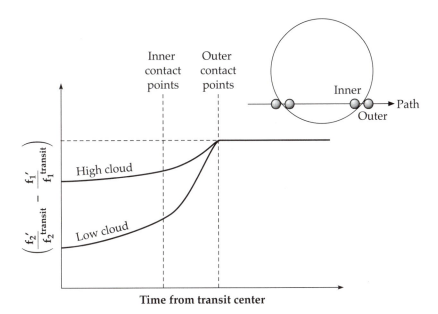

Figure 4.29. An illustration of the effect of opaque cloud structure in a transiting planet's atmosphere on transit measurements. Using the nomenclature of Figure 4.27 the relative flux *difference* on and off an absorption feature: $f_2'/f_2'^{\text{transit}} - f_1'/f_1'^{\text{transit}}$ is plotted versus time from the transit center. In this example both the ingress and egress transit data have been folded around the central transit time. The location of the inner and outer contact points is sketched in the upper right hand side of this figure. Since clouds are typically highly opaque (being condensates) the presence of *high clouds* actually reduces the net amount of atmosphere available to absorb photons, and thus $f_1'/f_1'^{\text{transit}}$ is smaller. Lower clouds block less of the atmosphere and so produce a distinctly different curve.

reduction of approximately one part in a billion. However, one known class of exoplanets—the hot Jupiters (Chapter 3 and above)—may contribute a substantially greater fraction of the infrared light of a system. The close proximity of these planets to the parent star suggests that their outer atmospheres will be significantly heated (c.f. HD 209458b in §4.2.18). Their emission temperature (assuming a blackbody) will be enhanced, and hence their net infrared luminosity can be quite large. We can make a rough calculation of this. Taking the physical parameters describing HD 209458b and assuming an atmospheric temperature of ~ 1000 K we arrive at a net luminosity (peaking in the infrared) of $L_p = 4\pi R_p^2 \sigma T^4 \sim 6.6 \times 10^{28}$ erg s^{-1}. Since a net solar luminosity (peaking in the optical) is $L_\odot = 3.8 \times 10^{33}$ erg s^{-1} we can see that a 1000 K hot Jupiter may represent *at least* 0.002% of the net system luminosity—and a correspondingly larger fraction of the infrared luminosity. This may not sound like much, but just such emission has indeed been detected in the infrared light of hot Jupiter systems. To give an example, the hot Jupiter in the system known as TrES-1 has been measured to contribute approximately 0.07% of the system flux at an infrared wavelength of 4.4 μm. This corresponds to a planet emission temperature of about 1100 K (Charbonneau et al. 2005). Such temperatures are similar to those of low-mass brown dwarfs (Chapters 1, 2), but spectral line widths are smaller.

With sufficient instrumental sensitivity and a suitable planet system, even more detailed information can be extracted by closely monitoring the emission from a planet as it goes into and emerges from secondary eclipse. In this case the limb of the star effectively "scans" across the face of the planet, providing information on the spatial distribution of observed emission. Again, the emission of the planet is deduced by careful measurement of the increase or decrease of net infrared emission from the system during eclipses.

The practical application of eclipse techniques relies heavily on having prior knowledge of the precise orbital period of the system. This enables both the targeting of observations as a function of time, and the "stacking" of data from individual eclipse ingress or egress episodes, in order to produce a statistically robust mean measurement of flux variation.

4.3   Direct Planet Detection and Imaging

The ultimate goal of exoplanet observations must surely be the direct detection and imaging of other worlds. As we have discussed above, there are considerable practical barriers to overcome. At the time of writing, however, the first glimpses of planetary characteristics beyond those of our own solar system have been obtained, and many of the necessary astronomical techniques are being developed and tested.

In Chapter 6 we discuss questions of how to interpret the light directly detected from planets; here we describe in outline some of the techniques that may be utilized to accomplish this detection.

As we described at the start of this chapter, the main challenge to detecting and imaging exoplanets is to separate the planet light from that of the parent star. If we had no terrestrial atmosphere to contend with for our ground-based observations, or could naturally manufacture huge, optimal telescopes, much of this would be moot. However, in reality we do face these obstacles—even if we move outside of the Earth's atmosphere we are still highly constrained by the size and precision of the optics required for this task. Here we will discuss some of the main approaches currently being pursued to solve these problems.

4.3.1 Phase Curve Photometry

The technique we refer to as **phase curve photometry** is perhaps intermediate between the partially indirect techniques of the previous sections and the direct observations described further below. Put simply, it involves the possible modulation of light from a stellar system due to the varying reflection and/or emission from exoplanets as a function of orbital phase, or position. Unlike the secondary eclipse method described in §4.2.20, it does not rely on the occurrence of a transit.

We will discuss the specifics of planetary reflectivity, emissivity, and optical phase more in Chapter 6, but here briefly outline phase curve photometry as applied specifically to hot Jupiter type planets. The modulation of reflected (optical wavelength) light can be observed in the different phases of the illuminated and nightside hemispheres of the planet (Figure 4.30). The modulation of *emitted* (infrared wavelength) light, due to the intrinsic temperature of the planet, may be rather more complex.

In the case of a hot Jupiter (as discussed for HD 209458b above) it is generally expected that tidal interaction between the giant planet and the parent star will, over time, result in **tidal locking** of the planet (see also Chapters 9 and 10). Thus the planet's spin period should eventually equal its orbital period, and the same side of the planet will always face the star. This is a form of resonance (see Chapter 3), and in the absence of external perturbations will also result in the circularization of the planetary orbit (see also §4.2.7). As briefly discussed in §4.2.18, the precise

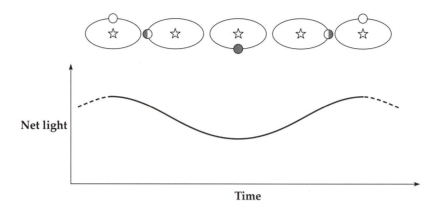

Figure 4.30. A schematic illustration of phase curve photometry. The upper pictorial sequence depicts the varying phases of a non-transiting exoplanet from the observer's point of view. The lower plot is a sketch of the apparent variation in net *reflected* light from the system, due to the varying planetary phases.

nature of the response of the planetary atmosphere to this highly non-uniform heating situation is unclear. Thus, if there is no redistribution of heat energy in the planet's atmosphere the day/night emission temperature may be dramatically different. If, on the other hand, the planet redistributes energy it may have a very uniform emission temperature. In this latter case the infrared phase curve may show little variation with time.

An example of the application of this technique is the case of the hot Jupiter orbiting the star Upsilon Andromedae (see §4.2.6) at a distance of 0.059 AU. Using prior knowledge of the orbital period and phase of the planet from radial velocity measurements, the tiny matching phase variation in infrared luminosity of the system can be searched for. Harrington et al. (2006) were able to make this measurement and deduced that the day/night emission variation of the planet is approximately $3 \times 10^{-3} L_*$ where L_* is the parent star's infrared luminosity. With careful modeling this can be translated into a day/night temperature *difference* of about 1400 K—with a nightside temperature below 273 K. In this case therefore the planet essentially has a "hot-spot" on its dayside, with relatively little redistribution of heat energy in its outer atmosphere.

Using this approach combined with secondary eclipse data the spatial variation in upper atmospheric temperature has also (for example) been determined for the transiting planet HD 189733b—a nearby (19 pc) hot Jupiter orbiting a K-dwarf star. In this case it has been found that there is indeed a likely "hotspot" on the dayside of the planet (which is tidally locked), peaking at some 1200 K and spanning some 20–30° on the planet (Knutson et al. 2007). This area of highest temperature also appears to be some 16° offset from the so-called substellar point (the closest part of the planet to the star).

4.3.2 Nulling Interferometry

A clever, but technically challenging approach to the direct imaging of planets involves exploiting the wave nature of light itself. The experimental setup is illustrated by the schematic in Figure 4.31.

If one considers wavefronts traveling from both a star and an associated planet then, although separated by a very small angle, the wavefronts will clearly approach an observer from slightly different directions (denoted by θ). If the star–planet system is observed by two physically separated telescopes (distance d apart) that are targeted on the star, then a wavefront from the planet will arrive at the telescopes at different times—or with a *phase* delay. The photons gathered by both telescopes can be combined to form a single image at a detector. An artificial *phase delay* is introduced in the light waves collected by one of the telescopes (number 2 in Figure 4.29) equal to exactly *half* the wavelength of the detected light (a delay of $\lambda/2$). In doing so then clearly the light from the *star* will cancel out, or *null*, when combined, or interfered, in the detector. Without the artificial delay, since the stellar wavefronts arrive at the telescopes at exactly the same time, they would otherwise be perfectly in phase at the detector. The light from the planet on the other hand is already out of phase between the two telescopes, and so a correctly configured interferometer will *bring the planet light into phase* at the detector.

Thus the starlight is nulled away, leaving only the light from the planet. This alleviates one of the principal challenges to planet detection—by simply removing the star! The actual requirement for

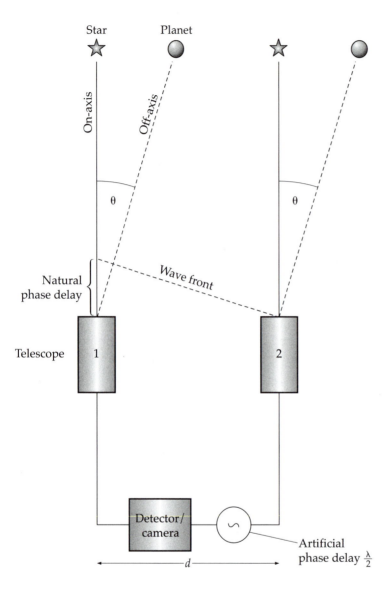

Figure 4.31. Schematic of a nulling interferometer designed to eliminate (null) the light from a star, but not that of an associated planet. Two telescopes observe the same system and are aligned so that the stellar light/wavefront arrives on axis. The planet is separated from the star by an angle θ and thus (in this depiction) the light from the planet arrives at telescope 1 slightly later than at telescope 2. The light from both telescopes is combined in a detector (e.g., a camera), but an artificial phase delay of $\lambda/2$ (where λ is the wavelength of photons detected) is added to the light from telescope 2. Upon combining, the light of the star is nulled out since the stellar light from the two telescopes is now out of phase by $\lambda/2$ (180°), but the planet's light has been brought *into* phase.

the setup of Figure 4.31 to successfully null out a star separated by θ from a planet is that the telescope separation must be

$$d = \frac{\lambda}{2\theta}. \tag{4.34}$$

If, for example, we consider a star some 14 pc distant (a criterion met by several dozen bright stars), an associated terrestrial planet at 1 AU subtends an angle of 0.07 arcseconds from the star. If the system is observed at a wavelength of $\lambda = 11\ \mu m$ (i.e., in the infrared, optimal for seeing warm dust or emitted radiation) then the required telescope separation is $d = 16$ meters. This is a reasonably practical requirement, and indeed close to the arrangement of current efforts to construct such systems (e.g., The Large Binocular Telescope [6]).

Much of the challenge on the Earth, as with all ground-based observation, is in dealing with atmospheric distortions of the wavefronts. Nulling interferometry and coronography (see below, §4.3.3) require integration with systems of so-called **adaptive optics** (AO) which attempt to remove the wavefront distortions by physically deforming the telescope mirrors and/or optical components based on direct measurements of the atmospheric effects as a function of time (e.g., using bright stars or artificial sources produced at altitude with lasers). To give a very crude indication of how difficult this is we introduce an important quantity in quantifying image quality. The **Strehl ratio**, S, is the ratio of the peak intensity of a real image to the peak intensity of the same image *if* there were no wavefront distortion (a purely diffraction-limited image, e.g., Figure 4.1). This can also be given in the following form:

$$S \equiv e^{-(2\pi\sigma)^2}, \tag{4.35}$$

where σ is the *root-mean-squared* deviation of the wavefront in units of wavelength—a measure of the difference of the wavefront from a perfectly flat plane. Equation 4.35 holds for most situations where S is greater than 10%. Thus, an optical instrument with a Strehl ratio of 100% is a perfect, diffraction-limited system. Typical present-day

---

6. http://medusa.as.arizona.edu/lbto/

adaptive optics systems can achieve Strehl ratios of between 30 and 70%. In order for the nulling and coronographic (see below) techniques to approach useful measurements for exoplanets, Strehl ratios of $\sim$ 90% are expected to be necessary. Ultimately the best strategy, albeit enormously technically and financially costly, may be to design space-based instruments which circumvent the problem of atmospheric distortion.

4.3.3   Coronography

Another approach which rather than exploiting tries to *reduce* the effects of the wave nature of light is **coronography**. In its very simplest form this quite literally involves blocking out the light of the star via some kind of optical barrier, or stop. In essence, an artificial eclipse is created.

However, even the basic demands of historical astronomical observation (before the explicit search for exo-planets) required a degree more sophistication. The fundamental problem is that any optical blocking device (for example, a simple disc) in the image plane of a focussing optic will itself produce **diffraction** of light, and so some of the stellar light will remain—in the form of a diffraction, or interference pattern (Figure 4.32). This problem is exacerbated by the finite size of a telescope's resolution (which is of course due to the diffraction pattern of light passing through the telescope aperture, Figure 4.1), and for Earth-based astronomy, due to the atmospheric distortion of incoming wavefronts (see above).

A traditional coronograph therefore includes a *further* optical blocking, or stop, to reduce off-axis diffracted light originating from the on-axis source (e.g., the star). The **Lyot stop** blocks some of the diffracted light before the image is reformed (Figure 4.32). If such a system is correctly designed it can (in the absence of atmospheric blurring) dramatically improve the contrast in the part of an image where exoplanets might lurk around bright stars. The success of this approach hinges on the quality of the wavefront entering the instrument (e.g., Sivaramakrishnan et al. 2001) and so is tightly correlated with the Strehl ratio and (for ground based work) the adaptive optics in a telescope system.

With both coronagraphic and interferometer approaches to reducing the light from the central star, a major difficulty in interpreting the remaining images arises from the phenomenon of **speckles**. Speckles

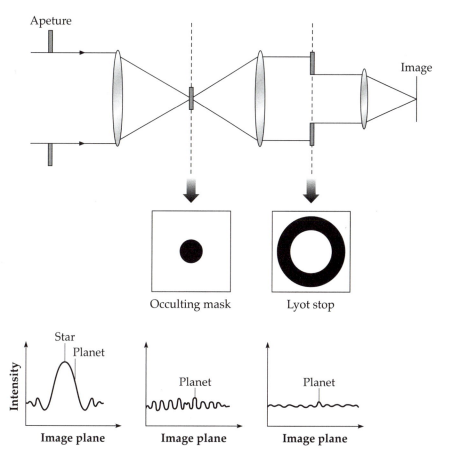

Figure 4.32. A schematic of a classical Lyot coronagraphic imager. Light enters the system from the left and in passing through an aperture (or in being reflected by a mirror) is diffracted as discussed above (see also Figure 4.1). The typical Airy diffraction pattern is illustrated on the bottom left, and the actual location of the center of a star and planet are indicated. The planet light is completely within the Airy disk and is therefore unresolvable from the much brighter starlight. Next the light passes through a circular occulting mask (other designs are possible, and in some cases superior). The resulting occulted image is shown as an intensity plot in the lower middle. Although the central peak of starlight has been almost entirely removed, there are new, additional diffraction features that still obscure the planet. Finally, the occulted light passes through a Lyot stop (annular stop) before being brought to a focus on the image plane. The Lyot stop blocks the majority of the secondary diffraction features of the occulted starlight and now (in sketch form) the light of the planet can be seen above the background light noise (lower right plot).

are a consequence of the inherently imperfect nature of optical surfaces (e.g. rough mirrors), and represent the results of the constructive and destructive interference of light in an optical system. One can experience speckles first hand by simply looking at the spot made by a laser pointer on a surface, or the image made upon reflection from a surface. Instead of appearing as a perfect circle or dot of light, the laser beam has a textured appearance, especially around its perimeter. In this case, a perfectly coherent wavefront is scattering off a surface with imperfections (even at the molecular or atomic level). There are countless interferences occurring between the scattered waves, and we see the resultant bright and dark spots. By distorting the wavefront entering a system, the atmosphere also creates speckles, which are typically time-varying.

For a coronagraph, speckles therefore create a multitude of apparent "sources," as illustrated in Figure 4.33 for the 55 Cancri system (§4.2.7). Some of these come and go as a function of time, and some are more persistent. It is immediately apparent that if one is to pick out the faint light of a planet amongst this mess, then additional strategies are required. First, the image of a planet will, over time, of course follow the motion of the planet in its orbit. Second, the spectrum of planetary light should be distinct from the spectrum of speckles—which are to first order just starlight. Based on this latter fact, it can be seen that a system that samples the spectrum of light at all parts of an image will be able to rapidly distinguish between planets and speckles. Such instruments are known as **integral field spectrometers**, and are of immense use in these efforts.

4.3.4 Planet Imaging and Interferometry

What might be the ultimate goal of exoplanet detection? In the context of comparative planetology it is the ability to precisely characterize a planet—its orbital path, composition, atmospheric structure, moon system, and so on. In the context of astrobiology the goal shifts to one of directly confirming the presence of a biosphere. We will return to this latter subject in Chapters 5 and 6. For both goals, it would be spectacularly useful to obtain a *spatially resolved image* of a planet— especially that of a terrestrial type world. With such data we would be able to look for continental structures, oceans, weather variability, and

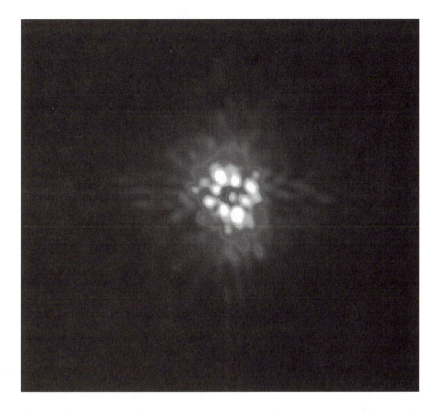

Figure 4.33. An actual Lyot coronagraph image (~2 arcsec across) of the star 55 Cancri. The central stellar light has been dramatically suppressed. What remains are numerous "speckle" patterns, as described in the main text. While this may appear discouraging for detecting faint exoplanets it is nonetheless a tremendous step towards accomplishing this (image courtesy of Ben Oppenheimer and the Lyot Project at the American Museum of Natural History, New York).

of course the direct visual signs of extensive life (e.g., forests or marine algae).

Is this possible? We have seen how the intrinsic diffraction limit of a telescope limits the resolution of any image. Consider the requirements for imaging an Earth-sized planet at a distance of 10 pc. The disk of such a planet would subtend $\sim 2 \times 10^{-7}$ arcseconds, or 0.2 μ arcsec. To actually see *details* of this disk in (for example) a 25 by 25 pixel image would require a resolution of better than $\sim 0.008\,\mu$ arcsec. If we look back to Equation 4.1 this would require (at a wavelength of 5000 Å) a telescope aperture of about 16,000 km!

However, what matters is not the size of a *single* instrument, but rather the total *baseline* over which simultaneous observation of a wavefront of light can be made. This is the basis of **interferometry** - whereby light received at physically separated locations is combined to produce an image with an effective resolution far greater than that of the individual receiving instruments. Thus, we can indeed conceive of a space-based array of telescopes that detect the wavefront from a given exoplanet and combine their measurements to produce an image with a resolution equivalent to that of a single instrument with the same size as the array. Whether such an instrument would ever be flown is another question entirely.

References

Agol, E., Steffen, J., Sari, R., & Clarkson, W. (2005). On detecting terrestrial planets with timing of giant planet transits, *Monthly Notices of the Royal Astronomical Society,* **359**, 567.

Beaulieu, J. P., et al. (2006). Discovery of a cool planet of 5.5 Earth masses through gravitational microlensing, *Nature,* **439**, 437–440.

Butler, R. P., et al. (1999). Evidence for Multiple Companions to Ups. Andromedae, *The Astrophysical Journal,* **526**, 916.

Butler, R. P., et al. (2006). Catalog of nearby exoplanets, *The Astrophysical Journal,* **646**, 505.

Butler, R. P., Marcy, G. W., Williams, E., McCarthy, C., Dosanjh, P., & Vogt, S. S. (1996). Attaining Doppler Precision of 3 m s^{-1}, *Publications of the Astronomical Society of the Pacific,* **108**, 500.

Charbonneau, D., Brown, T. M., Latham, D. W., & Mayor, M. (2000). Detection of planetary transits across a Sun-like star, *Astrophysical Journal Letters,* **529**, 45.

Charbonneau, D., et al. (2005). Detection of Thermal Emission from an Extrasolar Planet, *The Astrophysical Journal,* **626**, 523–529.

Einstein, A. (1915). Erklärung der Perihelionbewegung der Merkur aus der allgemeinen Relativitätstheorie, Sitzungsberichte Preussischen Akademie Wissenschaftten, **47**, 2, 831–839 (In this paper, Einstein includes the result of the deflection of light in the framework of General Relativity.)

Fischer, D. A., et al. (2007). Five Planets Orbiting 55 Cancri, astro-ph, arXiv:0712. 3917.

Gaudi, S., & Winn, J. (2007). Prospects for the characterization and confirmation of transiting exoplanets via the Rossiter-McLaughlin effect, *The Astrophysical Journal*, **655**, 550.

Gillon, M., et al. (2007). Detection of transits of the nearby hot Neptune GJ 4366, *Astronomy & Asprophysics*, **472**, L13.

Harrington, J., et al. (2006). The phase-dependent infrared brightness of the extrasolar planet Upsilon Andromedae b, *Science*, **314**, 623.

Henry, G. W., Marcy, G. W., Butler, R. P., & Vogt, S. (2000). A transiting "51 Peg-like" planet, *Astrophysical Journal*, **529**, L41.

Knutson, H. et al. (2007). A map of the day-night contrast of the extrasolar planet HD 189733b, *Nature*, **447**, 183.

Levine, H., Petters, A. O., & Wambsganss, J. (2001). *Singularity Theory and Gravitational Lensing*, Birkhauser.

Marcy, G., Fischer, D. A., Butler, R. P., & Vogt, S. S. (2006). "Properties of exoplanets: A Doppler study of 1330 stars." In *Planet Formation: Theory, Observations and Experiments*, eds. Klahr, H., & Brandner, W., Cambridge University Press.

Mayor, M., & Queloz, D. (1995). A Jupiter-mass companion to a solar-type star, *Nature*, **378**, 355.

Paczyński, B. (1996). Gravitational Microlensing in the Local Group, *Annual Reviews of Astronomy and Astrophysics*, **34**, 419.

Sackett, P. (1998). Searching for unseen planets via occultation and microlensing, arXiv:astro-ph/9811269, also in *Planets Outside the Solar System: Theory and Observations*, eds. J.-M. Mariotti & D. Alloin, NATO-ASI Series, Kluwer.

Sivaramakrishnan, A., et al. (2001). Ground-based coronagraphy with high-order adaptive optics, *The Astrophysical Journal*, **552**, 397–408.

Wang, Z., Chakrabarty, D., & Kaplan, D. L. (2006). A debris disk around an isolated young neutron star, *Nature*, **440**, 772.

Winn, J. N., et al. (2006). Management of the Spin-Orbit Alignment in the Exoplanetary System HD 189733, *The Astrophysical Journal*, **653**, L69.

Problems

4.1 Diffraction-limited, normal incidence (i.e., simple reflection) telescopes can be made in the extreme ultraviolet (e.g., operating at about 50 Å). What size mirror/aperture would be required to fully resolve the disk of a nearby $1R_\odot$ star at 10 pc distance in the extreme ultraviolet ? How much bigger would such a telescope need to be to resolve the disk of a Jupiter radius planet transiting such a star? Discuss what other potential difficulties might arise in observing star/planet systems at these wavelengths.

4.2   The expression for the stellar "wobble" velocity given by Equation 4.4 assumes a circular orbit. Can you generalize this to the case of an eccentric orbit to obtain an expression for v_* at periapsis? [*Note:* this requires a more detailed knowledge of orbital mechanics.]

4.3   In a high-resolution spectrometer most spectral lines used to measure velocity/Doppler shifts span a few pixels on a CCD detector. Since a spectrometer has a finite ability to spread out different wavelengths (its *dispersion*) on the detector this contributes to the sensitivity limits on velocities measured this way. For a required radial velocity precision of $1\,\mathrm{m\,s^{-1}}$ calculate $\Delta\lambda$ at 6000 Å (note in the non-relativistic regime $\Delta\lambda/\lambda \simeq v/c$ for pure radial motion). A typical dispersion of a good spectrometer is about 0.05 Å per pixel—what is the fractional pixel shift for $1\,\mathrm{m\,s^{-1}}$ Doppler precision?

4.4   Although terrestrial type planets orbiting solar mass stars are clearly of great interest it may be that terrestrial mass planets orbiting lower mass (and more abundant, Chapter 1) stars are equally important. Compute the reflex, or wobble velocity induced on a $0.4M_\odot$ and a $0.1M_\odot$ M-dwarf star by an Earth mass planet orbiting with a semi-major axis of 0.05 AU. Compare this to the wobble induced by the Earth on the Sun.

4.5   "Systemic" is a publicly available tool to fit and model radial velocity data and to search for the signature of exoplanets; its development is led by G. Laughlin (http://www.oklo.org/). A lengthy, but extremely useful set of exercises is available at http://www.oklo.org/wp-content/images/Systemichomework.doc. Download this file and complete the tutorials and problems. [*Note:* this exercise is suitable for a mid-term examination.]

4.6   Describe some of the currently observed trends and population characteristics of known exoplanets and their systems. You may want to refer to recent literature in order to be up to date. Discuss the apparent correlations with stellar metallicities, providing physical motivations.

4.7   Assuming a lens–source distance ratio of $\beta = 0.5$ estimate the optimal source distance for gravitational microlensing searches for giant planets at 5 AU around their parent stars when the lens stars are (a) $1M_\odot$ and (b) $0.2M_\odot$ in mass (recall that the planet needs to be close to the Einstein ring to produce a signal).

Compare the scales in (a) and (b) to the size of the Milky Way galaxy. What does this imply for microlensing searches for such planets?

4.8 Assuming an inclination of $90°$ compute the transit *duration* for (a) a Jupiter radius planet in a 10-day orbit around a $1M_\odot$, $1R_\odot$ star and (b) a Jupiter radius planet in a 10-day orbit around a $0.2M_\odot$, $0.2R_\odot$ star.

What will be the comparative effect of stellar limb darkening on the transit light curve shape between case (a) and case (b) (assuming limb darkening scales linearly with stellar radius)?

4.9 (a) What are the two fundamental challenges in trying to *directly* image exoplanets around other stars? What additional challenge is presented by ground-based optical telescopes?

(b) Describe the three major *indirect* techniques currently used for exoplanet detection and their relative merits and disadvantages. Pay particular attention to the motivating physics and quantitative differences in sensitivity/detection characteristics.

(c) Describe how a planetary atmosphere can be detected during transit. Explain and/or illustrate the impact of clouds/condensates on the transit characteristics.

4.10 Describe the techniques of secondary eclipse and phase curve photometry in detecting and characterizing planets.

Imagine an Earth-like planet undergoing eclipse by its parent star. If the star is a $0.2M_\odot$ M-dwarf with $L_* \sim 0.01L_\odot$, then estimate the net variation in luminosity seen during an eclipse. This can clearly be enhanced by narrowing the observational bandpass to the infrared.

4.11 Use Figure 4.28 to evaluate the potential composition of the ten most recently discovered transiting planets—you may obtain this list from the online exoplanet catalog: http://exoplanet.eu/

4.12 This exercise requires a literature search and some research. Coronographic stops can be much more complex than the simple Lyot stop. Describe some of the modern alternatives and discuss the specific characteristics that are chosen to help optimize imaging searches for planets.

CHAPTER 5

Life: a Brief History, and its Boundaries

5.1 Introduction

We have established a picture of the routes to star formation, and the routes to planet formation. As described at the outset, planets seem likely to offer the environments in the universe most suitable for the complex chemistry of life. That is not to say that all planets do. Gas and ice giants, and rocky worlds remote from their parent stars, may not satisfy the base requirements—namely, not too hot, not too cold, not too unstable, not too stable. But what dictates these requirements? As we have noted in Chapter 1, and elaborate on below, we are strongly compelled to base these constraints on the example of our one terrestrial biosphere. We have not, however, investigated what these constraints really are in detail.

In this chapter we will outline some basic biological facts, such as the mechanisms by which terrestrial life stores, expresses, and passes on information, and the apparent range in scale and complexity of organisms. We will also talk about the classification of terrestrial life—and some of the potential peculiarities of terrestrial life, such as the "handedness" of biological chemistry. In doing so we will also discuss the environmental ranges that life appears to occupy. Finally, we delve a little into the actual history of life, its corresponding geologic divisions, and some of the intriguing facts we do and don't know about life during the course of Earth's history. Much of this serves to set the stage for later chapters, where we will begin to integrate this biological knowledge into the description of planetary environments, and the potentially observable

195

characteristics that we might use to detect the presence of life on other worlds. This is a very brief survey of some of the more rudimentary knowledge about life, and is presented in a rather anecdotal style. Nonetheless, this serves to provide an overview of the sometimes unexpected diversity and complexity of life as we know it.

It is important to see how this discussion is integrated into the broader study of life in the universe. Other worlds, perhaps terrestrial in nature, will have had entirely distinct biological histories (if they have harbored life at all). The waxing and waning of very different biospheres throughout our own planet's past offers compelling evidence for this statement. Part of the challenge in astrobiology is to find ways to make use of this wealth of local knowledge when considering the expanded, alien territory of other planetary systems.

5.2 Two Histories

In our pursuit of the question of the origin and evolution of life in the Universe there are really two histories to consider—one is partially known, and one is total speculation—and it is crucial that we remain focussed on the differences. The first is the history of life on Earth. No less remarkable than the picture of life on this planet that we have currently is the story of how we have arrived at this picture. It is well beyond the scope of this text to even summarize this tale, and so instead we provide a number of important, or at very least entertaining, references (the extensive "Darwin Online" collection,[1] Gould 2002, Maynard Smith & Szathmáry 1999, Dawkins 2006). The second is the history of life in the Universe, about which we know absolutely nothing.[2]

There is a clear temptation, if not imperative, to use a knowledge of the history and characteristics of life on Earth as the template for both

---

1. http://darwin-online.org.uk/
2. This may be an unjust claim. Many people have thought long and hard about what the apparent lack of evidence for the signatures of "advanced" life actually imply about life in the universe as a whole. This is a case where an absence of knowledge can be exploited as an indicator for certain characteristics of a phenomenon. Nonetheless, until life beyond our world is discovered we simply cannot confirm or dispute any such arguments with certainty.

searching for and interpreting life elsewhere. It can be argued that this is a bad thing to do. Indeed, it seems that a favorite topic that comes up over cocktails is how "life elsewhere might be nothing like life on Earth," and how one is therefore faced with a seemingly hopeless task. While it is valid to see this as a potential pitfall of extrapolating what we know of life on Earth to beyond, and it should always be kept in mind, it is really in the best scientific tradition that we consider the history and characteristics of life on Earth as a starting point in astrobiology. Indeed, the former argument often presumes that we actually know all there is to know about life on Earth, but this is *not true*—life on Earth is constantly surprising us in its diversity and complexity. In Chapter 1 we discussed whether life considered as a "naturally emergent phenomenon" places it in the same league as the fundamental laws of physics. If we choose to try to test this conjecture, then we really need to understand one sample of life in as much detail as possible, since it can serve as our calibration for the entire enterprise.

Armed with this foundation we can then seek both the signs of comparable life, and have some preparation in seeking the signs of life that is somehow different, possibly radically so.

5.3   What is Terrestrial Life?

An elephant and a bacterium don't look much like each other, but we consider them to both be alive. A bacterium and a virus also don't look much like each other, and we might run into arguments about whether a virus (which is an entirely host-dependent entity) should be considered alive or not. What then are the criteria that separate "life" from "non-life"? This is not something we will spend much time on here, since there are easier, and more immediate, questions to answer. Nonetheless, many definitions of life have been considered throughout human history, and include factors such as metabolism, reactions to environment, ability to self-repair, and of course the ability to reproduce and pass on information from generation to generation. At some level there must also be a threshold of **complexity**, below which a molecular structure really operates only on the most basic, linear, physical/chemical laws, and does little that we would recognize in the way of complex self-organization. As in most of the rest of this book (and see Chapter 1, although see also

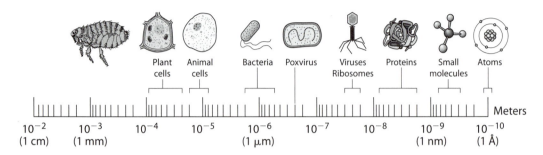

Figure 5.1. The scale of organisms and structures, from 1 cm to 1 Å. Plant and animal cells are significantly larger than those of microbial life; however there is considerable range amongst all types of organisms, forming a continuum of scales from greater than 1 cm all the way down to viruses and ultimately atoms.

Chapter 10) we prefer to sidestep this question, and simply state again that it seems most rational to not imbue life with any special significance at this stage in the game, and to just consider it as another (fascinating) phenomenon whose origin we have yet to fully understand.

5.3.1 Scales and Structures

Terrestrial organisms cover a remarkable range in physical scale. We illustrate this in Figure 5.1. Between the smallest bacteria (at several 10^{-7} meters) to the largest known single organisms (e.g., Giant Sequoia trees close to 100 meters in total length[3]) are some 10^9 orders of magnitude in scale. This fact is by itself quite informative. Any existing organism is a marvel of survival, and has typically been around for millions of years, if not longer. The very existence of such an enormous range in physical scale tells us that, in the broader scheme of things, there are opportunities for life on many, many levels.

As we will discuss in more detail below, there are however fundamental differences between organisms that go beyond just size. The most immediate is the division of life on Earth between organisms based on the structure of their cells; the "units" building multicellular life, and encapsulating entire organisms for microscopic life. Life

---

3. The record may in fact be held by a species of forest fungus *Armillaria ostoyae*, or "Honey mushroom," thought to span some 8.9 kilometers. However, this is perhaps not a single organism but rather a colony of clones.

on Earth is based upon either **Prokaryotic** or **Eukaryotic** cell structures. Figures 5.2 and 5.3 illustrate the basic nature of both cell types. To summarize the primary distinctions between them: Eukaryotic cells have their genetic material (§5.3.2) within a membrane-bound "capsule," or **nucleus**. Prokaryotic cells have (typically) a single, loop of genetic material within a non-membrane encapsulated region called the nucleiod, plus little circular scraps of genetic material called **plasmids** throughout the cell. Eukaryotic cells are also typically significantly larger than prokaryotic cells (Figure 2.27), and contain much more in the way of complex structure (e.g., organelles and cytoskeletons). Indeed, some of this structure appears to consist of symbiotically incorporated prokaryotic organisms, or at least prokaryotic-derived genetic structures. We will mention this a little further below, but both **mitochondria organelles** in most eukaryotes, and **plastids** in most plant or algae cells, play a critical role in metabolism and are almost certainly the product of **endosymbiosis**. This is to say that an earlier symbiotic relationship between the eukaryotic ancestor and a prokaryote (e.g., a proteobacteria) has evolved to completely subsume the prokaryote into the very cell structure itself.

An obvious question is why should two such fundamental cell types exist in all terrestrial organisms? The answer undoubtedly lies in the specific evolutionary pathways taken by life on this planet. As we will see later on, at various points in Earth's history, global environments have altered, and natural selection brought about by these changes can radically change life on the planet. Since all multi-cellular terrestrial organisms are eukaryotes, and multi-cellular life seems to have emerged some 1.5 billion years ago (well after single-celled life) it may be that global changes helped spur the development of these more complex biological structures (§5.4).

5.3.2 The Common Thread

All terrestrial life exploits the fantastically complex yet robust chemistry of carbon compounds. Many speculations have been made on the apparent uniqueness of carbon chemistry (organic chemistry), its connection to deep quantum mechanical physics, and whether it could be replaced by, for example, silicon-based chemistry. It is certainly true that silicon (with 14 protons, 4 valence electrons) is chemically similar to carbon (6 protons, 4 valence electrons). For example, both readily form molecules

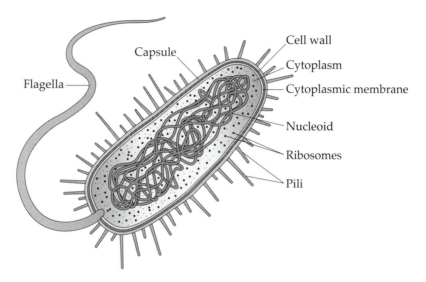

Figure 5.2. Prokaryotic cell structure. This drawing illustrates some of the basic features of a typical prokaryote—actual cell shape and details such as flagella can vary between species. The cytoplasm is a gel-like mix of water, enzymes, nutrients, wastes, and gases and contains cell structures such as ribosomes, a chromosome, and plasmids. The principal genetic material is contained within the nucleoid structure as a single chromosome (§5.3.2). Flagella are hair-like structures on some prokaryotes that enable them to move through a rotationally driven "beating." Some of the reasons for moving are to follow chemical gradients either towards nutrients or away from toxins, and in some species towards light. Ribosomes (common to all life) are essentially chemical factories central to the production of proteins using the information of the genetic code (§5.3.2). Pili are small hair-like protrusions found in many prokaryotes, which assist in attaching the organism to surfaces.

with four hydrogen atoms, and both can form long-chain polymers. However, the silicon atom is bigger than the carbon atom, and has a harder time forming strong (double or triple) covalent bonds—which are a central hallmark of carbon chemistry. Furthermore, very common silicon compounds, such as silicon dioxide (SiO_2) tend to be solid at terrestrial temperatures, and are also quite insoluble. By contrast, carbon compounds like carbon dioxide (CO_2) are highly soluble in a common solvent like water. It seems therefore that carbon-based chemistry offers far more flexibility and a far greater array of possible molecular

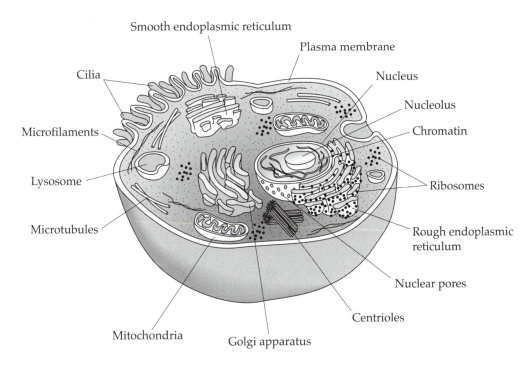

Figure 5.3. An example of a eukaryotic animal cell structure (plant cell structure differs in some ways). Key structures include the cell nucleus (a membrane-wrapped capsule containing the genetic material of the organism) that contrasts the much looser, less protected genetic material of prokaryotic nucleoids and plasmids. A eukaryotic cell contains many organelles—discrete structures with specific functions. Among these are mitochondria (§5.3.1, §5.6) providing energy in the cell. The Golgi apparatus is the distribution and shipping department for chemical products. It modifies proteins and fats and prepares them for export to the outside of the cell. The endoplasmic reticulum is a network of sac-like structures that make some of these proteins and fats and provide a pipeline between the nucleus and the rest of the cell for these products. As with prokaryotic cells, ribosomes are also present throughout the cell.

structures. These include the highly intricate ones that both store the information describing an organism (as we discuss below) and compose the myriad ways to feed, house, and reproduce it. This does not rule out life based around a different fundamental chemistry, but it does suggest that such life would appear profoundly different. In light of that, we here continue to focus on the nature of known life with an eye to more practical and tractable issues.

All organisms on Earth share a common information storage and
retrieval mechanism—namely that facilitated by the double-helix **de-
oxyribonucleic acid** (DNA) molecule. A section of this long molecule
is shown in Figure 5.4. Amongst its principal characteristics is the "lad-
der rung" structure of complementary **nucleic acids** (see Figure 5.5):
Adenine complements Thymine and Guanine complements Cytosine,
so that these molecules (acid bases) are referred to as **base pairs**. The
molecules attach through hydrogen bonds. Adenine and Thymine are
connected via two hydrogen bonds and Guanine and Cytosine via three.
This alphabet of four "letters"—A, T, G, C—encodes the information
describing all known terrestrial organisms. In higher eukaryotic organisms (such as ourselves) this information is not in a single lengthy
DNA molecule, but rather divided into sub-strands that are folded and
wrapped around proteins to make **chromosomes** which are encapsulated in cell nuclei (Figure 5.3). Chromosomes are a nice compact
piece of packaging, which undoubtedly helps protect the crucial DNA
molecules. DNA is very much the permanent storage device for life.
For its information to be translated into molecular structures that build,
power, and tend to the organism itself, DNA must be "read" out. In very
basic terms, a given part of the DNA will be transcribed into a shorter
molecule of **ribonucleic acid** or RNA by a **polymerase** RNA **enzyme**
that can "unzip" a bit of the DNA for reading. For a given section of
DNA, the transcribed RNA will consist of the complementary basepairs,
except that Thymine is replaced with another acid - Uracil. The raw materials to produce this copy are plucked from the surrounding cell. The
main information-carrying RNA is termed **messenger** or **m-RNA**, and
in eukaryotic cells the m-RNA is further processed, leaves a cell nucleus,
and floats off into the cell to find a place to do useful things—like make
proteins, which in turn make up the structure of cells and act as catalysts
for function.

DNA is a tricky molecule though: not all of it is of any immediately apparent use—indeed much appears to *not* translate to meaning-
ful structures such as proteins—although it all of course is involved
in determining its own wrapping and twisting structures. The modern
definition of a **gene** is a locatable region of genomic sequence, corresponding to a unit of **inheritance**, which is associated with regulatory
regions, transcribed regions, and/or other functional sequence regions.
In other words, a gene consists of a "useful" bit of DNA, which will

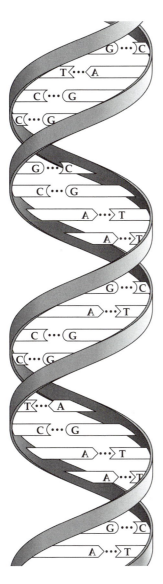

Figure 5.4. A representation of a section of the double-helix DNA molecule. Two complementary strands of nucleic acids (base pairs) are present, attached to two phosphate "backbones" and connected across the helix via hydrogen bonds between the nucleic acids (double bonds for Adenine–Thymine and triple bonds between Guanine and Cytosine, see Figure 5.5).

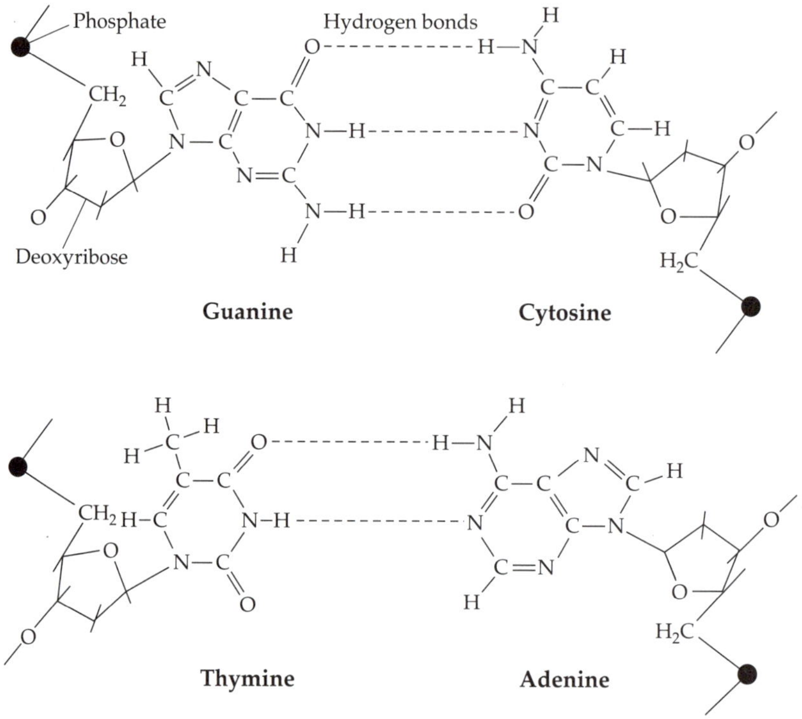

Figure 5.5. The basic structural form of the four nucleic acids in DNA: Adenine, Thymine, Guanine, and Cytosine. The location of the triple and double hydrogen bonds are indicated. The attachment points to the DNA "backbone" are also indicated schematically.

typically consist of a so-called **exon** region (see Figure 5.6) sandwiched between so-called **intron** regions. During transcription this whole piece gets unfurled and read, but only the exon parts of the sequence actually get copied and spliced together into the m-RNA. The intron regions do however clearly serve a purpose as part of the "structural language" of DNA, they may also represent "fossil" code for function, once used but now redundant or fragmented.

The actual information coding structure is well known. "Words" consisting of three base pairs in sequence code for particular **amino acids**—the building blocks of more complex molecules (Figure 5.7).

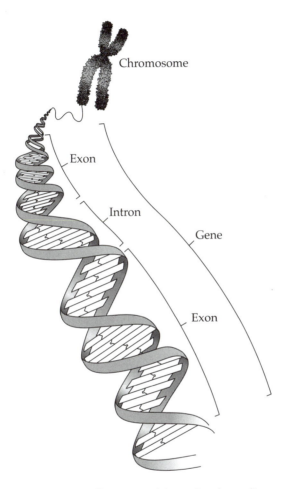

Figure 5.6. An illustration of the molecular realization of a gene. Within the DNA sequence, uncurled
from its wrapped form in a chromosome, are regions
consisting of transcribable code. Within these regions only the *exon* pieces are spliced together into
m-RNA and are eventually translated into structures
such as proteins within the cell.

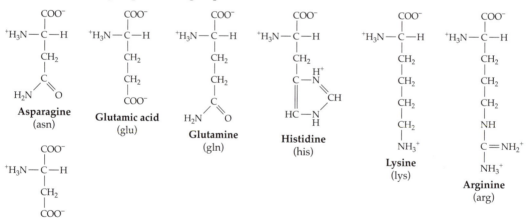

Figure 5.7. The general form of an amino acid is shown at the top of the figure. Below this are the structural forms of the 20 key amino acids utilized in all terrestrial life. There are three basic families; those with hydrophobic (water avoiding) side groups, those with hydrophilic (water loving) side groups, and those that are in between.

Amino acids that are in between

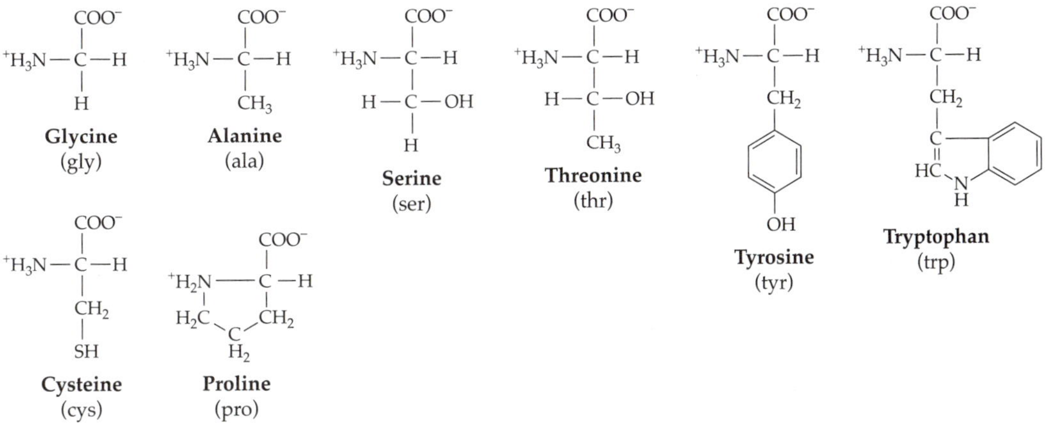

Figure 5.7 *(continued)*.

Although three-letter words with four unique letters suggests a total of 64 ($4 \times 4 \times 4$) unique words there is considerable redundancy so that only 20 amino acids are actually coded for, with different words encoding the same amino acid (see Table 5.1). Typically three of the possible words do not code for an amino acid, but are instead "stop" codons, for example TAG (or UAG in m-RNA). These appear to act as delimiters that indicate the end of a particular piece of transcribable DNA. Similarly, certain regions of the DNA with the sequence TATA correspond to the starting points for transcription.

Understanding the origins and evolution of the DNA code itself is a fascinating and tremendously important subject, well beyond our scope here. It is, however, worth re-emphasizing that DNA not only codes for molecules which are transcribed from it, but must also code for its own structural properties, which in turn influence the rates and pathways to particular m-RNA production.

It is also not sufficient to simply assume that the more base pairs in a given genome the more obviously complex the organism. While it is true that humans are coded by approximately three billion base

Table 5.1. A reverse codon table. The nucleic acid codons (words) that code each of the 20 biologically relevant amino acids are listed here. The "start" and "stop" codons are also listed—these assist in defining the reading frame (i.e., where the first letter of the first word begins and ends).

Ala	Arg	Asn	Asp	Cys	Gln	Glu	Gly	His	ILe	Leu
GCU	CGU	AAU	GAU	UGU	CAA	GAA	GGU	CAU	AUU	UUA
GCC	CGC	AAC	GAC	UGC	CAG	GAG	GGC	CAC	AUC	UUG
GCA	CGA						GGA		AUA	CUU
GCG	CGG						GGG			CUC
	AGA									CUA
	AGG									CUG

Lys	Met	Phe	Pro	Ser	Thr	Trp	Tyr	Val	Start	Stop
AAA	AUG	UUU	CCU	UCU	ACU	UGG	UAU	GUU	AUG	UAG
AAG		UUC	CCC	UCC	ACC		UAC	GUC		UGA
			CCA	UCA	ACA			GUA		UAA
			CCG	UCG	ACG			GUG		
				AGU						
				AGC						

pairs, chickens by some 1.2 billion base-pairs, and a bacterium like Escherichia coli by some 4.7 *million* base pairs, the pattern is (apparently) broken when one considers a "simple" plant such as wheat, which is encoded by some 16 *billion* base pairs! Of course it would be incorrect to assume that a plant such as wheat is actually "simpler" than an organism such as ourselves, but it is certainly true that it does not have the same multi-faceted metabolism and structural properties (e.g., brain, digestive system, musculature). Some strains of wheat are actually **hexaploid**, which means that they carry *six* sets of chromosomes in each cell. This is very different than humans, who usually carry only one.

5.3.3   The Tree of Life

Even prior to the revolution of genomics, the traditional way by which life has been classified and understood as a unified phenomenon has been by the study of **systematics**—based upon, amongst other things, structural characteristics (**taxonomy**). This notion—that current life is all related to a set of common ancestors, and possibly an ultimate, single, common ancestor—is almost implicit in these discussions. In many respects it makes good sense. It is natural to assume that whatever the origins of life, it didn't begin globally (or even universally) at precisely the same time. Indeed, if it is a rare occurrence, then it might be that there was a true, singular "seed" event from which all terrestrial life sprang. While it may be hard to establish this common ancestor through taxonomic measurements, it may be much more readily quantifiable through modern **phylogenetics**—the science of **phylogeny**; the connections between all groups of organisms as understood by ancestor/descendant relationships.

How can the true relationship of organisms be quantitatively determined ? There are specific, large molecules that play such a critical role in the functioning of an organism (e.g., those that play key roles in the synthesis of proteins within a cell; see also Chapter 10) that the DNA sequence that codes their design (i.e., the corresponding gene) simply *cannot* evolve very fast—or else the organism would be unlikely to function in subsequent generations. Furthermore, so fundamental are these genes that they are found in *all* terrestrial organisms.

However, over time (thousands to millions of years), these genes will gradually change or evolve. Some of the bases in the corresponding DNA sequence will change, or the sequence will itself grow. Remember, these genes code for complex molecules, with hundreds if not thousands of base pairs (and redundancy in the coding of amino acids: Table 5.1)—so there is plenty of room for tweaks and alterations without profound implications for the fundamental function. Since DNA is the unit of inheritance, this implies that a given species will carry its own set of unique changes to these key genes. To give a hypothetical example: imagine a species of insect that begins to diverge into two distinct families, one of which lives on the surface, and one of which burrows into the ground. Over evolutionary timescales these two types of insects—originally the same—essentially become two species, which do not interbreed. Nonetheless, the rate at which their functionally critical genes change will probably be very similar, and very slow. Thus, a genetic analysis will reveal that these crawling and burrowing insects are almost the same genetically—with only a short "distance" from a common ancestor. Since the rate at which such changes occur will vary enormously between species (and indeed between particular genes) the base-by-base comparison of such a gene in two different species of organism really yields an **evolutionary distance** rather than direct temporal information on when species perhaps had a common ancestor.

What this provides is a remarkable way to construct a **phylogenetic map**, or **tree of life** (Figure 5.8) for terrestrial organisms. The piece of DNA sequence (the gene) that is currently most often used to construct such a map codes for the molecule known as **16S small subunit ribosomal RNA (rRNA)** (and is some 1500 base pairs in length). This *ribosome* molecule is crucial for the synthesis of proteins in cells. Thus, in molecular biology parlance it is likely to be **highly conserved** (i.e., evolves slowly). The corresponding base-pair sequence is also coded for in all known cell types and contains smaller genetic regions which evolve at different rates.

Where the branches in this tree meet, we expect there to have been a common ancestor. Thus, organisms that appear to be closest to the major, or basal arms might be candidates for being the most ancient and unchanged forms of life. There are three major branches of terrestrial organisms **Eukarya, Bacteria**, and **Archaea**. Both bacteria and archaea are prokaryotes. The only things missing from this diagram are entities such as viruses and prions. Of these three subsets, eukarya and bacteria are the most familiar. Archaea represent a branch of apparently ancient, "bacteria-like" organisms that until relatively recently (Woese & Fox 1977) were not recognised as being genuinely distinct—although profoundly distinct they are. While archaea share much of the same basic cell structure and metabolic characteristics of bacteria, they differ in how the transcription of DNA into RNA occurs. In fact, they share many traits with eukaryotic life. Additional major differences between archaea and the other branches of terrrestrial life involve the detailed structure of their cell membranes. Some of these differences appear to be related to an ability to withstand extreme conditions, such as high temperatures (§5.5). It is also interesting to note that most archaea are harmless to other organisms and, unlike bacteria, do not cause disease. Together with their relatively short evolutionary distance from the base of the phylogenetic tree, this suggests (but by no means confirms) that archaea are ancient, and evolved in the absence of many, if any, "host" organisms.

The Tree of Life

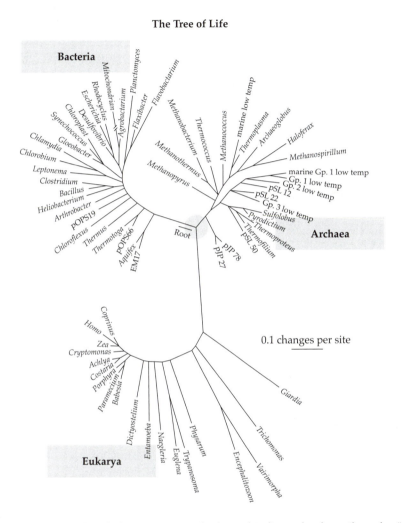

Figure 5.8. The Phylogenetic Tree of Life: In this figure the three "branches" of life on Earth are shown: Eukarya—organisms with complex cell structures and cell nuclei containing their genetic material; Bacteria—a class of prokaryote, without complex cell structures or nuclei; and Archaea—a class of prokaryote, considered among the most ancient of organisms, differing from bacteria primarily in their mechanisms of transcription and translation, which bear more resemblance to those of eukaryotes. Archaea are also often found in *extreme* enviroments. The length of the branches in this diagram indicates the genetic "distance" between organisms based on the highly conserved 16S rRNA gene. The true root of this tree is unknown, and it is not even clear that there should be a *single* common ancestor rather than a population of ancestor organisms.

So far, so good. However, there is a potentially tricky problem with some of the phylogenetic classification discussed above. This is a phenomenon known as **horizontal gene transfer**. Put simply, horizontal gene transfer occurs when *unrelated* organisms transfer a specific gene between themselves. Both bacteria and archaea are capable of this. The actual mechanisms seem to include direct exchange of "naked" DNA (i.e., whereby a fragment of the chromosome literally hops over), the transfer of plasmids (see above), or the transfer via a third party entity known as a **phage**. Now, this is wild and crazy stuff—it lets bacteria and archaea perform all kinds of experiments in genetic engineering. The debate on exactly how, and more importantly *why*, horizontal gene transfer occurs is still raging (e.g., Doolittle 2000). One of the major issues it raises is whether early life engaged in rampant gene exchange. Was this a way for the first structures we would term as organisms to rapidly (albeit blindly) search for viable survival strategies? Modern evolution takes place relatively slowly via (primarily) natural selection (although some bacteria can clearly evolve rapidly, for example in the case of developing resistance to drugs). If ancestral life was blindly and rapidly experimenting via gene exchange then the odds would have been greatly improved for hitting upon positive adaptations to the environment. As stable solutions (to use the mathematical terminology) were chanced upon, then more distinct groups of organisms would emerge. A second major issue that horizontal gene transfer raises is of course whether it destroys our attempts to construct a genetic tree of life. Gene transfer could render the concept of evolutionary distance somewhat moot. This is still actively debated. It does seem however that critical genes such as 16S rRNA, used in Figure 5.8, may be transferred relatively rarely. Thus, phylogenetic trees are likely to still be useful, if used with a little caution.

5.3.4   Chirality: The Handedness of Life

Before we discuss the relationship of terrestrial life to Earth's history, it is important to take a small step back and take note of a fundamental chemical characteristic of organisms. Life on Earth exhibits a "handedness" in its chemistry. Consider amino acids. For a given chemical composition (e.g., Adenine $C_5H_5N_5$) there exist *two* structural forms of the molecule. Each is the non-superposable mirror image of the other (Figure 5.9). Such molecules are **chiral**: they exhibit *left* or *right* chiral-

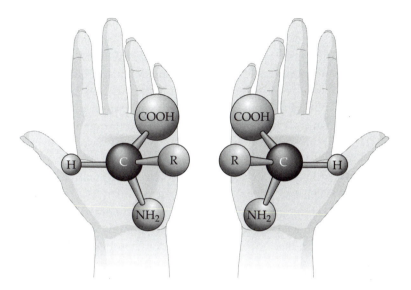

Figure 5.9. An illustration of the structural difference between the enantiomers of a simple organic molecule. One is the literal mirror image of the other, and can therefore never be rotated or translated to match the other, and has fundamentally different chemical behavior.

ity, with the two forms being known as **enantiomers**. The origin of the terminology of handedness is that the plane of polarization of light is rotated left or right in passing through a sample of either enantiomer.

What is remarkable and fascinating is that life on Earth is entirely *left-handed* in terms of amino acids, right-handed in terms of sugars, and so on—a characteristic known as **homochirality**. In fact, the entire set of left-handed chemistry that life employs is essentially *incompatible* with the equivalent right-handed chemistry. Right-handed amino acid enantiomers cannot be incorporated correctly into m-RNA, left-handed sugars cannot be metabolized by organisms, and so on. The only exceptions to this rule are archaea, which incorporate the "wrong" chirality molecules into their cell membranes—perhaps as a don't-eat-me defense mechanism, or a vestige of earlier diversity.

A test tube full of left-handed amino acids would, if left to itself, eventually turn into a test tube full of an *equal* mixture of left- and right-handed enantiomers—a process known as **racemization**. There is *no* evidence for any energetic bias towards one particular handedness in non-biotic chemistry. Indeed, this characteristic has long been used as a means to test the age of biologically produced chemical samples—as time goes by, a sample containing only a single enantiomer will racemize to a 50/50 mixture with a characteristic timescale for a given set of conditions.

This raises the question, *Why* is life on Earth left-handed (based on amino acids)? This is a vexing and intriguing problem. We can certainly come up with seemingly good arguments as to why life needs to choose one way or the other. For example, a world populated by two chemically incompatible strands of life might be far more competitive in terms of resources—the pyramid of larger things eating smaller things might be much narrower owing to the inedible "other" population, any regulatory feedback on environment might be greatly complicated by these two trees of life, and so on. However, it would be more elegant if we could say that there was indeed some advantage, no matter how tiny, to being left-handed rather than right-handed.

Many efforts have been made to solve this problem, ranging from appeals to minute theoretical differences in energy states of the left and right molecules due to fundamental physics, to far more mechanical reasons to do with inorganic chemistry and periodic structures that may have played a role in the formation of the first self-replicating molecules. It has also been argued that perhaps life arose in *both* forms initially, but then for simple reasons of population dynamics and stabilities one form died out, leaving the other to reign.

We will return, albeit briefly, to this issue in Chapter 8 when we discuss the origins of carbon chemistry in interstellar space and on the surface of young planets. For now, though, we note that regardless of its origins, if the homochirality of life on Earth is typical it provides an excellent test for the presence of life on other worlds. To give an example; if the Martian regolith is found to contain amino acids, and there is found to be an excess of left- (or right-) handed molecules, this would be a true smoking gun for the presence of extant life. Even a small bias towards one enantiomer over another would be intriguing.

5.4 The History of Life on Earth

The oldest, geophysically unaltered rocks we can access on Earth are about 4.0 billon years old: (the oldest minerals are some 4.4 billion years old, Chapter 9). The oldest known evidence of fossil organisms are prokaryotes of around 3.5 billion years in age, and there is some (controversial) evidence for prokaryotic life as old as 3.8 billion years. One of the most dramatic observations from the fossil record of life on Earth is that from these earliest times until some 1.5 billion years ago *only* prokaryotic - single-celled—life existed on the planet. Among the most interesting fossil structures of this single-celled early Earth are **stromatolites**. These are laminated (layered) sedimentary structures, likely formed by colonies of many different types of participating microbial organisms (including **cyanobacteria**, §5.4.1). Stromatolites flourish in warm aquatic environments. Although almost extinct today they can still be found in places of high salinity, such as Shark Bay in Australia (Figure 5.10). Microbes can form glue-like **biofilms**[4] and mats that can then trap and bind mineral grains—building up structures over long periods. These structures appear to have been quite diverse, from conical shapes to more spread-out mounds (for an excellent reference see Allwood et al. 2006), and likely occurred in a variety of environments.

The emergence of multi-cellular life on Earth seems to have coincided (over hundreds of millions of years) with another dramatic change— namely the oxygenation of the Earth's atmosphere (see below). In Figure 5.11 the basic major divisions of geologic time are illustrated. From the point of view of biology the major divisions are **Precambrian** (prior to approximately 550 million years ago) and **Phanerozoic** (within the last 550 million years). These divisions, and the many subdivisions therein, are based on the characteristics of the contents of the fossil record. This is also driven to a great extent by the phenomenon of extinction events. In describing the history of life on Earth a major theme is that at many times entire species and entire groups of species have quite literally vanished (see also Chapter 8). The time resolution of paleontological

---

4. A biofilm is a complex aggregation of microbes that can excrete a protective and adhesive material, including polymers. The ability to form biofilms is ubiquitous amongst bacteria and archaea.

(a)   (b)

Figure 5.10. Modern stromatolites at Shark Bay, Australia. Left panel (a): Aerial view from 700 ft. Shoreline is at top of image. Fields of stromatolites extend from tide zone out into bay at depths reaching 30 ft. (photograph M. Storrie–Lombardi, pilot Adrian Brown, McQuarie University, courtesy of Kinohi Institute, Inc.). Right panel (b): To illustrate scale. Dr. Michael Storrie-Lombardi, Director of the Kinohi Institute, inspects a Shark Bay stromatolite up close to take UV-vis-IR reflectance spectra (photography courtesy Kinohi Institute, Inc.).

studies is not sufficient to say whether such extinctions occured over a few days or weeks, or hundred to thousands of years. Nonetheless, in geological terms, these events were quite rapid. The actual causes of "mass" extinctions may be many, from catastrophic external events (asteroid collisions, supernovae, interplanetary dust blocking of stellar radiation), to catastrophic local events (supervocanoes, outgassing), and events due to changes in either global climate equilibria or disease and population dynamics. How these generally random events influence the pathways that life has taken on the Earth, and would take on other worlds, is open to a variety of interpretations. It is probably true, however, that *following* a large-scale extinction there arise new opportunities

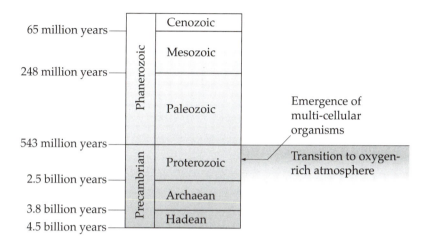

Figure 5.11. The basic divisions of geological time on the Earth, starting 4.5 billion years ago. Between approximately 2.5 billion and 550 million years ago the Earth's atmosphere underwent a transition from being rich in carbon dioxide (and possibly methane) to being rich in oxygen. This appears to have been driven by the oxygen production of the first photosynthetic organisms. In addition to being toxic to many other species, the oxygen atmosphere enabled aerobic organisms to flourish. Oxygen respiration enables a highly diverse range of biochemistry to take place and is likely to have played a role in the emergence of multi-cellular life some 1.5 billion years ago.

for surviving organisms. Furthermore, given the cross-phyla sharing of genetic coding (§5.3.3), it is also probably true that while a set of species may be eradicated, much of their genetic "knowledge" remains in the surviving organisms.

Irrespective of short timescale events, during the Proterozoic Era (between 550 million and 2.5 billion years ago) the Earth's atmosphere appears to have undergone a very significant transition from being carbon dioxide rich (and possibly methane rich) to being oxygen rich. This oxygenation was the direct result of photosynthetic life, with the main culprits likely being the cyanobacteria (see also §5.4.1). Prior to this time, the majority of life on Earth was **anaerobic**. As the atmospheric oxygen increased, the surface environment would have become more and more inhospitable to these organisms. It is appealing to think of organisms like cyanobacteria exploiting a form of planet-wide chemical warfare that improved the odds of their survival - although they of course were simply following normal evolutionary processes. Irrespectively, what *does* appear to have happened as a consequence is that organisms developed which utilized the remarkable network of chemical processes enabled via oxygen respiration. The emergence of multi-cellular life at the same time seems unlikely to be just a coincidence.

5.4.1 Implications for Astrobiology

Before we move on, it is also worth noting that cyanobacteria (Figure 5.12) are the most likely type of organism responsible for the existence of chloroplasts (§5.3.1) within eukaryotic cells. Cyanobacteria use photosynthesis (using stellar photons, atmospheric CO_2, and water to produce chemical energy) to produce oxygen (**oxygenic photosynthesis**). They still exist just about everywhere on Earth,[5] although large, permanent colonies are confined to just a few places (i.e., microbial mats and biofilms associated with stromatolites, §5.4). Among their many talents, some cyanobacteria can actually alternate between the pigments they contain for photosynthesis, switching from those best suited to redder light, or those best for bluer light (see also §5.5.3). Furthermore, some are capable of **nitrogen fixation**, taking strongly bound molecular nitrogen (N_2) directly from the atmosphere and converting it into ammonia (NH_3), nitrites (NO_2), and nitrates (NO_3), which are crucial for plant life. In fact, it is the extensive population of cyanobacteria in the rice paddies of Asia that ensures the success of the rice crops there, and which helps feed 75% of the worldwide human population. It would be hard to overstate the natural beauty, incredible success, and broad impact of cyanobacteria—they have survived for at least 2.7 billion years,[6] and are still going strong. Prior to cyanobacteria, **anoxygenic photosynthesis** was likely the main mechanism by which photon energy was exploited,

---

5. Much of the earthly slime you may encounter in life is likely to harbor cyanobacteria.

6. There is some controversy over the earliest fossil evidence for cyanobacteria, or similar organisms. Claims for 3.5 billion year old fossils are disputed and the issue is not yet resolved.

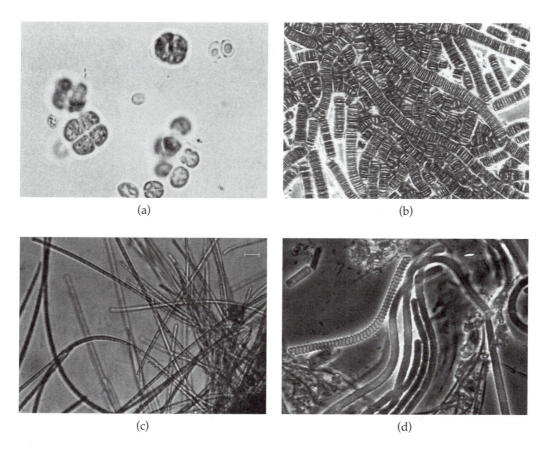

(a)

(b)

(c)

(d)

Figure 5.12. Microscopic images of four types of cyanobacteria. Cyanobacteria can exist in pure uni-cellular form (upper left), filamentous forms (lower left) and in large complex colonies such as in microbial mats (right hand images). Upper left: Andrea Kirkwood (University of Calgary), Mark Schneegurt (Wichita State University), and Cyanosite (www-cyanosite.bio.purdue.edu). Lower left: Roger Burks (University of California at Riverside, Mark Schneegurt (Wichita State University), and Cyanosite (www-cyanosite.bio.purdue.edu). Upper and lower right: Rolf Schauder (University of Frankfurt), Mark Schneegurt (Wichita State University), and Cyanosite (www-cyanosite.bio.purdue.edu).

by organisms such as purple and green bacteria. However, natural selection drove evolution to oxygenic photosynthesis (e.g. see Nisbet & Sleep 2001); once it was in place then microbes such as cyanobacteria—with the addition of nitrogen fixing abilities—could flourish almost anywhere that water, light, and CO_2 were present.

There is a tendency when talking about the longevity of species, and the implications for life beyond the Earth, to become fixated on our own brief rise as organisms, and the apparently catastrophic extinction of many others throughout Earth's history. The example of cyanobacteria and their precursors (and indeed archaea, see below) should give some pause for thought. Here are organisms that, by and large, have been resident on the planet effectively since the crust formed and liquid water existed. Asteroids have hit, the Sun has brightened, climate has changed, and these microbes have persisted through it all. This does suggest that if terrestrial life is anything like a fair sample life on exoplanets is much more likely to be like bacteria than it is to be like bananas, chickens, or spruce trees.

5.4.2   A Changing Planet

The geological and climate history of Earth are of course connected to the history of life in many ways. We will discuss some of the details relating to climate more in Chapter 9. In terms of geology, one critically important factor is the actual arrangement of dry land surface and oceans. In Figure 5.13 the estimated dry land/ocean configuration is shown at three different times in the Earth's history. Global and local climates are profoundly impacted by the variation in atmospheric and oceanic currents due to the shifting land masses. In Chapter 3 we discussed how the internal temperature of a rocky planet is set by a combination of the energy of formation (i.e. transfer of kinetic energy into thermal energy) and radiogenic heating. The outermost, solid shell of a rocky planet is known as the **lithosphere**. In the case of the Earth, the lithosphere consists of the **crust** and the **upper mantle** (c.f. Chapter 1). **Continental crust** and **oceanic crust** are very different on the modern Earth. This may not have been true very early on, during for example the Archaean 3.8 to 2.5 billion years ago (Figure 5.11).

Oceanic crust is some 5–10 km thick and is composed of **basalt** (overlaid by less dense sedimentary material), whereas continental crust is on average 20–70 km thick and is composed of a variety of lower density rocks. Convection of the molten interior of the planet drives the formation of **tectonic plates**, whereby these crusts are broken into zones (15 in the case of the Earth). At **mid-ocean ridges** (see also §5.5.3) mantle material is forcing its way up and forming new oceanic lithosphere. At the interface between oceanic and continental crusts the denser oceanic basalt is **subducted** underneath the lighter continental material and recycled into the molten mantle. In this way the topographical arrangement of the continental masses and the oceans is constantly changing over geological timescales.

Furthermore, as a consequence, while continental crust can be ancient (3.9 to 4.0 billion years old), and is rarely subducted today, the majority of oceanic crust is young (some 200 million years old). It also appears that the presence of liquid water oceans—and the presence of water throughout the lithosphere—probably acts to help "lubricate" the motion of tectonic plates. It can come as a surprise that the presence of substantial liquid water may have such a profound effect on the planetary geology—but this is really due to our tendency to underestimate the role of the oceans in general.

It is still unclear how long it took for continental crust to form on the young Earth. Some estimates suggest that about 3 to 3.5 billion years ago there may have been only 10–25% of the present-day surface area of continental crust, rising to about 60% by some 2.5 billion years ago (Taylor & McLennan 1995). This has some quite startling implications. Assuming that the Earth's water was in place by 3.5 billion years ago, then even more than the present 70% of the planet's surface could have been ocean (or at least water covered)—maybe even 97% of the surface at 3.5 billion years in the past. Stromatolite-producing microbial colonies may indeed have effectively had an aquatic planet to themselves.

We will raise this issue again (e.g., Chapter 10), but the clear message from the Earth's past is that a terrestrial type world, harboring successful life, need not exhibit any great resemblance to the *modern* Earth. Our discussion below also brings to bear on this important fact.

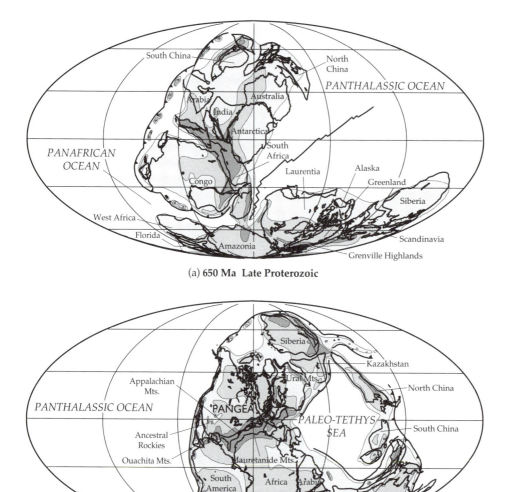

(a) 650 Ma Late Proterozoic

(b) Late Carboniferous 306 Ma

Figure 5.13. The extrapolated arrangement of the Earth's dry land and oceans at three periods: the Late Proterozoic (650 million years ago), the Carboniferous (306 million years ago), and the Cretaceous (94 million years ago). The environment experienced by life both on dry land and in the oceans would clearly have varied dramatically between these periods and the present day. Shaded regions indicate elevations above estimated sea-level. Outlines indicate modern land-masses. Heavy lines with triangles indicate direction of subduction zones. Plate Tectronic and Paleogeographic Maps by C. R. Scotese, © 2008, PALEOMAP Project (www.scotese.com).

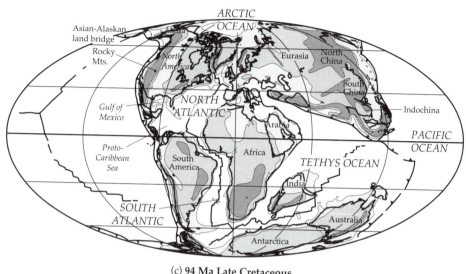

(c) **94 Ma Late Cretaceous**

Figure 5.13 *(continued)*.

5.5 Boundary Conditions and Habitability

The notion that certain environments in the universe are more suited to life than others is certainly a logical extrapolation from our day-to-day experience on the Earth. However, for astrobiology we need to be sure of the limits and ranges that life can occupy, and be able to quantify these. As briefly discussed in Chapter 1 we know that the true majority of life on Earth is almost certainly microbial (Whitman 1998), and occupies niches such as the terrestrial and oceanic subsurface (§5.6). Does such life care much about the stellar insolation at the planet's surface? It may, but much more indirectly than a daffodil or a human. In order to identify the potential niches for life, we need to know its physical limits. In order to get a handle on these we turn to a consideration of **extremophiles**— organisms that survive and thrive in conditions far removed from those optimal for ourselves. This encompasses both prokaryotic *and* eukaryotic life.

5.5.1 Overview of Extremophiles

An environment is considered extreme if some physical characteristic describing it differs strongly from the optimal conditions for organisms such as ourselves. Thus, temperature, acidity, alkalinity, salinity, water availability, toxicity, radiation environment, and nutrient availability (or a lack thereof) are all variables that may define an extreme environment. As organisms have been discovered, they have been categorized according to the dominant environmental characteristics of their habitat, some of which we list below. Of course some organisms may fall into more than one category—these are often termed **polyextremophiles**.

5.5.2 Types of Extremophiles

An organism that is *metabolically* active in a given environment may be placed in one or more of the following examples of extremophilic classes:

- **Acidophile**: Organism with optimal environmental pH level less than or equal to 3.
- **Alkalophile**: Organism with optimal environmental pH level greater than or equal to 9.
- **Barophile/Piezophile**: Organism with high optimal hydrostatic pressure (gas or liquid), typically considered to be above approximately 380 atmospheres, commensurate with deep ocean pressures. Obligate barophiles cannot survive outside of such environments (e.g., *Pseudomonas bathecetes* requires 1000 atmospheres and temperatures around 3 °C to survive), while "barotolerant" organisms can also deal with lower pressures.
- **Endolith**: Organism living in microscopic pores or fissures within rocks.
- **Halophile**: Organism requiring salt concentrations (NaCl) of at least 2 moles per liter of water.
- **Hyperthermophile** and **Thermophile**: An organism requiring ambient temperatures at or above 80 °C is termed a hyperthermophile. An organism requiring ambient temperatures between 60 ° and 80 °C is termed a thermophile (see below).
- **Hypolith**: Organism living inside rocks in cold deserts.

- **Lithoautotroph**: Microbial organism that "eats" rock. Derives energy from reduced compounds of mineral origin. Relies on **autotrophic** metabolic pathways, which produce organic compounds from carbon dioxide.

- **Metalotolerant**: Organism tolerant to high levels of dissolved heavy metals such as copper, cadmium, zinc, and arsenic.

- **Oligotroph**: Organism capable of growth in low nutrition environments.

- **Psychrophile/Cryophile**: Organism with optimal ambient temperatures below 15 °C, and as low as some −20 °C (see below), such as found in permafrost, polar ice, and cold oceans.

- **Radioresistant**: Organism resistant to high levels of ionizing radiation, including ultraviolet photons and nuclear decay products such as alpha particles.

- **Xerophile**: Organism capable of growth in extremely dry, dessicating conditions, such as high-altitude deserts.

Often the metabolic pathways of organisms are uniquely suited to their environment. For example, certain acidophiles are found in sulfuric acid springs emerging from mines associated with pyrite deposits (iron sulfide, or "fool's gold") and operate via an iron–sulfur metabolism at pH 0. It is also the case that organisms actively "defend" themselves against the environments in order to operate. For example, in alkalophiles, which are found in environments with pH as high as 11 (such as **soda–lakes**), there appears to be a mechanism at play within their cells which maintains an *internal* pH no higher than 8.

In the following short sections we discuss in a little more detail some of the above classes of extremophiles.

5.5.3 Hyperthermophiles to Psychrophiles

As stated above, thermophilic organisms are capable of tolerating much higher temperatures than others, and psychrophilic organisms operate at much lower temperatures.

In the past several decades, as the existence and diversity of extremophilic life have been appreciated, thermophilic organisms have perhaps been the most discussed and investigated. To place such organisms in context: in Table 5.2 we list some of the upper temperature

Table 5.2. The upper temperature limits for organism growth

Organism group	Upper temperature limit (°C)
Animals	
Fish	38
Insects	45–50
Ostracods (crustaceans)	49–50
Plants	
Vascular plants	45
Mosses	50
Eukaryotic microorganisms	
Protozoa	56
Algae	55–60
Fungi	60–62
Prokaryotes	
Bacteria	
Cyanobacteria	70–73
Other photosynthetic (non–oxygen-producing) bacteria	70–73
Heterotrophic bacteria (utilize organic nutrients)	90
Archaea	
Methane-producing	110
Sulfur-dependent	115

limits for growth for a range of phyla (not including the most extreme organisms).

As we can see from Table 5.2, if you are a fish you are out of luck, while if you are a microbe you stand a much better chance against high temperatures. Of course it is also true that low temperatures can stymie an organism, and many of the high-temperature bacteria and archaea do not do well in more temperate conditions (by our standards).

The full range of known temperature dependency of organisms is illustrated in Figure 5.14. It is quite likely that these ranges will increase somewhat as new organisms are discovered. At the uppermost end of the hyperthermophiles, the current record holder is an archaea known

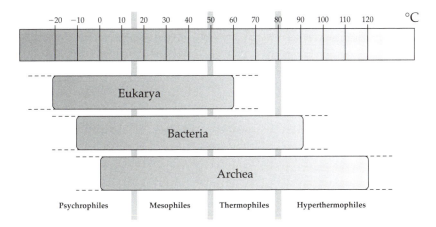

Figure 5.14. The currently known temperature ranges for metabolic activity of the major phylogenetic divisions of life on Earth. These ranges are broadly divided into those of **psychrophiles** (below 15 °C), **mesophiles** (between 15 and 50 °C), **thermophiles** (above 50 °C) and **hyperthermophiles** (above 80 °C). Organisms anywhere in the latter two categories are often just termed thermophiles.

as **Strain 121**. The name reflects the fact that it survives well to temperatures of 121 °C. This is a significant number because it corresponds to the temperature of medical autoclaves previously thought to destroy all living organisms. Strain 121 originates from a deep ocean hydrothermal vent system (see below) where it metabolizes by reducing iron oxide.

Having introduced the concept of thermal ranges for life on Earth, it seems appropriate to spend a little time considering the precise nature of the environments where thermophilic and psychrophilic organisms are found. This is especially important if we are to eventually extend this discussion to potential habitats beyond the Earth.

Much of the work thus far accomplished on thermophiles has stemmed from the realization that geothermal environments can harbor vast communities of organisms. One terrestrial location that epitomizes such a surface geothermal environment is Yellowstone National Park in North America. At Yellowstone, a prehistoric volcanic eruption some 600 million years ago left a **caldera** (a collapsed volcanic structure) some 50 by 70 kilometers across, sitting on top of a hotspot of molten mantle,

Figure 5.15. Photograph of a small hot water pool in Yellowstone park. Although close to boiling point, and mineral rich, a thick white microbial mat is growing in this liquid. It has a partially filamentous structure and is composed entirely of prokaryotic organisms. National Parks Service (NPS), photo by Canter, 1969.

or **magma** (see Chapter 1). The magma is sufficiently close to the surface that there is a very significant heat and chemical flux, especially as surface water interacts with the volcanic structures of the caldera. As a result there are many extreme environments at Yellowstone involving high temperatures and mineral-rich liquid environments of high acidity and alkalinity.

Among the many remarkable phenomena at Yellowstone are the presence of small and large microbial communities, often in the form of microbial mats and biofilms (§5.4) in hot and chemically rich water pools and lakes. An example is shown in Figure 5.15. Of particular interest to astrobiology is the very evident "zoning" of these ecosystems. An excellent example is seen in the Grand Prismatic Spring at Yellowstone, shown in Figure 5.16. Here a pool of boiling water is surrounded by thick microbial mats of vivid green and orange. The central water pool is too hot for the photosynthetic microbes forming these mats, hence they have

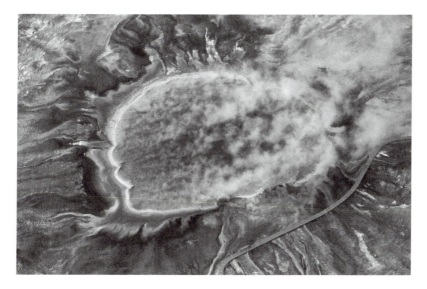

Figure 5.16. An aerial photograph of the Grand Prismatic Spring in Yellowstone park. It is approximately 90 meters across and some 50 meters deep. Steam can be seen drifting to the upper left from the hottest (87 °C) mineral rich water of the spring. The variegated shades of the surrounding features are due to the different pigmentation of the vast microbial mat communities living around the spring. An elevated path has been constructed (lower right), and for scale, what appear to be two human tourists can be seen. National Parks Service (NPS), photo by Jim Peaco, 2001.

populated the zone where the temperature drops to a level that they are more suited to (although still high by human standards). As a *community* they adjust their production of photosynthetic pigments depending on the season, producing more chorophyll (green) in the winter when the sunlight is limited and more carotenoids (red) during the summer (c.f. cyanobacteria). A little further away from the hot pool these thermophilic microbes do less well, although many others can occupy this terrain. Later, when we discuss definitions of habitability (Chapters 9 and 10) it will be informative to consider the basic principle that the Grand Prismatic Spring illustrates—namely that life, and indeed an entire ecosystem, can occupy a relatively narrow zone around a source of energy and nutrients.

On the continental surfaces of the Earth, thermophiles are found to exist in all manner of volcanic environments. The same is true in terrestrial oceans. What is particularly interesting is that in the deep oceans of the planet (i.e., thousands of meters below the water surface), the ecosystem differences between the ordinary "cold" environments and those associated with volcanic activity appear to be in much greater contrast than the equivalent environments on the dry continental surface.

At mid-ocean ridges (§5.4.2) magma is constantly rising and creating an interconnected, volcanically active underwater mountain range that spans the globe over some 60,000 km. Among many phenomena associated with this planet-wide system are **hydrothermal vents**. These are, in essence, similar to the hot springs and geysers of places like Yellowstone. They represent systems where liquid water has found its way into contact with the underlying hot, mineral rich, geologically young material, and is being superheated and returned to the surface (the ocean floor) through vents in the colder overlying rock (Figure 5.17).

At depths of over 1000 meters in the oceans, the ambient temperature is typically some 2–5 °C. The superheated water that emerges from hydrothermal vent systems can reach 400 °C and is rich in dissolved minerals. Due to the high pressure at these depths (e.g., in excess of 100 atmospheres) the water remains liquid (Chapter 9) and a remarkable temperature gradient is established between the superheated outflow and the surrounding ocean. As the vented liquid cools, many of the dissolved minerals begin to come out of solution and form deposits on the immediate ocean bed. Principal amongst these is **anhydrite** (anhydrous calcium sulfate), followed by sulfides of copper, iron, and zinc. These mineral deposits tend to form chimney-like structures around the hot vent, and can grow very rapidly and become very large (60 meter heights have been reported).

Perhaps the most commonly discussed type of submarine hydrothermal vents are known as **black smokers**. In these systems the venting liquid appears black and opaque due to a very high concentration of metal sulfides (see Figure 5.17). Black smokers also attain the highest temperatures, and the outflow is acidic, with pH typically of 2–3.

Remarkably, and in stark contrast to the surrounding deep ocean, these hydrothermal vent systems play host to complex and populous ecosystems—surrounding the vents over a distance of a few tens of me-

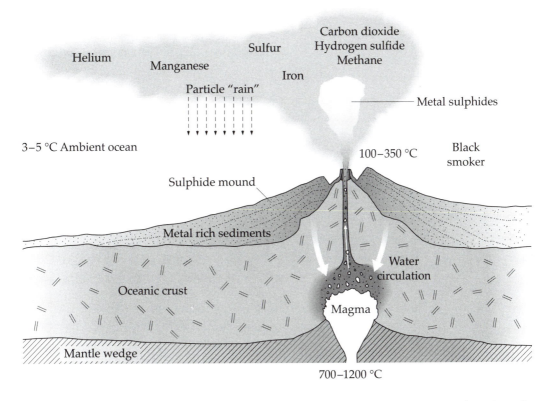

Helium

Manganese

Sulfur

Carbon dioxide
Hydrogen sulfide
Methane

Iron

Particle "rain"

Metal sulphides

3–5 °C Ambient ocean

100–350 °C

Black
smoker

Sulphide mound

Metal rich sediments

Water
circulation

Oceanic crust

Magma

Mantle wedge

700–1200 °C

Figure 5.17. A schematic of a mid-ocean ridge hydrothermal system. Ocean water circulates through pores and cracks in the oceanic crust at a ridge, becoming superheated at the interface with the mantle/magma. The superheated water, rich in dissolved minerals, returns to the seafloor through a vertical channel. As it enters the cold ocean, minerals and metals begin to precipitate out. In many cases these form a "collar" or tower around the exit point. The hottest examples of these systems are termed "black smokers," owing to the opaque, metal sulfide rich water venting into the ocean. As the vented water continues to cool, a plume of heavy elements (including metals) rains out over a few tens of meters, creating an extended mound of material, including large amounts of sulfides.

ters. In addition to multi-cellular organisms such as species of crab, shrimp, snails, and tube-worms (Figure 5.18) there is a rich and diverse microbial population—including, not surprisingly, thermophiles, hyperthermophiles, and acidophiles. These organisms occupy the temperature zones surrounding the vents according to their optimal condi-

Figure 5.18. Hydrothermal vents create oases for life in the deep ocean. Bacteria and archaea exploiting the mineral-rich water act as the base of the food chain. Large, eukaryotic life exists in abundance, including the adult giant "tube worms" (*Riftia pachyptila*) shown here. The outer, extended body of the tube worms acts to catch the compounds (e.g., hydrogen sulfide, carbon dioxide) in the vent water. Tube worms have no digestive tract, but half of their body mass is due to microbial populations that process these compounds into amino acids which the worm then uses. This species can grow to as much as 2.4 meters in length. Photo courtesy of Andrea D. Nussbaumer, Charles R. Fisher and Monika Bright.

tions, and also make use of the rich mineral content of the water through **chemosynthesis**, particularly of sulfur-bearing compounds (§5.6). Indeed, as in any terrestrial ecosystem, the small feed the large. Microbes not only produce generally usable nutrients for larger organisms, they also participate in symbiotic relationships (see also Figure 5.18). For example; the *Pompeii worm* (*Alvinella pompejana*) is perhaps the most heat-tolerant large organism, living happily at 80 °C. It feeds on vent microbes, but also secretes a mucus on its back which is food for hair-like colonies of bacteria. This covering of bacteria, maybe a centimeter thick, may not only provide direct thermal insulation for the worm, but may also produce a key enzyme which helps the worm's own cells resist the thermal environment.

Such deep-sea habitats are still poorly explored, since they are an enormous technological challenge, but they may provide fundamental clues as to both the boundaries of life and the possible origins of life—especially since many of the related ecosystems operate *independently* of the energy provided by the Sun.

At the opposite end of the temperature range shown in Figure 5.14 are the psychrophiles. These organisms have developed mechanisms to avoid cell destruction due to the freezing of water—including the ability to produce their own antifreeze. They can also withstand the effect of freezing/melting cycles on the environmental availability of liquid water. Among the habitats available to psychrophiles are those in the oceans. An amazing 7% of the Earth's surface area is covered by sea ice[7]. Within this ice, and in particular at the interface between the ice and the ocean, a remarkably diverse and extensive range of ecosystems exist (Figure 5.19). Truly psychrophilic organisms make up a large part of the population in these locations.

5.5.4 Biochemical Mechanisms for Survival

How do organisms such as thermophiles operate in environments where others cannot? This is still a matter of intense research and investigation, however some common traits and mechanisms are seen. The theoretical maximum temperature for *any* organism is 160 °C. At this temperature the molecule **adenosine triphosphate (ATP)** breaks down. ATP is *fundamental* to the process of energy transfer in all known living cells. In essence ATP transfers energy from chemical bonds to energy absorbing reactions in a cell—much like a rechargable battery. However, even prior to reaching such temperatures, biochemistry must exploit many different strategies in order for organisms to survive.

The key term here is **thermostability**—an organism and its biochemistry must be able to remain stable to the thermal environment. Conditions in cells must be modified; this appears to include changing the

---

7. As of the time of writing.

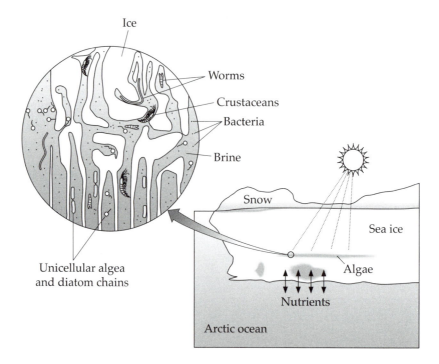

Figure 5.19. A schematic of the sea ice habitat for many psychrophilic organisms. The average freezing point of salt water (brine) is −1.9 °C. The ice at the interface with the ocean can (depending in part on the season) have a highly porous structure, consisting of an intricate network of pores and cracks, filled with brine. Depending on the ice thickness, snow cover (snow being more opaque), and season, this zone may receive sufficient light for photosynthetic organisms (e.g., algae) to flourish. Prokaryotic and eukaryotic organisms form a complex ecosystem in this subsurface environment, and there are many feedback (e.g. Chapter 6) mechanisms between these populations and the actual physical characteristics of the ice/brine environment itself. Adapted from C. Krembs and J. Deming, University of Washington and NOAA.

mix of intracellular salts, producing different solutes, and the synthesis of specific proteins (such as the so-called **heat-shock proteins, HSPs**, found in all hyperthermophiles). It seems that thermophiles tend to have more tightly coiled DNA, more tightly folded proteins, and special enzymes that help repair thermally damaged folds in the DNA. They also tend to have **mono-layer** cell membranes, as opposed to the bi-layer, lipid cell membranes of lower temperature organisms. This may make it harder for the cell walls to be disrupted.

Finally, the most extreme hyperthermophiles are of the phyla archaea. Since these are also considered to be among the most ancient type of organisms on Earth, this is suggestive, although not conclusive, that the environment on the Earth when the first life emerged was much more extreme. It may also point towards an origin for life far away from any temperate surface environment, in the deep earth, or close to volcanic systems such as hydrothermal vents (e.g., "The Deep, Hot Biosphere", Gold, 1992).

5.5.5 Radioresistance

Since the emphasis of this text is the astronomical end of astrobiology, it stands to reason that the ability of organisms to withstand ionizing radiation is of great interest, since many extraterrestrial environments will be subject to far greater exposure than the surface of the Earth. Among eukarya, humans rank quite low on the scale of ability to withstand ionizing (particle) radiation. Indeed, mice, fish, and insects all rank above us. As an example, the common fruit fly can survive at least sixty times more net radiation exposure than a typical human.

The true champion of terrestrial organisms is however a red bacterium called *Deinococcus radiodurans*. It can withstand an instantaneous (as opposed to prolonged) radiation dose some 500 times more than a human, and with some loss of functionality it can even survive an astonishing 1500 times larger dose. *D. radiodurans* accomplishes this remarkable feat through several strategies: it maintains multiple copies of its genome in a cell, its DNA is tightly packed in toroidal forms, and it has an amazing ability to self-repair damaged DNA with specialized proteins. It is also able to effectively switch off cell processes until it is safe to begin repairs, and may possibly use manganese to directly protect itself against radiation.

What is puzzling, and extremely interesting, is *why D. radiodurans* has evolved this capacity for radioresistance. It has been posited that the ability to withstand radiation is actually just a side effect of an ability to withstand extreme *desiccation* (Mattimore and Battista 1996, but see also Exercise 5.5 below). What is curious about this interpretation is that *D.*

radiodurans is found all over the place, including around black smokers, so there appears to be no shortage of niches for it. Based on its place in the phylogenetic tree of life *D. radiodurans* is certainly related to the pure thermophiles; however, it also shares genetic traits with both archaea *and* eukarya. It is possible that horizontal gene transfer has resulted in organisms like *D. radiodurans* becoming amazing *generalists* (i.e., good at many things), and storing many of the most desirable genes. So, although highly speculative, it is possible that *D. radiodurans* acquired the ability to resist radiation from another organism (either long gone, or yet to be discovered)—which may or may not have developed that capability to deal with desiccation.

5.5.6 Spores and Survival

When faced with adverse environmental conditions, some bacteria are capable of forming **endo-spores**—often simply termed **spores**. A bacterial spore is series of tough layers of material encapsulating the bacterial DNA in a suspended state. In this form the bacterial DNA can lie dormant for extended periods until conditions are favorable, at which point it may activate and germinate to produce a new population of bacteria. Although not meeting the primary criteria for extremophiles (i.e., metabolic activity in extreme environments), bacterial spores *can* provide a means for the parent organism to survive through highly adverse conditions, well beyond the normal range for function. It is for this reason that we include them in our brief survey of some of life's boundary conditions.

As discussed in Chapter 1, this ability of bacteria has provided considerable fodder for the concept of *panspermia*—the wide-scale distribution of life through interplanetary and interstellar space. Indeed, one of the best documented examples of the possible tolerance of spores comes from the Apollo lunar missions. During the Apollo 12 mission, samples of plastic and other material were retrieved from the Surveyor 3 robotic lander which, at that time, had spent some three years on the lunar surface. Despite exposure to hard vacuum, solar radiation, and extreme desiccation, spores of the common terrestrial bacterium *Streptococcus mitis* were found on the material (having stowed away on the Surveyor probe from Earth) and successfully revived. It should be noted, however, that there is some disagreement about the result. Surveyor 3 was purposely *not* sterilized before being sent to the Moon, but revivable spores were found only in one of the few dozen tested samples of the Apollo retrieval. There is some suggestion therefore that the spores were simply the result of later contamination.

Even more remarkable claims have been made about spores recovered from the guts of insects (e.g., bees) encased in amber (fossilized tree resin). Some work has suggested that 20 to 130 million year old spores can be successfully revived. Dealing with contamination by more modern microbes is a major difficulty in such studies, compounded by our relative ignorance of the true diversity of currently living microbes. Perhaps the most convincing examples of long-term survival of bacteria and other organisms comes from the study of the subsurface permafrost in locations such as Sibera and Alaska. Mosses likely to have been in a dormant state for some 40,000 years have been discovered, as well as apparently dormant bacteria such as *Carnobacterium pleistocenium.*

In brief: some bacteria do indeed form highly resilient spores. These may play an important role in the survival and propagation of life within a biosphere—and conceivably beyond—however we do not yet know to what extent.

5.6   Deep Life

In discussing the dominant life on Earth (the subsurface microbes) and the extreme environments that some of that life inhabits (the hydrothermal vent systems) we have only just begun to touch upon the "deep" habitats of the planet. The **subseafloor** of the world's oceans (which we recall cover 70% of the surface of the modern Earth) is now known to harbor microbial cells to average depths of *at least* 1000 meters. This subseafloor is a layered environment of biologically produced sediment at the top, down to the basalt of the Earth's oceanic crust (§5.4.2). Furthermore, it is now understood that water actively permeates, and flows in and out of, this subseafloor—driven by pressure from above and from geothermal energy from below. In fact, current estimates suggest that the *entire world's oceans* are effectively filtered through the subseafloor approximately every *million* years or so (e.g., Davis & Elderfield 2004).

Much of the subseafloor microbial population is known to consist of species of **Methanogens** and **Sulfate-reducing bacteria** (SRB). The methanogens are archaea (Figure 5.8), and produce methane as a product of metabolism. They are **anaerobic** (literally "without air," or more specifically without oxygen), meaning that oxygen is actually deadly for them. Many take up carbon dioxide as a source of carbon and use hydrogen as a reducing agent to produce energy (the production of methane naturally follows). Sulfate-reducing bacteria on the other hand exploit the oxidizing power of sulfate (i.e., containing sulfate ions SO_4^{2-}), and reduce it to sulfide. They too are anaerobic in nature, and play a major role in bio-corrosion—for example during earth drilling processes.

These microbes are therefore **chemosynthetic** in nature—they derive energy from chemical reactions. By contrast we are, of course, much more familiar with **photosynthetic** organisms, which exploit the energy of UV and visible photons to perform much of their chemistry. We ourselves might be considered as aerobic chemosynthesizers—however, we rely on our microbial passengers to do the work for us. Both our digestion of food and our basic cell function are actually largely a result of microbial processes. In the digestion of food we rely on a flora of microbes (including both bacteria and archaea—such as the methanogens), and in the running of our cells we rely on the presence of **mitochondria** (Figure 5.3), which convert food molecules (e.g., amino acids) into ATP (see §5.5.4). Mitochondria are almost certainly an example of **endosymbiosis**—whereby a microbial organism is incorporated directly into a eukaryotic cell in the form of an organelle (§5.3.1). This symbiosis ensures the survival of both the microbe and the host cell.

Returning to the subseafloor: the combination of a remarkable system of water circulation and a seemingly vast microbial population (as well as the more localized mid-ocean ridge communities—§5.5.3) points towards a deep oceanic biosphere that more than rivals the one we see on the planet's surface. This "deep life" does not end in the oceans. In fact, even in the apparently harsh conditions at depths of 3–4 km in South African gold mines, healthy populations of microbes are found. Some of these organisms are sulfate-reducing bacteria. Remarkably, they obtain the raw hydrogen needed for their metabolism as a byproduct of the natural radioactivity at these depths. Uranium ores and radioactive potassium emit alpha particles that can readily ionize molecular hydrogen and provide the protons needed by these bacteria. What is particularly fascinating is that these bacteria seem to show a deep genetic linkage on the tree of life (Figure 5.8) to **phototrophic** (photosynthesizer) organisms. Indeed, there are examples of subsurface microbes that are at least genetically equipped to deal with both chemosynthetic *and* photosynthetic metabolism.

As we have already alluded to, such discoveries must profoundly effect our most fundamental theories of both the origin and evolution of life on the Earth. From the astrobiological perspective they also throw open a bewildering array of possibilities and questions. If we are to know where to look for life "by hand" on other worlds (e.g., Mars), or where, and how to look for life remotely (e.g., exoplanets), it is clear that we must incorporate this information and seek knowledge of how these deep biospheres interact with the external characteristics of a world.

References

Allwood, A. C. et al. (2006). Stromatolite reef from the early Archaean era of Australia, *Nature*, **441**, 714.

Doolittle, W. F. (2000). Uprooting the tree of life, *Scientific American*, February 2000.

Davis, E. E., & Elderfield, H., (Eds.) (2004). *Hydrogeology of the Oceanic Lithosphere*, Cambridge University Press.

Dawkins, R. (2006). *The Selfish Gene: 30th anniversary edition*, Oxford University Press.

Gold, T. (1992). The deep, hot biosphere, *Proceedings of the National Academy of Sciences*, **89**, 6045–6049.

Gould, S. J., (2002). The Structure of Evolutionary Theory, Harvard University Press.

Mattimore, V., & Battista, J. R. (1996). Radioresistance of *Deinococcus radiodurans*: Functions necessary to survive ionizing radiation are also necessary to survive prolonged desiccation, Journal of Bacteriology, **178** (3), 633–637.

Maynard Smith, J., & Szathmáry, E. (1999). *The Origins of Life: From the Birth of Life to the Origin of Language*, Oxford University Press.

Nisbet, E. G. & Sleep, N. H. (2001). The habitat and nature of early life, *Nature*, **409**, 1083.

Taylor S. R., & McLennan, S. M. (1995). The geochemical evolution of the continental crust, *Reviews of Geophysics*, **33**, 241–265.

Whitman, W. P., Coleman, D. C., & Wiebe, W. J., (1998). Prokaryotes: The unseen majority, *Proceedings of the National Academy of Sciences.* **95**, 6578–6583.

Woese, C. R., & Fox, G. E. (1977). Phylogenetic structure of the prokaryotic domain: The primary kingdoms, *Proceedings of the National Academy of Sciences of the United States of America*, **74**, (11), 5088–5090.

Problems

5.1 At the beginning of this chapter we mentioned how the apparent absence of the signature of advanced life can be used to say something about its characteristics. Within 25 pc of the Earth there are currently known to be some 2029 stellar systems. Assuming that the absence of any signature of advanced life indicates that there is *no more* than one such population in this volume then determine (trivially) the upper limit for the number of examples of advanced life per cubic parsec. Assuming the Milky Way Galaxy is a cylinder of radius 7.5 kpc and total height 0.6 kpc then estimate the *upper* limit for the number of stellar systems with advanced life in the Galaxy—describe any further assumptions you make.

5.2 Describe the fundamental differences between Prokaryotic and Eukaryotic life. Can you speculate on the evolutionary advantages/disadvantages of each?

5.3 Proteins consist of a linear chain of amino acids (a *polypeptide*) coded for by DNA (e.g., Table 5.1) that can fold into a three-dimensional, tertiary, structure. It is this final structure that performs the protein function. Protein folding is typically a highly complex process, and is very hard to correctly predict from the simple base-pair code that a given protein corresponds to.

From Table 5.1 compute the *mean* number of codons that correspond to any given biologically important amino acid. Now consider a protein that consists of 100 amino acids. How many ways (as a statistical average) can this protein be coded for? Can you comment on what this number may imply ? (*Suggestion:* Consider how such molecules may have evolved as their parent organisms evolved.)

5.4 Describe the techniques of modern phylogenetics and the construction of a "tree of life" for terrestrial organisms. Explain why horizontal (or lateral) gene transfer may confuse phylogenetic measurements.

5.5 Describe the circumstances and nature of thermophilic organisms on the Earth. Can you think of additional potential environments (natural and human made) to those mentioned in this chapter that thermophilic organisms might inhabit?

5.6 In the paper "Was Earth ever infected by Martian biota? Clues from radioresistant bacteria" (Pavlov, A. K., et al. (2006). *Astrobiology*, **6**. 911) Pavlov et al. argue that the radioresistance characteristics seen in a bacterium like *D. Radiodurans* (see §5.5.5) could have arisen from natural selection in an environment such as that of Mars. Subsequent transfer of material from Mars to the Earth (e.g., meteorites) could have then brought these traits here, rather than their arising terrestrially. Review this paper and attempt to critique its arguments.

5.7 A typical value for the solar energy flux at the Earth's *surface* at any given time/latitude is of the order 1×10^9 erg s^{-1}m^{-2} (about 100 W m^{-2}). (Note that this allows for atmospheric absorption and scattering.) An average *geothermal* energy flux close to the surface on Earth's continents is about 6×10^5 erg s^{-1} m^{-2}. Taken at face value, what does this suggest for the nature of surface life versus subsurface life? (e.g., consider metabolism).

In reality, geothermal energy flux is much less evenly distributed than solar flux. How might this effect the evolutionary specialization of "deep biosphere" organisms?

CHAPTER 6

Planetary Radiation, Comparative Planetology, Biosignatures, and Daisyworld

6.1 Introduction

In this chapter we return to questions surrounding the observation and evaluation of planetary bodies. This differs from our earlier brief foray into such matters (Chapter 4) in that we will now begin to consider some of the issues surrounding biological systems (Chapter 5). Although it is certainly true that the quantitative description of the external characteristics of planets is relevant for all planet types—for the science of astrobiology it is naturally the evidence for life on rocky worlds that is of most concern. The arrangement of this discussion may seem unusual— going from astrophysics to biology, and now back to astrophysics and planetary science. However, a strictly linear approach suffers the danger of becoming little more than a disconnected history of subjects. We are interested in seeing the *inter*connected discussion across disciplines, and perhaps more importantly, conveying the nature of the scientific strategies that need to be employed to address the complex questions at the heart of astrobiology.

Our only available probes of the characteristics of exoplanets are their orbital dynamics and the electromagnetic radiation that they reflect, emit, and sometimes block. We may never travel to worlds beyond our Solar System but we are capable of seeing their light, and we must be prepared to use this as one of our principal diagnostics of the physical characteristics of planets—and ultimately of the life they may harbor. As we discussed in Chapter 5, the perception that terrestrial life (and therefore by extrapolation, extraterrestrial life) is a "visible" or surface phenomenon is in many regards demonstrably false. Nonetheless, it is still true that although life may not universally (or even typically) occupy and influence a planetary surface and atmosphere, it clearly has the capacity to do so. We will return to the question of the impact of widespread life on the external characteristics of a planet (specifically terrestrial type worlds) in later discussions. Our first step must, however, be to develop the analytic tools and strategies that allow us to investigate a planet's external characteristics regardless of its nature.

We also present a rudimentary investigation into the potential interplay between life and the external characteristics of a host planet through the use of a relatively simple, analytic model. This is fun to do, and it also offers a glimpse of the ways in which future observations of other worlds will need to be evaluated in order to search for the signs of life. Seeking evidence for surface environments that are held out of equilibrium (either thermodynamically or chemically) may be a critical indicator of the presence of a biosphere.

6.2   The Reflectivity of a Planet: Albedo

As a class of astrophysical objects, planets both absorb, emit, and reflect radiation. These processes are intimately connected with each other and have the potential to be complex functions of both radiation wavelength and location on the planet itself. For now we will focus primarily on the spatially and spectrally "integrated" radiation properties.

The classical approach to quantifying the reflectivity of an object is to define its **albedo** (from the Latin for white, "albus"). If an object reflects 100% of radiation at all frequencies then its albedo, usually denoted as A, is unity (as we will see below this is often termed the **Bond albedo**). For example, the Earth's albedo (time averaged, to allow for seasonal and weather variations) is about 0.31, which implies that about 69% of the incident solar radiation is absorbed and will both heat the planetary surface and atmosphere and be processed via other mechanisms (e.g., photon absorption that drives chemistry). Ultimately this energy will go into powering phenomena ranging from atmospheric and oceanographic currents (Chapter 9) to inorganic chemistry, as well as the biosphere. The slight eccentricity of the Earth's orbit ($e = 0.0167$) results in an annual variation of input energy of about ±3.4%, which,

Table 6.1. Examples of the
albedo of Solar
System bodies

Object	Albedo (A)
Comets	< 0.005
Moon	0.07
Mars	0.16
Earth	0.31
Jupiter	0.51
Venus	0.76
Enceladus	~ 1.00

together with time variation of the albedo (due to clouds, biology, etc.), results in a more complex variation in the input. This should, of course, not be confused with the additional *latitudinal* variation in radiation received at the terrestrial surface due to the inclination, or obliquity, of the Earth's spin-axis and the finite optical opacity of the atmosphere (see also Chapter 9).

Before we look at albedo in a more mathematically rigorous way it is illustrative to see that there is a wide variation in its basic value for different solar system objects; a few are listed in Table 6.1

Comets, with a large complement of dark surface dust and organic chemicals (Chapter 8) typically have extremely low albedos. It is perhaps surprising that the Moon has such a low albedo: after all, we are used to it shining brightly in both the day and night skies, whereas Venus is ten times more reflective. This is simply a result of our limited perception. The Moon produces a greater flux incident on our eyes or instruments, but that is due to its proximity. In fact the lunar regolith is well worn by large and small impacts as well as solar radiation and is typically dark and absorbing. Enceladus, orbiting Saturn, has a remarkable, little cratered (therefore youthful, Chapter 8) water ice surface, which accounts for its spectacular reflectivity. These numbers vividly demonstrate that the albedo is a strong function of the precise composition and nature of an object—and therefore a powerful tool with which to understand objects.

6.3   The Thermal Characteristics of a Planet

Armed with the albedo we can begin to assess the thermodynamic balance between a planet and the radiation field it is bathed in. The cross-sectional area of a planet absorbing/reflecting radiation at any given instant from a single (distant) source is just πR_p^2. The incident flux from a parent star is just $L_*/(4\pi d^2)$, where d is the star–planet separation. Thus, the rate of **absorption** of energy is the product of these quantities:

$$\frac{(1-A)L_*}{4}\left(\frac{R_p}{d}\right)^2. \tag{6.1}$$

If the planet is in thermal equilibrium with the external radiation field then clearly the energy absorbed will equal the energy emitted (after being thermalized by the planet). If we assume that the planet can be treated as a blackbody then its bolometric luminosity is just $L_p = 4\pi R_p^2 \sigma T_p^4$, where T_p is its **effective** blackbody temperature. Thus, in equilibrium we can equate this luminosity to the rate of energy absorption and rearrange for T_p:

$$T_p = \left[\frac{(1-A)L_*}{16\pi d^2 \sigma}\right]^{1/4}, \tag{6.2}$$

which is *independent* of the planetary radius R_p. We have made three big assumptions in arriving at Equation 6.2. The first is generally known as the "fast rotation" approximation. We have assumed that the absorbed radiation is instantaneously thermalized, and that this energy is redistributed from the illuminated hemisphere to the entire sphere of the planet. If this assumption were relaxed to the opposite extreme - "slow rotation"—T_p would be a factor $2^{1/4}$ larger. The second assumption is that the planet does *not* have an atmosphere and is a perfect blackbody. In Chapter 9 we will deal with this issue in some detail. The third assumption is that there are no significant *internal* energy sources contributing to the effective planetary temperature—such as radiogenic driven heating, latent heat of formation, tidal energy dissipation and so on.

Although this is highly simplified we now have at our disposal a first-order means to estimate the blackbody temperature of a planet around a star, and therefore a means to estimate one of its observable characteristics. For the Earth $T_p \simeq 255$ K is calculated assuming an albedo of 0.31 (Table 6.1). This is clearly lower than we might expect—which is the subject of more extensive investigation in Chapter 9. However, from Wein's Law it does immediately tell us that the thermal *emission* from the Earth will be found in the infrared, while the peak of the *reflected* sunlight will still be found in the optical. We will return to this discussion later (§6.6.3), but for now we emphasize that this temperature need not equal the actual surface temperature of a planet at a given location; it is an integrated *effective* radiative temperature.

6.4 Interpreting Reflected Light from Planets

Having looked at the energy absorbed by a planet let us turn our attention to the reflected component of incident radiation. It should be noted that detailed calculations of reflection quickly become complex (as do those for absorption); however, they are generally tractable. This discussion is naturally organized by considering the three most commonly used definitions of albedo.

6.4.1 Geometric Albedo

The **phase** of an object is defined as the angle (ϕ) between the lines connecting the source of illumination (e.g., the star) to the object and the observer to the object (Figure 6.1). When $\phi = 0$ an object is said to be in **full phase**: the observer sees a fully illuminated hemisphere/disk of a planetary body. The **geometric albedo** (A_g) of a body is then defined as its reflectivity (the fraction of light reflected in the direction of the observer) at full phase relative to a **Lambert** sphere (generally at a specific radiation wavelength, although sometimes integrated over a band or all wavelengths). A Lambert sphere is a **perfect, diffuse reflector**. In other words, such an object will reflect all incident radiation isotropically.

For a spherical object, such as a planet, the net geometric albedo is therefore an integrated property over the visible hemisphere. Figure 6.2 illustrates the situation. The geometric albedo (at a given wavelength) is obtained as an integral over the illuminated hemisphere at full phase:

$$A_g = \frac{1}{\pi I_{inc}} \int_{\theta=-\pi/2}^{\pi/2} \int_{\alpha=0}^{\pi} I(\theta, \alpha, \phi = 0) \cos \theta \sin \alpha \, d\Omega, \quad (6.3)$$

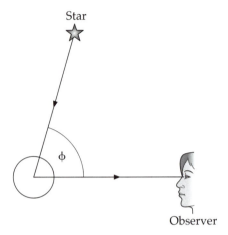

Figure 6.1. The definition of *phase* as the angle ϕ between the line connecting the source of illumination (e.g., a star) to an object (e.g. a planet) and the line connecting the object to the observer.

where I_{inc} is the incident specific radiation intensity (so that πI_{inc} is the incident flux), and I is the emergent/scattered radiation intensity.

6.4.2 Spherical Albedo

The **spherical albedo** (A_s) is defined as the fraction of incident radiation reflected by a sphere *at all angles*. In other words it corresponds to the reflected flux integrated over all phases—or all possible observers.

For a unit radius sphere the reflected flux F at a given phase angle can be given as

$$F(\phi) = \int_{\theta=\phi-\pi/2}^{\pi/2} \int_{\alpha=0}^{\pi} I(\theta, \alpha, \phi; \theta_0, \alpha_0) \cos \theta \sin \alpha \, d\Omega, \quad (6.4)$$

where θ_0 and α_0 correspond to the *incident* angles of radiation (which are no longer fixed as in the case of full phase). Thus,

$$A_s = \frac{1}{\pi I_{\text{inc}}} \int_{4\pi} F(\phi) \, d\Omega = \frac{2}{I_{\text{inc}}} \int_0^{\pi} F(\phi) \sin \phi \, d\phi, \quad (6.5)$$

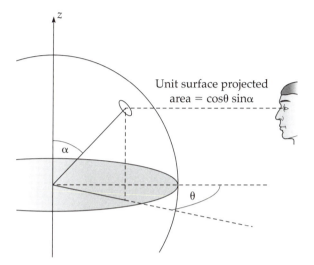

Figure 6.2. Illustration of the geometry used in considering the geometric albedo. A unit surface at an angle α from the vertical axis ($\hat{z}$) of an object (e.g., a planet) and a longitude θ measured from the line connecting the object and the observer has a *projected* area to the observer of $\cos\theta \sin\alpha$. The net observed intensity is therefore an integral over the visible hemisphere of this projected area.

assuming a uniform sphere. We can then see that the spherical albedo is directly related to the geometric albedo by

$$A_s = A_g \cdot \frac{2}{F(\phi = 0)} \int_0^\pi F(\phi) \sin\phi \, d\phi, \qquad (6.6)$$

where the second set of terms on the right-hand side of Equation 6.6 is more generally known as the **phase integral**, q, so that $A_s = A_g \cdot q$. Thus we see that while the geometric albedo refers to the reflected radiation in a *specific direction*, namely that of the source itself, the spherical albedo refers to the *net* reflection of radiation in all directions.

6.4.3   Bond Albedo

Given the definition of the spherical albedo we can now provide a formal definition of the most often used form of albedo (which we used in introducing the concept in §6.2). The **Bond Albedo** (A_b) is the ratio of the *total* power reflected to the *total* incident power, and is therefore explicitly defined as an integral over all wavelengths λ:

$$A_b = \frac{\int_0^\infty A_{s,\lambda} I_{inc,\lambda}\, d\lambda}{\int_0^\infty I_{inc,\lambda}\, d\lambda}. \qquad (6.7)$$

Thus, Bond albedo should be used with some caution since it depends on the spectrum of incident power.

6.4.4   Application

We have already seen one simple application of the Bond albedo in estimating the equilibrium effective temperature of a body (Equation 6.2). However, actually measuring A_b may not always be straightforward. Consider the situation for the observation of reflected light from an unknown exo-planet. We recall that in most cases, with current technology, the light of the planet is neither resolved from that of the parent star, nor is the disk of the planet ever seen (Chapter 4). For an arbitrary, and possibly unknown geometry the observer will, in general, *never* observe the planet at full phase, and must therefore deduce A_b by modeling the system and recognizing that $A_b \propto A_s$ and $A_s = A_g \cdot q$. Furthermore, the detailed radiation scattering properties of planet atmospheres and surfaces must be incorporated, as in Equation 6.4. In this case we can certainly make use of the properties of planets in our own solar system as a means to calibrate the calculations.

Given a series of observations of the reflected light from an unknown planet that cover an entire orbital period, and given some knowledge of the orbital configuration (e.g., eccentricity, orientation), it is possible to constrain both A_g and q. With further modeling it is, in principle, possible to constrain much more, such as the planet's inclination (obliquity) - although issues of non-uniformity (e.g., clouds, features such as the Great Red Spot on Jupiter etc.) become important (see also Chapter 4, §4.3.1).

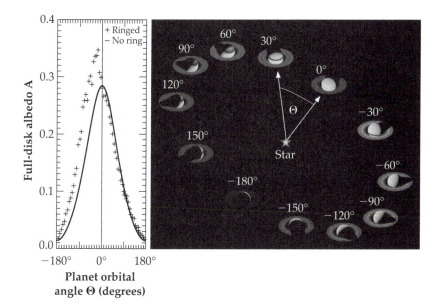

Figure 6.3. An illustration of the information content of albedo measured as a function of phase (adapted from Dyudina et al. 2005). In this case a hypothetical distant observer of the Saturn system is positioned some 35° above Saturn's orbital plane (i.e., the apparent system inclination is $i = 55°$). The left-hand panel shows the expected phase curve for the full-disk albedo (the ratio of the observed power to a perfect Lambertian sphere of the same size in full phase). Without ring structures (solid curve) the full-disk albedo is a simple symmetric function about the zero point of the orbital angle (shown in the right panel). With rings included the estimated full-disk albedo is more highly peaked (owing to an increased area of reflection) and is asymmetric. The right panel provides a visual guide to the actual illumination and shadowing arrangement at the different phases.

As a great example, Figure 6.3 illustrates the situation for a hypo-thetical observer of the Saturnian planet–ring system positioned some 35° above the orbital plane of Saturn about the Sun. The left-hand plot in this instance shows the so-called "full-disk" albedo, which is simply the ratio of the *observed* reflected power to that of a perfect (Lamber-tian) sphere of the same geometric size at full phase, as a function of Saturn's position in its orbit. In this case the effect of Saturn's rings has been explicitly modeled, and demonstrates that simple observations of reflected light from such a planet might lead one to deduce the presence of these structures—even without direct imaging of the planet itself.

6.5   Total Observed Flux

The total observed flux, f_{planet} from a planet is of course a sum of reflected radiation and emitted radiation due to the planet's finite temperature:

$$f_{planet} = f_{emitted} + f_{reflected}. \qquad (6.8)$$

Thus, if we assume no additional internal source of thermal energy,

$$f_{emitted} = \frac{L_{emitted}}{4\pi D^2} = \frac{R_p^2 \sigma T_p^4}{D^2} \qquad (6.9)$$

where D is the distance of the observer from the planet. For the reflected component

$$f_{reflected} = \frac{1}{4\pi D^2} A_g \pi R_p^2 \frac{L_*}{d^2} q(\phi). \qquad (6.10)$$

Simplifying we find

$$f_{reflected} = \frac{A_g}{4} \left(\frac{R_*}{D}\right)^2 \left(\frac{R_p}{d}\right)^2 f_* q(\phi), \qquad (6.11)$$

where f_* is the flux at the stellar surface, πI_*, and $q(\phi)$ is the instantaneous phase (removing the integral over ϕ), assuming again a uniform planet surface.

We also note that Equation 6.8 can be re-written to obtain:

$$\frac{L_{planet}}{L_*} = \frac{R_p^2}{4D^2}, \qquad (6.12)$$

where L_{planet} is the total emitted and reflected luminosity of the planet (integrated over all frequencies, a **bolometric** luminosity) when it is in perfect thermodynamic equilibrium as a blackbody. Thus, Equation 6.12

provides a first-order estimate of the relative total luminosity of a planet to its parent star, *independent* of albedo or temperature.

The peak wavelengths of the emitted and reflected fluxes will of course be very different. We discuss this and other issues of planetary spectra in more detail in the following sections.

6.6 Comparative Planetology

We have mentioned before (Chapters 1, 3, and 4) the notion of **comparative planetology**. Planets represent a phenomenon that is likely more complex than that of stars. The bulk characteristics of a star, especially on the Main Sequence, are governed to quite high precision by only a few basic parameters (e.g., mass, composition, age). Planets on the other hand, with their potentially complex and diverse formation pathways (Chapters 2 and 3), and wide range in composition and internal and external structure, are seldom adequately described by simple physical modeling. Nonetheless, the assumption is that there must be underlying physical parameters that ultimately determine planetary characteristics, albeit in a nonlinear fashion (e.g., mass may not be linearly related to surface composition). Faced with this difficulty in "decoding" a planet's nature the best course of action is to develop a methodology for comparative studies. For example; we might consider the differences and similarities between cold gas giants and the hot Jupiters (Chapter 4), or the differences between terrestrial-type worlds in young systems and those in old systems (based on the stellar parent ages). By looking for correlations and patterns we can hope to gain insight to the more fundamental physical characteristics of planets.

The use of albedo and phase measurements, as described above, provides a starting point for such investigations (although it should be noted that for exoplanets, albedo will be very often assumed, not measured). The study of atmospheric structure, composition, and dynamical state provides a rich set of data to extend this.

6.6.1 Planetary Atmospheres

A major part of the puzzle in deciphering the external (and sometimes internal) characteristics of planets is the structure and composition of

Table 6.2. Approximate pressure scale heights (km) for planets in the solar system

	Venus	Earth	Mars	Jupiter	Saturn	Uranus	Neptune
Scale height (km)	16	8.5	18	18	35	20	19

their atmospheres. We will return to this subject in discussing the habitability of terrestrial planets (Chapter 9), but introduce it here by investigating the globally averaged vertical pressure and temperature of planetary atmospheres in our own solar system.

If one makes the approximation that the atmosphere of a planet is in hydrostatic equilibrium then we can appeal to the equations describing this (Chapter 2) and derive the gas pressure as a function of altitude (z):

$$P(z) = P(0)e^{\frac{-z}{H(z)}}, \tag{6.13}$$

where we define the **pressure scale height** H as

$$H(z) = \frac{kT(z)}{g(z)\mu(z)}, \tag{6.14}$$

where $T(z)$ is the temperature at altitude z, k is Boltzmann's constant, $g(z)$ is the gravitational acceleration at altitude z, and $\mu(z)$ is the mean molecular *mass* at altitude z. Thus H is the distance over which the atmospheric pressure decreases by a factor e. In Table 6.2 we list the approximate pressure scale heights for the major planets with significant atmospheres (i.e., excluding Mercury and Pluto).

At first glance this may seem counterintuitive—the gas giants have scale heights comparable to those of the terrestrial planets. However, this is simply due to the variation in the parameters of Equation 6.14. By comparison, the tenuous atmosphere of Mercury (not listed) can have a scale height on the illuminated side of about 95 km, due to a combination of high temperatures (725 K) and low gravitational acceleration. Similarly, despite a surface temperature of 90 K, the Saturnian moon Titan has a scale height of some 20 km, due to its low gravitational acceleration.

Clearly this parameterization must be strongly effected by what $T(z)$ and $\mu(z)$ are in detail. The nature of $T(z)$ for a planetary atmosphere is mainly governed by *vertical* energy transport, which is in turn largely governed by the transmission of photons (radiative transfer), which is itself a function of the **opacity** of the atmosphere and the chemical and physical processes taking place (§6.6.3; see also Chapter 9 for discussion of planet-wide energy transport). The actual globally averaged temperature profiles for the terrestrial planets used in Table 6.2 are illustrated in Figure 6.4. For comparison, the pressure–temperature profiles for the giant planets are shown in Figure 6.5. In the lowest altitude zones of all the atmospheres shown in Figures 6.4 and 6.5, the temperature initially decreases with increasing altitude within the region known as the **troposphere**. Within the troposphere cloud formation typically occurs. For the Earth the condensing species is H_2O, on Mars it is CO_2, Venus H_2SO_4, and on the giant planets it can be NH_3, H_2S, and CH_4. At a certain minimum temperature at the **tropopause** the temperature then begins to *increase* again with altitude in a region known as the **stratosphere**. In the case of the Earth this temperature inversion is governed by ozone (O_3) which absorbs UV photons. It is notable that Mars does not appear to exhibit this temperature inversion structure. Above this is the **mesosphere**, where the temperature often remains relatively constant with altitude. Above this layer the stellar irradiation (the UV in particular) causes the temperature to increase with increasing altitude. This next layer of the atmosphere is generally termed the **thermosphere** (for the Earth this begins at about 85 km altitude). Within the thermosphere is the **ionosphere**, which as its name suggests, is a region where dissociation of molecules and atoms takes place at a higher rate and can be treated more as a plasma. Above the thermosphere is the **exosphere**, which extends out into space. The density here is sufficiently low that the atoms and molecules of the atmosphere collide much less frequently and begin to follow essentially "ballistic" trajectories that can result in their escaping altogether and being lost to interplanetary space (see below). The exosphere of the Earth starts at around 500 km altitude at the **exobase** which is defined as the point where the mean free path of a high-velocity atmospheric particle is roughly equal to the depth of the remaining atmosphere above it (see also optical depth discussion in Chapter 9).

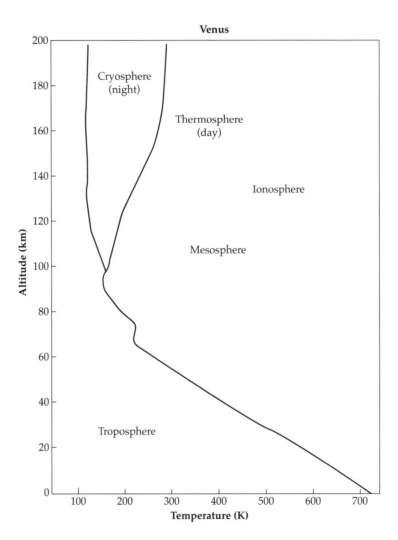

Figure 6.4. The atmospheric temperature structure is plotted for Venus (panel (a)), Earth (panel (b)) and Mars (panel (c)) (adapted from Lissauer & de Pater (2004) and references therein). For all three worlds the atmospheric temperature initially declines with increasing altitude above the planetary surface (the troposphere). Both Venus and the Earth exhibit a temperature inversion above this layer. In the outer atmospheres (e.g., panels (a) and (b)) there is a dramatic shift in temperature between day and night. A strong variation is also seen (e.g., panels (b) and (c)) due to changes in solar activity, as high-energy particles bombard the outer atmospheres and heat it up during episodes of increased activity.

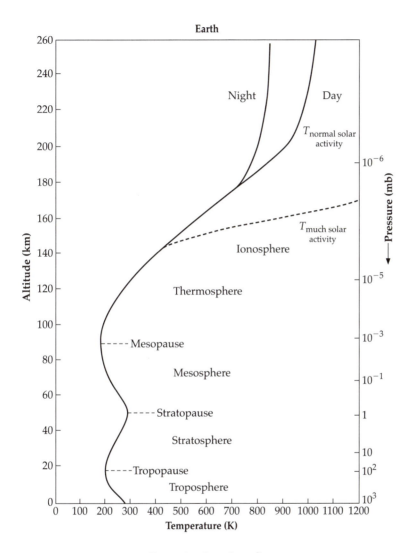

Figure 6.4 *(continued).*

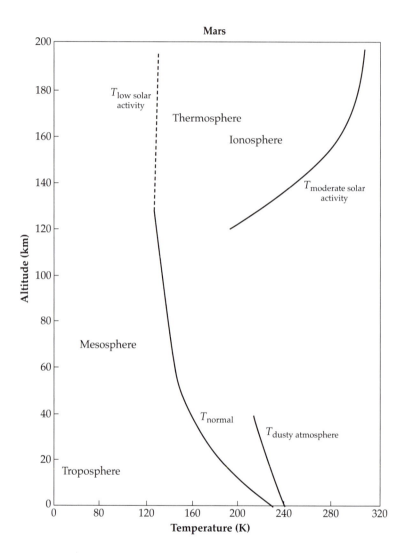

Figure 6.4 *(continued)*.

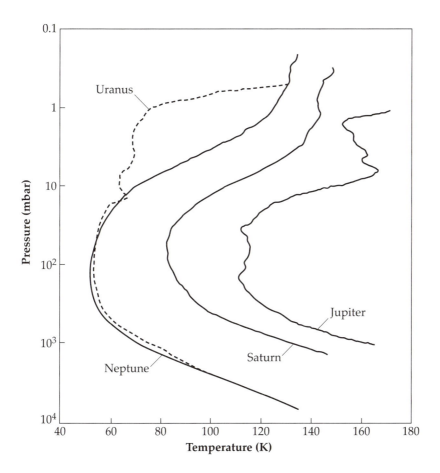

Figure 6.5. The atmospheric temperature structure is plotted for the giant planets Jupiter, Saturn, Uranus, and Neptune. In this case the temperature is plotted versus atmospheric pressure, which varies with altitude according to Equation 6.13. Much as for the rocky worlds (Figure 6.4) the giant planets exhibit a decline in temperature with increasing altitude within the tropopause before it increases again in the outer atmosphere. Adapted from Lissauer & de Pater (2004)

It is readily apparent that planetary atmospheric structure can be complex. Indeed, energy transport in an atmosphere varies with altitude (and location, Chapter 9). In the troposphere of the Earth, for example, energy transport is actually dominated by **convection** (i.e., the physical vertical movement of atmosphere) rather than pure radiative transfer.

6.6.2 Atmospheric Loss

As described above, a planetary atmosphere will be continually "leaking" away above the exobase. The rate at which this occurs for different atmospheric components is ultimately a complex function of many physical mechanisms. These include:

- **Diffusion**: molecular (thermal) diffusion and **eddy diffusion** (the vertical motion of packets of atmosphere) transport atmospheric components upwards to the exobase. The rate of vertical diffusion is therefore a fundamental limit on how rapidly atmosphere can be lost to space. This limiting flux of atoms or molecules is determined by the pressure scale height of the atmosphere and the **mixing ratio** of a given component. As an example, in the case of the Earth the hydrogen limiting flux is some $\sim 2 \times 10^8 \ \mathrm{cm}^{-2} \, \mathrm{s}^{-1}$.

- **Jeans Escape**: Given the definition of the exosphere as the region of the atmosphere of the same depth as the mean free path of atmospheric particles, then the criterion for a particle to escape (since statistically it will, on average, not collide with anything) is just that its kinetic energy must exceed the gravitational potential energy at that altitude. Assuming its kinetic energy is entirely due to a thermal distribution of particle velocities (the Maxwell–Boltzmann distribution, see Chapter 7) the *most probable* particle velocity is just $v_0 = \sqrt{2kT/m}$ where T is temperature and m is particle mass. An **escape parameter** can then be defined as the ratio of the potential to kinetic energy at a given altitude. This ratio is then $X = (v_e/v_0)^2$, where v_e is the escape velocity at this altitude. If $X > 3$, then by integrating the Maxwellian distribution of particle velocities from v_0 to ∞ the Jeans formula for the upwards flux of lost particles (F) is obtained (e.g., Lammer et al. 2003):

$$F = \frac{N_{\text{exobase}} v_0}{2\sqrt{\pi}} (1 + X_{\text{exobase}}) e^{-X_{\text{exobase}}}, \qquad (6.15)$$

where N_{exobase} is the number density of a given particle at the exobase. For the Earth $N_{\text{exobase}} \sim 10^5$ cm^{-3}, and the temperature is about 900 K. For atomic hydrogen then $X_{\text{exobase}} \sim 8$ resulting in $F \sim 6 \times 10^7$ cm^{-2} s^{-1}. If on the other hand $X < 3$ at any point in the atmosphere then the thermal energy exceeds the gravitational potential energy and gas will begin to flow from the planet in **winds**. These winds are only limited by external factors such as the ambient radiation pressure. Such conditions may exist for hot Jupiters (Chapter 4) with exospheres exceeding 10,000 K in temperature.

- **Non-thermal escape:** Non-thermal processes can in fact dominate the rate at which atmosphere is lost. These include **dissociation** whereby UV photons (or even electrons) dissociate a molecule and one of the products gains enough kinetic energy to escape the potential well (this is particularly important for water loss, see below). A variety of other ion reactions and charge exchange mechanisms can also result in boosted kinetic energy. **Sputtering** is an important mechanism involving the high-energy impact of particles from interplanetary space (including cosmic rays). Collisions can impart significant kinetic energy and cause particle escape. It is particularly important in thick atmospheres and in environments with intense incoming particle fluxes—such as those experienced by the Jovian moons due to Jupiter's powerful magnetosphere. Planets without substantial magnetic fields may also be subject to **solar wind sweeping**, where direct interaction with stellar winds can result in both the loss of atmosphere and the temporary entrapment of stellar wind particles (e.g., Mercury).

- **Impact erosion:** During the impact of large bodies on a planet (see for example Chapter 8) a significant amount of atmospheric loss can occur. Objects larger than the scale height of an atmosphere can result in shock-heated gas being "blown off" the planet.

Atmospheric loss is clearly an important mechanism to account for in assessing the nature of a given planet. It is one reason why the smaller planets in our solar system do not have primordial hydrogen or helium in their atmospheres—these lighter elements have been lost due to the lower planetary mass. It also appears particularly critical in the case of terrestrial-type worlds and their potential for life. An example that we will return to (Chapter 9) is that of water loss. If water molecules are able to enter the exosphere then they are more readily dissociated by UV stellar photons. The lighter hydrogen atoms are then more likely to suffer Jeans escape (see above). The net result is that these water molecules are forever lost and the planet becomes dryer. As an example; both the Earth and Mars suffer water loss through atmospheric processes. If we take the present loss rates due to Jeans escape alone and integrate these over the ages of the planets we find that the Earth has lost enough water to represent a global layer some 6 meters deep. In the case of Mars the water loss due to Jeans escape represents a global layer some 10–20 meters deep (which is in fact a smaller amount of water loss than that of the Earth—due to the smaller radius of Mars). If we allow for mechanisms such as impact erosion (during earlier, heavier impact periods, Chapters 3 and 8) then it appears that Mars could have even had a global water layer 1 kilometer or more deep some 4 Gyr ago (although not necessarily as exposed liquid, e.g., it could be incorporated into the regolith much like groundwater on Earth). Since non-thermal escape (see above) can be very important, the presence or absence of a significant planetary magnetic field can also play a role in controlling the loss of lighter elements. A magnetic field can reduce the flux of high-energy particles (e.g., solar wind protons and electrons) entering the upper atmosphere.

The loss of more volatile species can also serve as a diagnostic in other situations. If we expect a given atomic or molecular species to have long ago been lost from a planetary atmosphere, but observe its presence, then we can safely assume that it is being replenished via some mechanism on the planet itself. A good example is the extremely tenuous neutral oxygen atmosphere observed around the Jovian moon Europa. Europa has a mass of only $0.008 M_\odot$ and is subject to intense sputtering (see above) in the high radiation environment surrounding Jupiter. The presence of any measurable atmosphere indicates that it must be actively replenished. The surface of Europa is water ice, and the same particles that sputter away its atmosphere must also dissociate water molecules on its surface, thereby providing (temporarily) its oxygen atmosphere. The presence of unexpected gaseous components in an atmosphere is particularly relevant to the presence of a biosphere (see §6.7 below).

6.6.3   Planetary Spectra

In the absence of in-situ measurements, our only opportunity for determining the precise composition and structure of a planetary atmosphere (or surface) is the detailed investigation of its spectrum of emitted and reflected radiation. Figure 6.6 shows a sketch of the basic components of the spectrum of a terrestrial-type planet. The task of understanding the details of such data ultimately involves many factors. These include the structure of a planetary atmosphere (e.g., temperature as a function

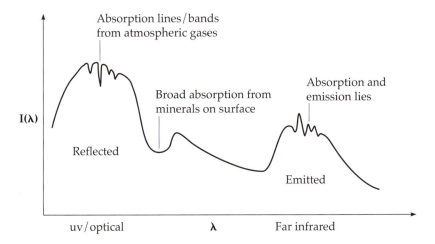

Figure 6.6. Sketch of the basic features of a planetary spectrum (adapted from Lissauer & de Pater (2004)). In the ultraviolet, visible, and near infrared wavelength range (approximately 0.2 to 1 μm) the spectrum is dominated by the reflected stellar light—modified by atmospheric absorption, and in the case of a rocky world with a thin atmosphere, further modified by absorption due to minerals and structures on the surface. In general (e.g. for a planet such as the Earth with a temperature of just a few hundred kelvins) the spectrum in the mid and far infrared (approximately 1–30 μm) is dominated by the planet's own emission, modified by both absorption and emission processes in the atmosphere.

of altitude), its chemical composition (e.g., molecular species which absorb and emit radiation), its surface characteristics (if applicable, as for terrestrial-type worlds), and the spectrum of incident (stellar) radiation. In no particular order, here are some of the specific processes and quantities involved in the detailed interpretation of planetary spectra (see also Marley et al. 2007):

- Absorption and emission of gaseous species: The type and abundance (or column density) of absorbing and emitting species dictate the likelihood of photons of a given wavelength originating from, or penetrating to, a particular depth in an atmosphere. Thus the **opacity** (see Chapter 9) at a given wavelength is important. Since an atmosphere will have a vertical structure, this likelihood will also depend on the degree to which a species is mixed with other species, what the pressure scale height (and hence gravity) is, and whether there are cloud layers (see below). The temperature structure also comes in to play. For example, at some infrared wavelengths where the atmosphere has low opacity we will see to greater depth. If, as in the case of giant planets, temperatures are significantly higher deeper in the atmosphere then radiation emitted by these layers may escape more easily and such worlds can appear as bright infrared sources.

- Opaque atmospheric structure: Clouds or condensates can dramatically alter planetary albedo—especially at longer (infrared) wavelengths, depending on the relevant molecular species (e.g., H_2O), and on wavelength-dependent scattering and absorption (see also Chapter 4). For cloudless planets, the thermal emission may dominate in the infrared (Figure 6.6) compared to cloudy worlds. This is a particularly important issue for water-rich terrestrial worlds at $\sim 300\,K$ where H_2O condensates may begin to form (see below for more discussion of clouds).

- Surface albedos: For planets with thin atmospheres and well defined solid or liquid surfaces, the detailed surface albedo is a major factor in determining their spectrum. As discussed above, albedo can be a complex function of wavelength. Figure 6.7 shows some examples of the wavelength dependence of spherical albedo for a range of surface compositions on the Earth. These include two categories that are implicitly a function of biota (tundra and forest), in particular

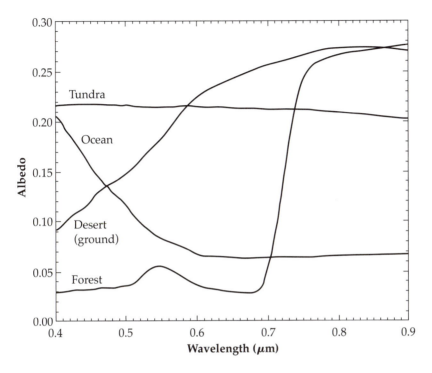

Figure 6.7. A sketch of the UV to near-infrared geometric albedos computed using remote sensing data of the Earth for a variety of surface compositions (adapted from Tinetti et al. 2006). The albedo of tundra, which includes biological structure (i.e., plant life), is relatively flat with wavelength. A sterile (life free) ocean shows relatively high UV/optical reflectance. Desert—or sterile ground/soil—increases in reflectivity with wavelength due to mineral absorption characteristics. Forest (or photosynthetic surface life in general) shows a dramatic increase in reflectivity beyond 0.7μ m, corresponding to a feature associated with chlorophyll and cell structures (see §6.7.2).

an Earth-centric photosynthesizing population of organisms. We discuss this more in §6.7.2, but note that the sudden increase in albedo for "forest" at about 0.7 μm is directly related to the photosynthetic nature of terrestrial plant life. The curves for ocean and desert (ground) assume sterile conditions (i.e., no marine or desert life).

- Chemistry and photochemistry: Unlike a test tube of ingredients, the atmosphere of a planet can be hard to quantify in terms of chemical equilibria. For example, the nature of hydrostatic equilibrium (Chapter 2) results in the settling (or sedimentation) of heavier condensates. Thus, potentially reactive elements or compounds may be removed from certain zones, allowing other species to exist where we might otherwise not expect them. Furthermore, the absorption of stellar radiation in a planetary atmosphere (especially at UV wavelengths and shorter) can initiate a wide range of chemistry (**photochemistry**) that in turn alters the composition of the atmosphere. For a steady-state system a chemical equilibrium should be reached—however this equilibrium need not be that expected from thermal conditions alone.

- Rayleigh scattering: Particles smaller than about 1/10th of the wavelength of incident radiation will scatter photons in a wavelength-dependent way—due to their interaction with the bound electrons in the scattering particles. The scattered light intensity due to a *single* particle is:

$$I = I_0 \frac{(1 + \cos^2 \theta)}{2R^2} \left(\frac{2\pi}{\lambda}\right)^4 \left(\frac{n^2 - 1}{n^2 + 2}\right)^2 \left(\frac{d}{2}\right)^6, \qquad (6.16)$$

where I_0 is the intensity incident on the scattering particles, I is the observed intensity at an angle θ from the forward direction of the incident radiation, R is the distance of the observer from the particle, n is the refractive index of the particle material, and d is the particle diameter. Thus, the scattering increases rapidly with decreasing λ (i.e., as $1/\lambda^4$) and is the dominant mechanism in the "blue" optical spectral range. Rayleigh scattering also redistributes the photons equally in *both* the forward and backward directions (via the $1 + \cos^2 \theta$ term in Equation 6.16). On Earth we see Rayleigh scattering by atmospheric molecules manifested as the blue daytime sky. Similar effects will be at play in exoplanet atmospheres, and can influence the phase dependency of albedo.

- **Mie scattering**: For particles larger than the wavelength of the incident radiation the scattering is described by the **Mie solution** to Maxwell's Equations. The intensity of scattered light is only a weak function of wavelength in this regime; however, the scattering is asymmetric with most photons being scattered in the *forward* direction—increasingly so as the particle size increases. On Earth, atmospheric water droplets produce Mie scattering and can give rise to (among other optical phenomena) apparent white "haloes" around the Sun. As with Rayleigh scattering, such effects will operate in exoplanet atmospheres.

- Together with clouds, **photochemical hazes** can play an important role in determining planetary spectra. For example, ultraviolet photolysis of high concentrations of methane in giant planets and terrestrial worlds produces chains of complex hydrocarbons that in turn condense into particles.[1] Indeed, it has been suggested that methanogenic organisms on the early Earth could have produced sufficient methane that the resultant haze affected global climate (see also Chapter 9). At high altitudes hazes will tend to reflect more incident radiation.

- Atmospheric dynamics: Both horizontal and vertical transport in planetary atmospheres can affect temperature structures (e.g., the redistribution of incident energy) and the way in which atmospheric components are mixed. This in turn affects atmospheric chemistry—especially if the chemical equilibrium timescales are long compared to dynamical timescales.

One of the most difficult (and important) aspects of understanding the observable characteristics of planetary atmospheres is the presence of *clouds* (see above and Chapter 4). In this context clouds need not be the familiar water vapor condensations that we see on Earth. For example, Jupiter has significant and complex cloud structures, but these are clouds of NH_3, H_2S, and CH_4 condensates. In higher temperature

---

1. Terrestrial **smog** is a good example of this type of organic haze.

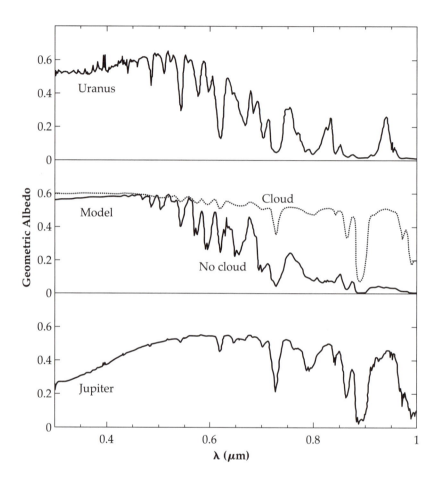

Figure 6.8. An example of the UV to near infrared geometric albedo of Uranus and Jupiter compared to a model of a gas giant planet both with and without high-altitude cloud (condensate) structures (adapted from Marley et al. 1999). These results suggest that Jupiter contains significantly more high-altitude condensate structures than Uranus.

environments ($\sim 1000-1600$ K), such as those of hot Jupiters, the condensing species are expected to be silicates, such as enstatite ($MgSiO_3$) or even iron. Why are clouds so important? Clouds introduce high-opacity, complex structures to a gaseous atmosphere—dramatically altering the passage of radiation. In Figure 6.8 the geometric albedo is plotted for Uranus, Jupiter, and a model for a general gas-giant type atmosphere—with and without cloud condensates. The difference is striking: the presence of clouds greatly increases the geometric albedo from visible to near infrared wavelengths. Such models indicate that while Jupiter does indeed have substantial cloud condensates in its upper atmosphere, such structure is largely absent from a planet like Uranus. Clouds can also introduce strong time variability to a planetary albedo. For example, the Earth, with both a highly variable surface albedo (oceans versus land) and significant but varying cloud cover, turns out to be the most photometrically variable planet in the solar system—with the net albedo deduced by an external observer changing by 10–20% about the mean! By comparison, Venus has a perpetual 100% cloud cover, a much higher Bond albedo (76%), and almost no variation in albedo with time. Thus, *partial* cloud cover can cause great time dependence on the albedo of a planet—especially if the clouds come and go.

In Figure 6.9, spectra are shown for a model that attempts to reproduce the characteristics of the present-day Earth. In this case the model is shown both with cloud structure (water condensates) and without. In the visible to near infrared (0.5–1.7 μm) wavelength range the effect of clouds is similar to that seen in Figure 6.8 for giant planets—the reflectivity is greatly enhanced at almost all wavelengths and the reflected flux greatly dominates any planetary emission. The least enhancement occurs at those wavelength bands corresponding to particular molecular H_2O spectral absorption features. In the mid- to far-infrared (5–25 μm) the emission of the Earth (Chapter 9, and see Figure 6.6) begins to dominate over the reflected flux and in fact cloud cover tends to *reduce* the intensity of radiation escaping.

The external characteristics of terrestrial type worlds are also a complex function of the planet's age. Although the Earth today clearly exhibits different global characteristics from those a few hundred million

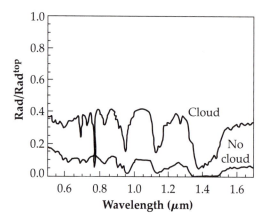
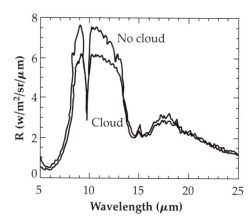

Figure 6.9. Examples of the spectrum predicted for the Earth from the visible and near infrared (left panel) to the mid to far infrared (right panel) both with and without cloud cover (water condensates). In the left panel the effective albedo is plotted since reflected radiation dominates the planetary spectrum in this wavelength range. The addition of cloud structures significantly increases albedo, although less so over bands corresponding to strong molecular water spectral absorption features. In the right panel the expected net radiation intensity from the planet is plotted in the mid to far infrared. Emission from the planetary surface dominates over reflected light and in this case water cloud cover tends to actually trap radiation (Chapter 9) and results in a reduction in intensity at certain wavelengths (plots adapted from Tinetti et al. 2006).

years ago (e.g., global temperature, atmospheric composition; see Chapters 9, 10) this is largely due to natural fluctuations in climate system equilibria, continental rearrangement due to plate tectonics, and the effects of the biosphere rather than some steady aging process. By contrast the bulk characteristics of gas giant planets are largely governed by the rate at which they *cool*. In very crude terms, giant planets cool and shrink with time as they radiate away energy while maintaining hydrostatic equilibrium (Chapter 2). Thus, young giant planets are hotter and larger (see also Chapter 3), and therefore (from the blackbody law) more luminous. Modeling this evolution requires some sophistication, and is not unlike the modeling of stellar interior structures and evolution. Chemistry, energy transport (radiation and convection), and condensates must be accounted for (see e.g., Burrows et al. 1997). Figures

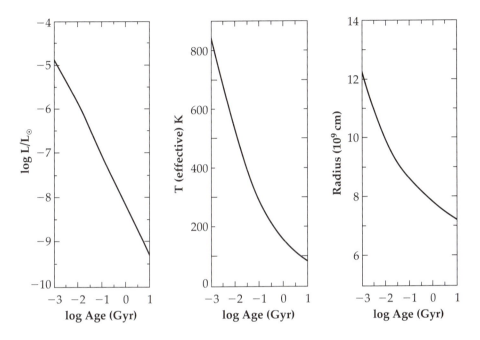

Figure 6.10. The evolution with time of the luminosity (left panel), effective (i.e., apparent blackbody) temperature (middle panel), and radius (right panel) of a $1\,M_J$ planet, from 10^6 years after formation until approximately 9.6×10^9 years after formation. Data is taken from the non-gray atmosphere models of Burrows et al. (1997).

6.10 and 6.11 summarize the evolution of luminosity, effective temperature, and radius for planets the mass of Jupiter, and 5 times the mass of Jupiter.

6.7   The Impact of Life: Biosignatures

Based on the example of the Earth, a substantial biosphere will profoundly effect the atmospheric and surface (if not subsurface) conditions of a terrestrial type world. As we have seen in Chapter 5, the precise composition of the Earth's atmosphere is a result of the actions of organisms. If all oxygen-producing life were to vanish then, over time the atmospheric oxygen would combine with the carbon-silicate rocks on

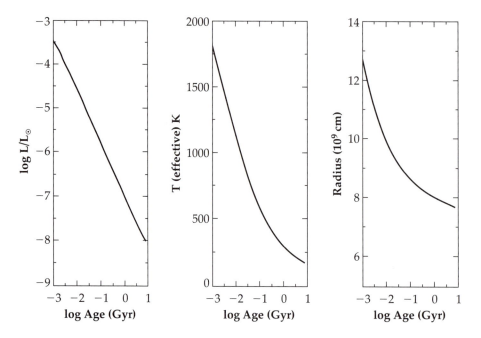

Figure 6.11. The evolution with time of a 5 M_J planet, following the same models as in Figure 6.10.

the surface. Similarly, if all methane-producing bacteria were to suddenly vanish then atmospheric methane would quite rapidly disappear as it oxidized into water and carbon dioxide. The albedo of the Earth's land masses and oceans is also strongly effected by the pigmentation of the organisms living there (e.g., plants, algae, diatoms, etc.). Even the **hydrological** cycle of the planet, and hence cloud formation, is effected by the presence of life.[2] All of these effects push and sustain the planetary environment away from its default chemical and energetic equilibrium. Seeking **biosignatures** therefore involves seeking evidence

---

2. There is even evidence that plankton can influence local climate by producing the gas dimethyl sulphide (DMS). DMS is a source of small sulphur particles in the atmosphere, which act as condensation nuclei for the water vapor. In effect, large plankton colonies may be able to encourage their own cloud cover.

for out-of-equilibrium conditions which are difficult to explain through abiotic mechanisms (e.g., volcanism). Biosignatures may also simply be physical characteristics of a planet (see below) which are hard to explain through any other means than the presence of life.

One of the fundamental characteristics of the Earth's biosphere is how it changes with the planet's seasons. The varying solar input between northern and southern hemispheres has a dramatic effect on local conditions, and life responds accordingly. Thus, not only will life help maintain out-of-equilibrium conditions on a planet, it will almost certainly react to variations in stellar input in ways that are more complex than abiotic phenomena. An excellent example of this can be seen the atmospheric concentration of terrestrial CO_2 as a function of time. From 1958 until the present day, CO_2 measurements have been taken from the Mauna Loa Observatory at approximately 3 km above sea level, located on the side of the Mauna Loa volcanic mountain in Hawaii. The resulting plot is known as the "Keeling Curve" (after C. D. Keeling) and is shown in Figure 6.12. Two features are immediately apparent: (1) there is an annual variation in CO_2 concentration (the wiggles) and (2) there has been a consistent increase from 1958 to the present. Both characteristics can be directly attributed to biological systems. The annual variation is a consequence of the respiration and photosynthesis of organisms. During summer, photosynthesis is dominant and carbon is removed from the atmosphere; during winter this is reversed as foliage (and organisms, such as oceanic photosynthesizers) dies off and decays. The longer term trend is almost without doubt attributable to industrialized human activity, which is arguably still the effect of a biological system, namely us.

As a biosignature, the long-term trend is less unique—a planet could be undergoing increased volcanism, or experiencing some abiotic shift in atmospheric chemical equilibrium. The annual cycle is, on the other hand, much harder to explain without invoking a large biosphere. It is worth noting that Mauna Loa is situated in the Northern terrestrial hemisphere—CO_2 measurements taken there are therefore strictly speaking just sampling the Northern atmosphere. Since the Southern hemisphere experiences the opposite summer–winter cycle one might expect that a *globally averaged* CO_2 concentration would therefore show little annual variation as the two hemispheres cancel out. However, in

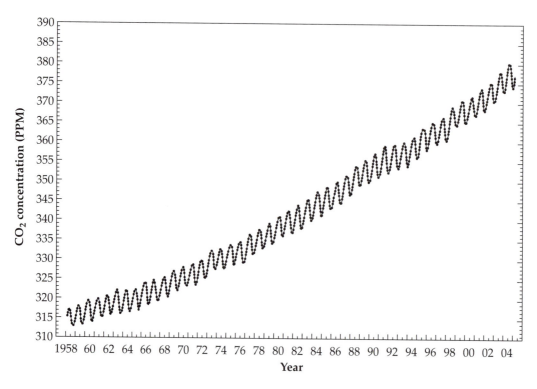

Figure 6.12. The "Keeling Curve" showing the measured variation with time of atmospheric CO_2 concentration in the northern terrestrial hemisphere. Data has been taken at the Mauna Kea Observatory in Hawaii, 3 km above sea level, since 1958. Two clear trends are seen: a general increase with time, attributable to human activity, and an annual variation attributable to the planet-wide seasonal variation in biological photosynthesis, respiration, and decay.

the case of the Earth the Northern hemisphere contains a significantly greater ratio of land to ocean surface area, and most of that is at high latitudes, where seasonal change is largest. Thus, the globally averaged concentration of CO_2 would still exhibit annual variation. This is significant for the long-term prospects of studying variations in other terrestrial planets. As a hypothetical example, consider an Earth-type planet that we suspect experiences regular variation in stellar flux—either due to an eccentric orbit or a measured spin-axis tilt (Chapter 9). If we monitored its atmospheric composition via spectroscopy and detected a seasonal variation in a species such as CO_2 then not only might this indicate an active biosphere but it could provide clues as to the *surface arrangement* of biota (and therefore potential geography).

6.7.1 Atmospheric Signatures

An excellent, and now classic, example of the application of atmospheric biosignature measurement is an experiment carried out by the Galileo spacecraft during its 1990 flyby of the Earth en route to Jupiter. Galileo was equipped with an array of sensitive detectors, including imaging and spectroscopic instruments spanning radio, infrared, optical, and ultraviolet wavelengths. During close passage to the Earth, Galileo was able to amass a range of data, making in effect an unbiased examination of the possible existence of a biosphere on this blue–green world (Sagan et al. 1993, Sagan 1994). This was a beautifully conceived and executed control experiment: after all, if life could not be detected by a well-equipped spacecraft passing only some 1000 km away then efforts further afield might prove impossible. The Galileo data were able to show the likely presence of both solid, liquid, and gaseous water through infrared spectral absorption features (c.f. those in Figure 6.9) and the observation of large areas of very low infrared albedo (0.04), but high optical **specular** (mirror-like, as opposed to diffuse) reflectivity— consistent with liquid water oceans. A high abundance of atmospheric oxygen was measured that would be hard (although not necessarily impossible) to explain by non-biotic means. The detection of methane in this oxygen-rich atmosphere indicates an active replenishment mechanism, since methane is rapidly oxidized to CO_2 and H_2O.[3] It is therefore strongly out of equilibrium, but could readily be produced by methanogenic organisms (Chapter 5). Nitrous oxide was also detected in the atmosphere, where it would otherwise be expected to be quite rapidly photodissociated. Again, atmospheric N_2O could be readily produced by nitrogen-fixing organisms (Chapter 5).

Finally, although not atmospheric signatures, infrared imaging of the planet surface indicated some features that could be well explained by

---

3. The entire terrestrial atmospheric CH_4 content can be removed in ~ 10 years via this route.

photosynthetic life (see section below). Furthermore, structured narrow-band radio emissions were detected that suggested an artificial (i.e., life-produced) origin. Taken altogether, the Galileo data provided good evidence for the presence of a substantial biosphere on the Earth.

It is however interesting (and sobering) to consider what the conclusion might have been in the absence of, for example, the structured radio emissions—or any one of the other major indicators for a biosphere. In this respect the Galileo experiment—arguably an "ideal" experiment—demonstrated that future efforts to infer life solely on the basis of planetary spectra are going to be open to considerable interpretation.

6.7.2   The Vegetation Red Edge

Terrestrial plant life (and some prokaryotic life, e.g. cyanobacteria, Chapter 5) employs the pigment **chlorophyll**[4] to absorb photons that are then utilized to drive the synthesis of sugars (specifically **glucose**) from the raw ingredients of carbon dioxide and water, producing oxygen as a waste product. This process of photosynthesis using the chlorophyll molecule involves a complex series of chemical reactions, and has dependencies on both temperature and the atmospheric concentration of carbon dioxide. The familiar green color of plant life on the Earth is due to the preference of chlorophyll to absorb photons of slightly shorter and longer wavelenghs than 500 nm. This is illustrated in Figure 6.13, where chlorophyll absorption is very strong at wavelengths shorter than 450 nm (and into the UV) and has a second peak around 680 nm.

However, something quite remarkable happens to the reflectivity of terrestrial vegetation at wavelengths slightly longer than 700 nm (the near infrared). As illustrated in Figure 6.14, the reflectivity increases by about a factor of 10 between 700 and 800 nm and continues to be high to about 1400 nm. This spectral feature is known as the **vegetation red edge**. In addition, the *transmission* of radiation through leaf structures is also very high at these wavelengths (making up the remaining 50–60% of the photon budget incident on plants). In other words, terrestrial plant life seems to have as little absorption of these infrared photons as

---

4. The primary molecules are chlorophyll a and chlorophyll b, with slightly different peak absorption wavelengths, Figure 6.13.

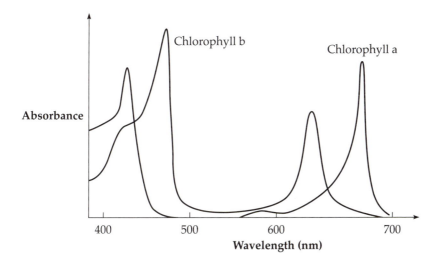

Figure 6.13. The absorption efficiency (absorbance) of the two primary chlorophyll molecules (a and b) in terrestrial photosynthetic life is plotted as a function of wavelength. The net result is two absorption peaks, one around 450 nm and one around 680 nm, i.e., the visible/red part of the spectrum.

it can. If this weren't enough, then an inspection of the wavelength of solar photons actually reaching the surface of the Earth (on average) indicates that the majority of solar *energy* is deposited between 600 and 1100 nm (e.g., Seager et al. 2005). Thus, vegetation appears to be strongly avoiding the bulk of stellar energy deposited on the planet's surface. Rather, it selectively "harvests" only those photons of direct biochemical use.

Before considering the utility of the vegetation red edge as a biosignature it is worth briefly discussing the mechanisms, and possible reasons, for this behavior. The leaf structure of terrestrial vegetation consists of layers of water-containing cells with air gaps between them. This results in highly efficient **scattering** of light—photons reflect off cell walls, and the strong changes in refractive index between the water–air boundaries result in efficient internal reflection. At a secondary level, the size of the organelles (Chapter 5) within the cells is comparable to the wavelength of the light, and Mie or Rayleigh scattering (§6.6.3) can occur. Thus, photons entering a leaf experience multiple scatters but only those that

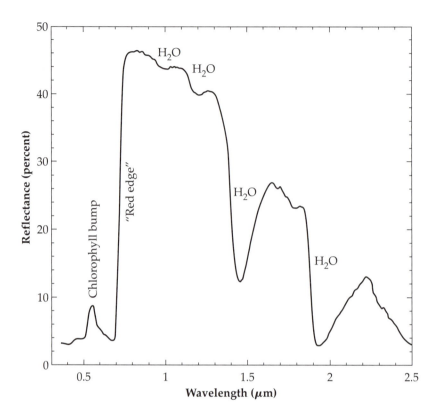

Figure 6.14. The reflectivity (in percentage) of terrestrial vegetation as a function of wavelength. The chlorophyll "bump" at around 0.55 μm is due to the lack of absorption between the two peaks shown in Figure 6.13. At approximately 0.7 μm (700 nm) there is a dramatic increase in reflectivity known as the vegetation red edge. At longer wavelengths are several features attributable to molecular water absorption.

encounter an absorbing pigment molecule tuned to their wavelength will be absorbed; essentially all others will be scattered back out. The leaf structure is a highly efficient "sieve," which optimizes the chances of a suitable photon interacting with a pigment molecule, while rejecting unsuitable photons. There may also be a good reason for discarding the infrared photons longward of 800 nm—if these were instead absorbed then the plant might become too hot (e.g., see discussion in Seager et al. 2005). This is, however, quite speculative, and there may be other reasons for plants to evolve this way. It is a fascinating subject, which is actively researched.

As a biosignature, the vegetation red edge presents what seems to be a very clear spectroscopic indicator of the presence of photosynthetic life. If the reflected surface spectrum of a planet could be isolated (i.e., accounting for atmospheric albedo and inorganic surface albedo) then the 700–800 nm reflectance edge (and to a lesser degree the chlorophyll green bump) could indicate the presence of widespread vegetation. However, it must be treated with caution. There are mineral species (e.g., cinnabar (HgS)) that also exhibit quite strong reflection edges at different wavelengths. Furthermore—and most importantly— it is not clear that photosynthetic organisms on other worlds would have evolved precisely the same set of biochemical and biomechanical strategies as terrestrial organisms. A good example of this is the situation for planets orbiting lower mass stars (e.g., M-dwarf stars). The peak photon output of such as star (to first order a blackbody) will occur at longer wavelengths. Not only would this require the use of different pigment molecules by organisms, but also different photosynthetic steps—potentially involving 3 or 4 photons rather than the two photons generally utilized on the Earth. This would alter the pigmentation spectral signature, and it is unclear whether photosynthetic organisms would evolve the same light scattering structures responsible for the terrestrial red edge. It is also the case that the terrestrial red edge is due to multi-cellular structures—something that the Earth has possessed only in the last 1 to 1.5 billion years of its existence.

Probably the most useful things that we can learn from the terrestrial vegetation red edge are these:

- The exploitation of stellar photons by organisms can be a highly successful evolutionary strategy. Thus, the presence of life that modifies the surface albedo of a planet should be carefully considered.

- There may be both weak and strong planetary spectral features associated with photosynthetic life, likely correlated with both the peak output of a stellar spectrum and the modifying influence of the planetary atmosphere.

- Spectral features due to life should vary according to local conditions and any seasonal/climatic variation (for example, the annual cycle of chlorophyll seen on Earth), and such variation may offer a route to separating these features from those due to inorganic material.

The practical search for a biosignature such as the red edge requires quite extensive modeling and measurement of those other planetary characteristics that contribute to the albedo. In Figure 6.15 a sketch is shown of some of the features that need to be "de-convolved" from terrestrial data in order to pinpoint an edge-like spectral feature.

6.8   The Impact of Life: Feedback

In the previous sections we began to consider the possible impact of life on planetary observables, both in terms of explicit bio-signatures, and in terms of modifications to the gross surface and atmospheric properties of a world. This is an extraordinarily rich and complex topic with application to our own planet as well as others. It would be fun to delve a little deeper into this, and in particular to try to come up with more *quantitative* descriptions of the possible impact of life on the gross properties of a planet. As we shall see later, perhaps one of the best clues we will have available to guide us in the quest for life on other worlds will come from the tendency for complex, organized structures and systems to emerge from previously unstructured environments. As a first step in accomplishing this we will appeal to a highly simplified, artificial model. This serves as an excellent introduction to the notion of mathematical and numerical modeling of real systems, which is one of the greatest legacies of the computer revolution of the twentieth century. It also provides a specific (albeit hypothetical) example of **feedback** processes. Feedback in systems is incredibly important for our own planet and will likely be so for others as well.

6.8.1   Daisyworld

The history of Daisyworld is an interesting one. It was developed by the proponents of the so-called Gaia hypothesis. This, in short, is the

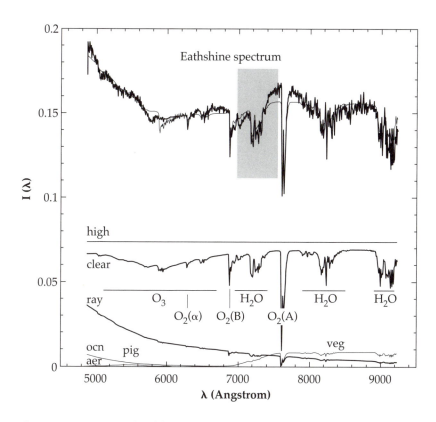

Figure 6.15. An example of the reflectivity spectrum of "Earthshine" (uppermost curve)—obtained by observing the light reflected from the dark side of the Moon (adapted from Woolf et al. 2002). Units of reflectivity are arbitrary. Some evidence is seen (shaded box centered on ~7300 Å) for the vegetation red edge although the face of the Earth responsible at the time the data was taken consisted primarily of the Pacific ocean. Interpreting the spectrum is however a complex problem. This is illustrated by the *seven* component spectra (lower curves) required to attempt to model the observed data. These are a model for high cloud cover (labeled high), the clear atmospheric transmission (clear), Rayleigh scattering (ray), the vegetation spectrum (veg), the subsurface or sterile ocean spectrum (ocn), the aerosol or haze spectrum (aer—negligible here), and the green-pigmented phytoplankton spectrum from the upper/surface ocean (pig—negligible here).

idea that the entire Earth can be considered to be a self-regulating living organism with all that follows in terms of equilibria, stability, and response to external forcing factors. Central to this is the notion that life itself helps maintain a favorable environment through feedback. Andrew Watson and James Lovelock (e.g., Watson & Lovelock 1983, Lovelock 1972) came up with Daisyworld as a means to demonstrate this idea in a somewhat more rigorous and quantitative way. Depending on who you talk to, reference to Gaia will either raise eyebrows or garner appreciative nods—since it has on occasion been appropriated by one scientific or political camp or another to bolster arguments about how the Earth environment and biosphere function and even how it is impacted by human activity. In truth, one can answer such questions in far more direct ways. Here however we will escape any possible controversy by referring the interested reader to some of the original articles and focus instead on the rather neat piece of mathematical modeling that was developed.

Daisyworld is a hypothetical planet inhabited (or seeded) by only *two* species of lifeform; black daisies and white daisies, which are identical in all properties *except* for color, or albedo. For the sake of simplicity we assume that the daisies grow in large patches (as opposed to being highly intermingled). Between the patches is just bare earth. *Both* species grow best at a single, unique temperature, which in the classical Daisyworld model corresponds to 22.5 °C. Above or below this temperature the daisies grow less well, and don't grow *at all* if the temperature is below 5 °C or above 40 °C.

They are also affected by crowding–as less space is available they grow more slowly. In other words, space is the common resource, and is finite on Daisyworld. Their lifespan is also finite and both species of daisy die off at a fixed rate.

The property of daisies that represents their influence on the planet is albedo. Black daisies have lower albedo than bare earth and white daisies have a higher albedo than bare earth. From what we have learnt in previous sections this implies that a planet covered in black daisies will reach thermal equilibrium with the radiation from its parent star at a mean *effective* temperature higher than that of bare earth, or a white daisy–covered world. Furthermore, locally, a patch of black daisies will wind up being warmer than bare earth, which would in turn be warmer

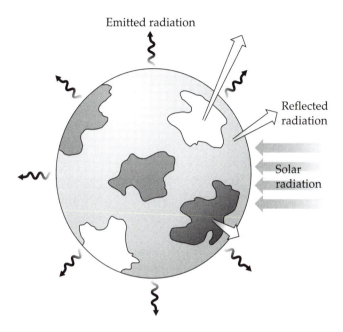

Figure 6.16. Schematic illustration of the Daisyworld system as modeled here. A planet harbors two types of daisies, black and white. These grow in large patches (avoiding real intermingling) on a bare earth surface. The albedo of white daisies is higher than that of the bare earth, which is in turn higher than that of the black daisy patches. Both species grow optimally at a single temperature (22.5 °C in the model used here) and less well away from this temperature. Neither species grows at all below a minimum or above a maximum temperature (5 °C and 40 °C respectively). No atmosphere is included in the basic model. Solar radiation is absorbed and reflected and the planet emits as a blackbody according to its global mean surface temperature.

than a patch of white daisies. Figure 6.16 provides a schematic representation of this Daisyworld system.

6.8.2 The Model Equations

Now, we need to set up the mathematical model that will describe the above scenario.

First we define the *areas* covered by black (a_b) and white (a_w) daisies such that the *bare earth* area is

$$x = 1 - a_b - a_w, \tag{6.17}$$

in other words the total planet area is just unity.

The *dynamics* of daisy growth is governed by a pair of coupled equations expressing the rate of change of area covered by black and white daisies:

$$\frac{da_b}{dt} = a_b(x\beta_b - \gamma), \tag{6.18}$$

$$\frac{da_w}{dt} = a_w(x\beta_w - \gamma), \tag{6.19}$$

where γ is a constant death rate (number per unit time) and in what follows will be set to a value of 0.3. $\beta_{b,w}$ is a growth rate per unit area which is a function of *local temperature* $T_{b,w}$ and obeys

$$\beta_{b,w} = 1 - 0.003265(295.5 - T_{b,w})^2 \tag{6.20}$$

for $278 < T_{b,w} < 313$ K, and $\beta_{b,w} = 0$ for all other temperatures. In other words there is no daisy growth outside of the temperature range 5–40 °C. This set of equations (6.18, 6.19) is an example of **Verhulst Dynamics**—a way of describing population dynamics (see also Saunders (2004)).

Now, if Daisyworld is in thermal equilibrium with the radiation field of its parent star then, as we have seen before, the total *emitted* radiation flux is

$$\sigma T_p^4 = SL(1 - A) \tag{6.21}$$

where T_p is the mean *effective* temperature of the planet (treated as a blackbody), A is the *total* (Bond) albedo, L is the stellar luminosity, and S is just a constant that depends on the star-planet separation d: $S = 1/(16\pi d^2)$.

The total planetary albedo can be written as an area-weighted average of the bare earth or ground albedo (A_g), the black daisy albedo (A_b), and white daisy albedo (A_w):

$$A = A_g x + A_w a_w + A_b a_b. \tag{6.22}$$

According to our model $A_b < A_g < A_w$, and in this case we choose $A_b = 0.25$, $A_g = 0.5$, and $A_w = 0.75$ for convenience.

Returning to the coupled equations (6.18, 6.19), we note that these are defined in terms of a *local* temperature $T_{b,w}$, while the thermal equilibrium of the planet is in terms of a net, planet-wide emission temperature T_p. In order to relate these we therefore need to allow for **heat flow** between different regions (ignoring any other possible temperature variations such as latitude dependence). You might well ask why we can't just take a simple area weighted mean temperature? This is a valid question, but in this case we must be careful to both allow for the redistribution of thermal energy and ensure that the *energy* budget of the planet is preserved, in other words the following must always hold true:

$$\sigma x T_g^4 + \sum_{b,w} \sigma a_{b,w} T_{b,w}^4 = \sigma T_p^4. \tag{6.23}$$

This can be accomplished by introducing another parameter, which is a measure of the heat flow; we label this q and then

$$T_{b,g,w}^4 = q(A - A_{b,g,w}) + T_p^4. \tag{6.24}$$

If $q = 0$ then all $T_{b,g,w} = T_p$ and there is no heat flow. Alternatively, if $q > SL/\sigma$ then heat (meaning thermal energy) would actually flow *against* any local temperature gradient. q must therefore take values between these extremes. In our case a suitable value turns out to be about $2.06 \times 10^9\,K^4$. The way we have constructed the problem therefore implies explicitly that *all* local daisy temperatures are *always different* from T_p.

6.8.3 Putting the Model to Work

Now we have set up the Daisyworld model how does it perform? In thermal equilibrium the planet and its daisy population will clearly be in a state of dynamic equilibrium in terms of daisy populations. Daisies will be born at a fixed rate, but they will also die at a fixed rate, and the finite planet surface area will constrain the total population. The only external force in this system is the parent star's luminosity. If the star is on the Main Sequence then its luminosity will gradually *increase* with time as its core gets denser and hotter (Chapter 1)—so we'll use this as an external change. The Main Sequence luminosity evolution of our own Sun can be parameterized as

$$L(t) \approx \left[1 + \frac{2}{5}\left(1 - \frac{t}{t_\odot}\right)\right]^{-1} L_\odot, \tag{6.25}$$

where $L_\odot$ and $t_\odot$ correspond to the *present* luminosity and age of the Sun, and t is zero when the Sun first becomes a Main Sequence star.

Let us consider three cases for Daisyworld, in the first there are no daisies.

Zero daisy case

Trivially in this case (assuming $A_g = 1/2$)

$$T_p^4 = \frac{SL}{2\sigma}, \tag{6.26}$$

and the emission temperature of the planet T_p simply grows slowly (Figure 6.17) as the star gets more luminous according to Equation 6.25.

Black (or white) daisies only

Now consider the case where only one species of daisy exists. From Equation 6.18 then we see that *in equilibrium* the rate of change of the daisy population must be zero; $da_b/dt = 0$, which implies that $a_b(x\beta_b - \gamma) = 0$ and $x = 1 - a_b$, and thus $\beta_b = \gamma/x = \gamma/(1 - a_b)$. Then, Equation 6.20 yields

$$T_b = 295.5 \pm 17.5\sqrt{\frac{a_b + \gamma - 1}{a_b - 1}} \tag{6.27}$$

and $A = 0.25a_b + 0.5(1 - a_b) = 0.5 - 0.25a_b$. Thus, $T_b^4 = 0.25q(1 - a_b) + T_p^4$ and we can now eliminate T_p using the relation $\sigma T_p^4 = SL(1 - A)$, producing

$$\sigma T_b^4 = 0.25\sigma q(1 - a_b) + SL(0.5 + 0.25a_b). \tag{6.28}$$

Combining this with Equation 6.27 we can now solve for T_b and a_b and subsequently find the equilibrium value of T_p as a function of L.

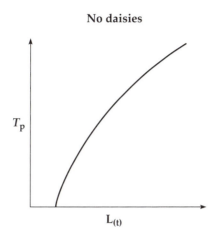

Figure 6.17. The emission temperature of the planet (T_p) as a function of the stellar luminosity ($L(t)$) as it varies due to normal Main Sequence evolution. In this case there are *no* daisies and T_p is given simply by Equation 6.26.

The question is now, what happens to the planet temperature as the star evolves? It is implicitly assumed that the planet is seeded with black daisy seeds, so if conditions are favorable then black daisies will appear. Initially let us suppose that the planet is too cool for daisies to appear (i.e., $T_p < 5\,°C$). As L evolves then T_p will grow (according to Equation 6.26) until the magic $5\,°$ C is reached and black daisies begin to grow. At this point Equation 6.27 becomes valid, and provides two possible solutions for T_b, which translates into two possible values of T_p. The lowermost of these represents an unstable state and so as L continues to slowly grow T_p will rapidly jump to the upper, stable solution (Figure 6.18).

Thus, when L reaches a critical value the daisies *profoundly* affect the planetary temperature by lowering the net albedo. However, the planet's surface area is *finite* and the black daisy population will rapidly

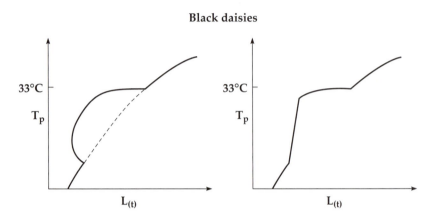

Figure 6.18. The emission temperature of the planet (T_p) as a function of the stellar luminosity ($L(t)$) as it varies due to normal Main Sequence evolution in the case of a black daisy populated world. The left panel illustrates the results from the solutions to Equations 6.27 and 6.28. Stable solutions for T_p are shown as solid curves. In the right-hand panel the actual values taken by T_p are illustrated. As L increases a critical point is reached where T_p jumps to the stable solution for the system, resulting in a sharp global temperature increase. This curve is followed until the black daisies die off and T_p continues to evolve on the bare earth curve.

saturate and then begin to reduce as L continues to grow and local temperatures increase. For a significant period though a relatively "steady-state" is achieved where T_p is almost constant as the slowly diminishing black daisy population offsets the slowly growing stellar luminosity L, until all the daisies die off. Beyond this point (at about $T_p = 33\,°C$ in this example) the planet's temperature again grows according to Equation 6.26.

If we had instead started with a planet seeded by white daisies the converse would have happened (Figure 6.19). The daisy population will act to initially *lower* T_p and keep it almost constant—until the daisy population runs out of space and then experiences *rapid* death beyond a critical L.

An interesting and important point is that if one took the black (or white) daisy model at some stage and somehow *reduced* L then the popu-

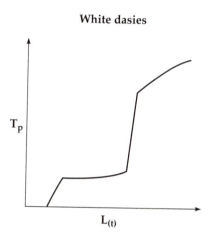

White dasies

Figure 6.19. The equivalent plot as that of Figure 6.18, but in this case for an entirely white daisy population. The global mean emission temperature T_p is stabilized at lower L but then rapidly jumps to an upper solution as L increases.

lation changes will also be reversed—so, for example, the black daisies will appear slowly and disappear quickly. However, they will *not* all disappear at the same L that they would appear at for the forward (increasing L) case given above. In fact L has to go to a rather lower value. This is an example of **hysteresis**, and is such that it gives Daisyworld an inherent stability to external changes in L, since while (black) daisies can appear rapidly as L increases, if L then decreased, they would die off slowly and persist beyond their original point of appearance.

Black and white daisies

Now let us consider the situation we've been aiming at understanding— where the planet has been seeded by *both* black and white daisy species. Since $\beta_w = \beta_b$ (the same growth dependency) and $T_b > T_w$ (the local black temperature *must* be higher than that of white), then

$$T_b - 295.5 = 295.5 - T_w \tag{6.29}$$

so,

$$T_b + T_w = 591. \tag{6.30}$$

We also have the relations

$$T_b^4 = q(A - A_b) + T_p^4, \tag{6.31}$$

$$T_w^4 = q(A - A_w) + T_p^4. \tag{6.32}$$

Subtracting 6.32 from 6.31 yields

$$T_b^4 - T_w^4 = q(A_w - A_b), \tag{6.33}$$

and eliminating T_w we obtain

$$T_b^4 - (591 - T_b)^4 = q(A_w - A_b). \tag{6.34}$$

This may look innocent, but it is in fact truly shocking! T_b (or T_w) depends *only* on A_w, A_b and q, *not* on either the initial conditions *or* (more remarkably) on the stellar luminosity L! It further implies that the *total* area covered by *all* daisies is a constant—as long as both species exist (Figure 6.20).

We can see this explicitly by noting that $\beta_b = \beta_w = 0.918375$ in this case. In *equilibrium* (i.e., at a given stellar luminosity) then clearly $da_b/dt = 0$ and $da_w/dt = 0$ and we find that $x = \gamma/\beta_{b,w}$. Since $a_b + a_w = 1 - x$ we determine that

$$a_b + a_w = 0.673. \tag{6.35}$$

Thus, 67.3% of Daisyworld is covered by daisies at *all times* when both species are present and the system is in equilibrium. This is not an intuitively simple result!

Further manipulation leads one to solve a cubic equation (since the 4th-order terms in Equation 6.34 cancel out in this case) and it can be shown that when finite numbers of both species are present;

$$T_p = 299[(a_b + 0.668)L]^{1/4}. \tag{6.36}$$

Thus, the planet temperature, T_p, has a very *weak* (actually declining as a_b shrinks) dependence on the stellar luminosity L in the two-daisy

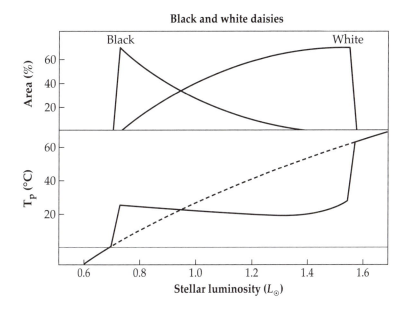

Figure 6.20. The results of a two-daisy (black and white) model with an evolving (increasing) stellar luminosity. The lower panel shows the variation of T_p, which is stabilized over a wide range of stellar input. The upper panel shows the area of the planet covered by black and white daisies as a function of stellar input. The sum of these two curves is always 67.3% for the model computed here. Initially black daisies dominate and are able to raise T_p while the stellar input is low. As L increases the white daisy population begins to grow to counter this energy input. Much later the black daisies essentially die off, leaving the white daisies to force T_p to stay low compared to the bare earth solution. Finally, even the white daisies are unable to cool Daisyworld sufficiently and they die off as T_p reaches 40 °C.

world! This is shown explicitly in Figure 6.20, and is true only within the regime where significant numbers of both black and white daisies exist.

There are several very profound implications of this finding. First, life (or at least some simplified surrogate for one of the most basic properties of life) can, in principle, dramatically influence the environmental conditions of a planet. Maybe this doesn't sound like news, but remember—we've just demonstrated this *quantitatively*. Second, conditions on a planet can be *stabilized* in such a situation, thereby *increasing* the window of opportunity for a favorable environment for life, even as the parent star evolves. In the case of Daisyworld, Figure 6.20 reveals that over a factor of about three in L (~ 0.5–1.5) the temperature T_p can be stabilized to be close to $\sim 22\,°C$. Without the daisies then T_p would vary between approximately $0\,°C$ and $60\,°C$ with this variation in L. Thirdly, and perhaps most disturbingly, this could be considered as an example of evolution *without* natural selection! Remember, the fact that the daisies grow optimally at $22.5°C$ was a hard-wired initial condition. *The daisies have not adapted to the planet, instead they have forced the planet to "adapt"*! Of course it is not entirely fair to describe the situation in these terms—nothing has really adapted in the true sense used in biological evolution, where favorable characteristics to external pressures are adopted through natural selection. In this case the daisies are creating a regulating feedback loop between their two species and the underlying stellar driven changes in the energy budget of the planet. Cooperation between the black and white daisies, born out of necessity due to the conditions imposed on the system, wins the day by stabilizing the planet's environment.

This property of self-regulation is however a fairly universal phenomenon seen in terrestrial organisms, and is referred to as **homeostasis**. In effect then, the situation could be used to argue that a coupled system, such as a life-bearing planet, could indeed be modeled as an organism. It is here that the connection with the Gaia hypothesis is made. Before taking this further it is worth pausing to consider the implications for studying terrestrial-type exoplanets. At face value, Daisyworld implies that a planetary surface maintained in a physical state far from its default, static equilibrium may be a smoking gun for the presence of a large biosphere, with all of its attendant feedback processes.

6.9   Variations

At the outset of our consideration of Daisyworld we enforced strict and highly unrealistic rules about this planet and its daisies. Clearly additional parameters and phenomena can be added to our model.

For example, the detailed spatial distribution of daisies could be explicitly modeled to allow for complex 2-D arrangements (i.e., relaxing the earlier "patch" requirement). Such numerical exercises enter into the realm of **cellular autonoma**. In such models the same self-regulation/homeostasis can still occur, but only while the degree of fragmentation of the daisy distributions remains within some limit. In other words, the daisies lose their stabilizing ability when the species become too interspersed. In part this is due to the role of heat/energy flow in the regulation, which is effected as the daisies blend together more and more.

Other variants include the addition of a population of "herbivores" that eat daisies and have their own population dynamics, but do not themselves contribute to the surface albedo. In this case it most typically appears that the herbivores do best in a *narrower* range of environment (e.g., temperature) than the daisies. If the herbivores are too successful they wipe out the daisies and then die off themselves, so the situation is most stable when the herbivores have a somewhat limited impact on the daisies.

6.9.1 Evolving Daisies

An obvious variation to test is one in which the daisies *are* allowed to evolve—Specifically, if they are allowed to vary their optimal growing temperature. In this case black daisies would evolve to grow best at a higher local temperature and white daises at a lower local temperature than the previous uniform 22.5 °C.

Initially one would suppose that this would allow the daisies to become even more successful at stabilizing their environment. However, this is not the case—depending on how this additional freedom is modeled it is found that at best the evolving daisies produce a marginal improvement in regulating T_p, and at worse they evolve themselves into a niche that leaves them vunerable to further change.

Let us look at this a little more explicitly, because it is a surprising result. The first step is to determine what the *local* equilibrium temperature should be for black and white daisies (T_b, T_w). Consider again Equation 6.34. For the values of albedo and heat flow we have specified we find that $q(A_w - A_b) = 1.03 \times 10^9$, by trial we can then determine a value for T_b which satisfies Equation 6.33. From Equation 6.30 we can then immediately find T_w. In this case $T_b = 300.5$ K and $T_w = 290.5$ K.

We now rewrite Equation 6.20 and replace the optimal growing temperature with these local temperatures:

$$\beta_b = 1 - 0.003265(300.5 - T_b)^2, \qquad (6.37)$$

$$\beta_w = 1 - 0.003265(290.5 - T_w)^2. \qquad (6.38)$$

This serves as a crude description of "evolved" black and white daisies. Now, as for our original case, we can evaluate the *total area* of the planet that is covered by daisies when both species are present. Since the optimal growth temperature is now the actual local temperature of either daisy species then clearly in equilibrium $\beta_b = \beta_w = 1$ and consequently

$$a_b + a_w = 0.7. \qquad (6.39)$$

Thus, in this case—perhaps rather surprisingly—the total areal coverage of daisies is now 70% compared to 67.3% for the un-evolving case, a gain of only 2.7%! Furthermore, the actual ability to regulate the planet's temperature is not altered at all—in this case $T_p = 299[(a_b + 0.65)L]^{1/4}$—so the L dependency is the same as for the non-evolving case. In addition, the actual *range* of L over which T_p is stabilized (Figure 6.20) is actually about 16% *smaller* than in the non-evolving case. The reason for this is simply that at lower L (lower T_p) the black daisies, which would previously have helped boost the temperature, have now evolved to do better at slightly higher temperatures, and so cannot respond as well. Similarly, at high L, the white daisies, which would previously have helped lower the temperature, are actually now further from their optimal growth temperature, and so provide less regulation. In this case it would appear that over the Main Sequence lifetime of the parent star the planet environment is better stabilized by daisies which have *not* evolved to grow optimally at their equilibrium local temperature.

This can all come as a bit of a shock. We are used to thinking of evolution as a process of optimization—albeit somewhat haphazard. However, we have left out many factors that may mitigate the apparent ineffectiveness of our "evolved" daisies. In real ecosystems there is **diversity** even within a given species—thus there will be a spread in optimal conditions. Furthermore, if timescales are long enough (such as for Main Sequence evolution of solar mass type stars) then the population will shift and respond.

One could of course take this further using our simple model. Suppose that instead of evolving to grow best at their natural local temperature the black and white daisies actually begin to evolve their *albedos*. Black daisies become lighter and white become darker—depending on what L is. For example, early on, with low L it would make sense for the white daisies to become darker to boost their local temperature. Somewhere in the midpoint of the stellar evolution one might imagine a "grey" daisy population. Whether or not this actually improves the overall situation we leave as a mental exercise for the reader.

6.10   Concluding Thoughts

Finally, the basic Verhulst dynamics of Equations 6.18 and 6.19 is unlikely to be truly representative of the full spectrum of interactions and responses contained in a dynamic system of living organisms and environments. It is a big jump to go from such a simple model of feedback and regulation to a more realistic picture where there are *many* interrelated feedback loops operating on both small, local scales and on global scales. Determining the response of such a system to external forcings (such as the changing stellar input) is enormously difficult, as demonstrated by the remarkable but complex efforts to model the Earth's past, present, and future climate (Chapter 9). In reality it becomes much harder to make predictive statements about how life influences its environment, although it clearly does in many ways. Nonetheless, in a highly simplified and artificial case we *have* shown how life has the potential to profoundly impact a planet's physical characteristics. We have also shown how powerful mathematical or numerical modeling can be. Furthermore, we have shown in quite a general way how generic "biofeedback" processes can maintain a planet system in a seemingly stable state that is very far from its default, barren, equilibrium state. While there may be planetary phenomena (e.g., tectonics, geochemistry) that also provide feedback processes, it seems at least plausible that these would be less flexible or responsive than those of a biological nature. Chapters 9 and 10 pick up this tale from a different angle—namely the investigation of what it takes to make an environment "habitable" for life as we know it.

References

Burrows, A., et al. (1997). A nongray theory of extrasolar giant planets and brown dwarfs, *Astrophysical Journal*, **491**, 856.

Dyudina, U., et al. (2005). Phase light curves for extrasolar Jupiters and Saturns, *Astrophysical Journal*, **618**, 973.

Lammer, H., et al. (2003). Hydrodynamic escape of exoplanetary atmospheres. Proceedings 2nd Eddington workshop "Stellar structure and Habitable Planet Finding," ESA SP-538, Eds. Favata & Agrain.

Lissauer, J. J. & de Pater, I. (2004). Planetary Sciences, Cambridge University Press.

Lovelock, J. E. (1972). Gaia as seen through the atmosphere. *Atmospheric Environment*, **6**, 579.

Marley, M. S., Gelino, C. S., Stephens, D., Cunine, J. I., Freedman, R. (1999). Reflected spectra and albedos of extrasolar giant planets. I. Clear and cloudy atmosphers, *The Astrophysical Journal*, **513**, 879.

Marley, M. S., Fortney, J., Seager, S., & Barman, T. (2007). Atmospheres of extrasolar giant planets. In *Protostars and Planets V*, ed. B. Reipurth, D. Jewitt, & K. Keil, University of Arizona Press, 733.

Sagan, C., et al. (1993). A search for life on earth from the Galileo spacecraft. *Nature*, **365**, 715.

Sagan, C. (1994). The search for extraterrestrial Life. *Scientific American*, October 1994, 93.

Saunders, P. T. (1994). Evolution without natural selection: Further implications of the daisyworld parable *Journal of Theoretical Biology*, **166**, 365.

Seager, S., Turner, E., Schafer, J., & Ford, E. (2005). Vegetation's red edge: A possible spectroscopic biosignature of extraterrestrial plants, *Astrobiology*, **5**, 372.

Tinetti, G., et al. (2006). Detectability of planetary characteristics in disk-averaged spectra II: Synthetic spectra and light-curves of Earth, *Astrobiology*, **6**, 88.

Watson, A. J., and Lovelock, J. E. (1983). Biological homeostasis of the global environment: The parable of Daisyworld, *Tellus*, **35B**, 284.

Woolf, N. J., Smith, P. S., Traub, W. A., & Jucks, K. W. (2002). The spectrum of earthshine: A pale blue dot observed from the ground. *Astrophysical Journal*, **574**, 430.

Problems

6.1 Describe the basic concept of albedo. Describe the definitions of albedo based on geometric arguments and how astronomical observations may be used to place constraints on their nature.

6.2 (a) The equation we gave in the text (Equation 6.2) for the simple equilibrium temperature of a spherical body was

$$T_{\mathrm{p}} = \left[\frac{(1-A)L_*}{16\pi d^2 \sigma} \right]^{1/4}.$$

This assumed *fast* rotation of the body. Determine the factor by which the equilibrium temperature of an extremely *slowly* rotating body is larger, showing your reasoning (remember, this does not include any physics due to energy transport by an atmosphere or ocean).

(b) Now let us assume that some fraction (β) of the energy received on one side of a *slowly* rotating planet of uniform albedo is transported/redistributed (e.g., by an atmosphere or by conduction) to the "night" side of the planet, where it is absorbed. Determine what value of β is required to maintain a dayside temperature of 373 K and a nightside temperature of 273 K.

6.3 Equation 6.12 suggests a technique for deducing the physical size of planetary bodies. What observations would be needed of (for example) a Kuiper belt dwarf planet in order to apply this? [*Note:* pay careful attention to the definition of luminosity used here.]

6.4 (a) Equation 6.15 provides an estimate of the thermal escape rate for a given species from a planetary atmosphere. Using this equation and Figure 6.4 (c), estimate the *total* hydrogen loss rate (atoms per second) from Mars when the sun is at low and moderate activity levels (this refers to the particle flux from the sun, which heats the exosphere). Assume that the density of atomic hydrogen at the Martian exobase at ~ 200 km is $\sim 2 \times 10^4$ cm^{-3}, and that the escape parameter X is ~ 8.

(b) If the loss rate during moderate/high solar activity were sustained for 10^6 years, and assuming each hydrogen atom was originally in a water molecule, then estimate the total *volume* of liquid water (at 1 atmosphere pressure and room temperature) lost by Mars over this time period through thermal atmospheric hydrogen escape.

6.5 Discuss the major factors that must be considered when interpreting a detailed planetary spectrum. How do condensation structures in particular complicate any such interpretations?

6.6 A 100-Myr-old $1M_{\mathrm{J}}$ planet orbits at 0.1 AU from a solar-luminosity star. What would the observed effective temperature of this planet be?

6.7 Read the journal article by Carl Sagan (*Nature*, 1993, see references) and describe briefly the various "bio-signatures" that were extracted from the Galileo Earth flyby data.

Using the data in this *Nature* article, find out which terrestrial atmospheric molecule exhibits the *greatest* deviation in abundance from its thermodynamic equilibrium value. Why might this be *so* far out of equilibrium in comparison to other potential bio-signature molecules?

6.8 Use Wien's Law to compute the peak wavelength of the Solar blackbody spectrum—assume a temperature of 5770 K. Lower mass stars such as M-dwarfs are of great interest for exoplanet studies. A $0.25 M_\odot$ main sequence star has $L_* \sim 0.013 L_\odot$ and $T_{eff} \sim 3200$ K.

Comment on the potential viability of terrestrial-type plant photosynthesis on a planet orbiting an M-dwarf star. In what ways might such life have evolved to deal with this situation in comparison to life on Earth (you may want to refer to Figure 6.14)?

6.9 Describe how much of the so-called "vegetation red edge" is a unique product of the structure of terrestrial plant life. How likely do you think it is that a terrestrial-like exoplanet harboring extensive surface life would exhibit similar spectral features? In particular, consider the variations with different stellar parents, and different evolutionary trajectories. This question could be expanded to a larger project.

6.10 Describe the Daisyworld numerical experiment. The parameter q controls the heat flow between patches of daisies and bare earth. Discuss the possible implications of a heat flow parameter that varied depending on the combination of neighboring surface types (e.g., white/black has higher q than white/bare earth).

6.11 Suppose that the Daisyworld sun undergoes periodic up-and-down fluctuations in its output (L_*) over several millions of years. Can you speculate as to how the non-evolving daisies would respond? And how would the evolving daisies, as described above, cope? Will they fare better or worse than the non-evolving ones in this case? A truly complete answer requires some quantitative reasoning—for example, consider a sinusoidally varying L_*. Recall also the discussion of *hysteresis* given in this chapter.

CHAPTER 7

Cosmochemistry, Dust, and Prebiotic Molecules

7.1 Introduction

The precise circumstances that lead to the complexity of chemistry necessary to initiate terrestrial life are unknown, and may well occur in many different situations and follow different pathways. What is more certain is that an enormous amount of chemistry (carbon chemistry in particular) can, and does, take place *beyond* the confines of planetary surfaces or subsurfaces. This is not a trivial statement to make. It is in fact quite astonishing that interstellar space contains both simple and complex molecules. Far from there being a clean boundary between the chemistry on, and in, planetary bodies, and the rest of the universe, there is really a continuum between this "local" chemistry and that of the largest structures of a galaxy. In general in astronomy, because the majority of the normal matter in the universe is still in the form of hydrogen and helium (Chapter 1), the heavier elements are treated as pollutants. Nonetheless, they play critical roles in controlling the pathways by which material dissipates and absorbs energy over the largest scales (for example, galaxy clusters, galaxies, and molecular clouds, Chapter 2). This is because their diversity of electronic structure allows a much wider range in energy, and efficiency, of photon absorption and emission than that of just hydrogen and helium. Indeed, heavier elements such as oxygen and iron act as some of the principal "coolants" in the gaseous interstellar and intergalactic media.

In biology these cosmic pollutants are of course an indispensable part of the fundamental components of life (i.e., the CHON elements). In a rather abstract sense, at some point these two views must merge—pollutants become incorporated into more biologically relevant molecules, and the vast sea of matter in the universe recedes largely into irrelevance for life on a local scale. The same is true for the types of planets we consider as potential harbors for life—the gravitational behavior of interstellar matter and the eventual coagulation of heavier elements form them. However, it is the chemistry just prior to coagulation, and following the final stages of formation, that is critical in determining their initial surface (and subsurface) environments.

In this chapter we are going to dig deeper into the question of where solid material comes from in the cosmos—the same material that we invoked in our discussions of the formation of planetesimals and planets. At the same time we will begin to survey the chemistry that is capable of building complex molecules in interstellar and interplanetary space. These two things are, in general, intimately linked. This is an enormously complex set of phenomena, for which we have quite incomplete knowledge. Nonetheless, it is possible, and useful, to gain an appreciation of some of the physics behind the major processes.

7.2   Elements and Materials

The main physical processes that must be understood in order to pursue questions of the origins of solids, and the chemistry of interstellar and circumstellar space, are

- The origin of the elements
- The formation of dust (structures large enough to no longer be well described in molecular terms)
- The formation of complex molecules

It will become clear that the second and third items on this list are intimately linked.

We have already briefly discussed the origin of elements, both in the Big Bang and in stellar nucleosynthesis (Chapter 1). The heavy elements (essentially anything of higher atomic mass than helium, which we term metals in astronomy) are dispersed via late-stage stellar evolutionary processes: envelope mass loss, nova, and supernova. We have also invoked "dust" in discussing planet formation, but up until now we have neither specified its precise physical characteristics nor origins. Furthermore, we have both implicitly and explicitly invoked the formation of complex molecules in interplanetary or interstellar space. However, as with dust, we have not delved into the details of how, where, and when such molecules might form.

7.3   The Origin of Dust

We know that dust exists in space because we see it. It exists in both high and low density regions of the interstellar medium, throughout our Galaxy and other galaxies (Chapter 2). The interstellar medium represents about 20–30% of the mass of the Milky Way galaxy, and approximately 1% of the mass of the interstellar medium is in the form of microscopic dust grains. These grains exist in varied density and temperature environments—from diffuse clouds of gas with particle density (hydrogen density) $\sim 100 - 300 \, \mathrm{cm^{-3}}$ and temperatures $\sim 100 \, \mathrm{K}$, to dense clouds of $\sim 10^{4-8} \, \mathrm{cm^{-3}}$ (hydrogen atom density) and temperatures $\sim 10 - 30 \, \mathrm{K}$. The size of the dust grains is quite varied, but an average value is a particle radius of some $\sim 0.1 \, \mu\mathrm{m}$. The closest example of extraterrestrial dust is somewhat different and is in the form of the **zodiacal dust** in our own Solar system. Roughly speaking this is a disk-like cloud of dust consisting of $1 - 300 \, \mu\mathrm{m}$ particles that lie in the ecliptic plane between the Sun and a radius of about 5 AU. The dust is most likely strewn by comets as they are evaporated by the Sun and from asteroid collisions grinding down rocky material. If you are lucky you can see it forming a faintly glowing band in the sky towards the Sun immediately after sunset, similar in apparent size and brightness to the Milky Way, but with a different orientation. This is the **zodiacal light**, sunlight forward scattered by the interplanetary zodiacal dust. In the anti-solar direction of the sky one might see **gegenschein**, which is the backscattering of sunlight by the interplanetary dust. Like most dust of this grain size and temperature, it emits radiation predominantly in the infrared, and scatters optical light. An example of an interplanetary dust particle is shown in Figure 7.1. In amongst these grains there will also be smaller particles of true interstellar dust: an example of what is thought to be such a particle is shown in Figure 7.2.

The Earth actually accretes some 4×10^7 kg a year of this interplanetary dust mixture (which is approximately equivalent to a block of rock

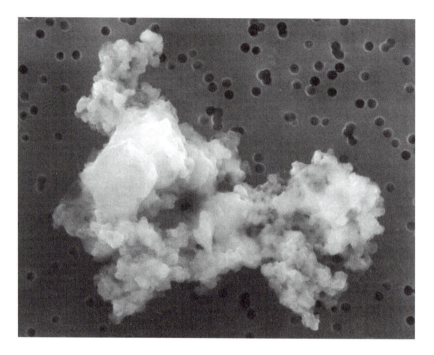

Figure 7.1. An image of a 10 μm interplanetary dust particle (IDP) collected in the Earth's stratosphere by a U2 aircraft. This IDP is predominantly silicon and carbon in composition (image: NASA/JPL).

some 23 meters on a side, or over a million years, about 2 km on a side). Indeed, so-called **Brownlee particles**, which are micro-meteorites or interplanetary dust, have been retrieved from the stratosphere by high-altitude aircraft (Figure 7.1).

However, the dust that is of primary interest to us in the present discussion is that out in interstellar space—and the origin and history of this can be different from that of the majority of the zodiacal dust.

In order to begin to track down the environment in which dust might form we can consider the growth of dust grains, specifically the growth *timescales*. Among the physics that we will assume is the notion that dust is the result of the condensation of gas. A gaseous phase provides the raw atoms or molecules that will condense out to make dust grains.

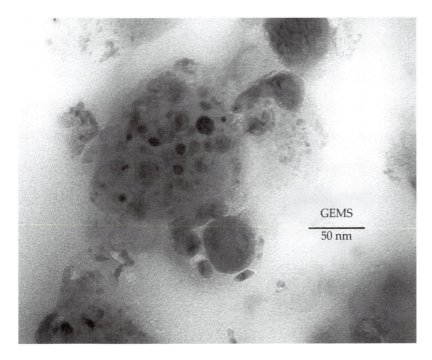

Figure 7.2. An image of a so-called GEMS (glass with embedded metals and sulfides) particle. This may be an example of a true interstellar dust grain which has become incorporated (and perhaps chemically processed beyond its original state) into a larger, interplanetary particle. At 0.1 μm across it is much smaller than the IDP shown in Figure 7.1 (image: NASA/JPL).

7.3.1 Order-of-magnitude Estimate for Grain Growth Rate

What follows here is a rather pedagogical discussion involving some quite gross assumptions about how gas condensation works. Nonetheless, it provides a logical framework and introduces the basic physical concepts. Ultimately we can best learn the origins of dust, and mechanisms of dust grain growth, through direct observation and experimentation.

Consider a spherical grain of material X, which has zero electrical charge (which is something we must be wary of for such small particles) and an initial radius a_0. The grain is within a gaseous environment of temperature T_{gas} and a space number density n_X of atoms of element X. Thus, atoms of X will condense onto the initial grain (assuming the temperature is low enough to allow this). We will temporarily sidestep the issue of where that first grain came from.

If we assume that the gas is well described by Maxwell–Boltzmann statistics then the root-mean-squared velocity of the atoms is given by

$$v_{rms} = \left(\frac{3kT_{gas}}{2m_X}\right)^{1/2},$$
(7.1)

where m_X is the mass of an atom of element X.

Suppose that the *cross section* of interaction between the grain and the atoms in the gas is σ (recall the situation for planetesimals, Chapter 3). The rate of collision between the grain and the gas atoms is then roughly

$$R \simeq n_X \sigma \left(\frac{3kT_{gas}}{2m_X}\right)^{1/2}.$$
(7.2)

Given a rate of collision and an assumption about how "sticky" the grain and gas atoms are (i.e., the likelihood of the atom becoming incorporated into the grain) then we could immediately calculate the growth rate of the grain. However, we do not know what σ is, and we have not allowed for the fact that in a real situation there will be many grains, and as the grains grow then n_X will shrink (assuming the gas is not replenished). In the section below we expand this analysis to deal with the first issue; we will deal with the second later on. Much of this discussion is directly related to the work we did in Chapter 3 on planet formation and the time evolution of planetesimal populations, except gravity is not (directly) governing the processes of dust condensation.

7.3.2   Growth Rate without Gas Depletion

Consider the full Maxwell–Boltzmann distribution of gas atom velocities:

$$f(v)dv = \left(\frac{2}{\pi}\right)^{1/2} \left(\frac{m_X}{kT_{gas}}\right)^{3/2} v^2 exp\left(\frac{-m_X v^2}{2kT_{gas}}\right) dv.$$
(7.3)

$f(v)dv$ is the fraction of atoms with velocities between v and $v + dv$. Thus the number of atoms with velocities between v and $v + dv$ is $n_X f(v)dv$. The total rate at which atoms collide with the grain is then given by

$$R = \pi a^2 n_X \int_0^\infty v f(v)dv, \tag{7.4}$$

where πa^2 is the geometric cross section of a grain of radius a. The integration yields

$$R = 4\pi a^2 n_X \left(\frac{kT_{gas}}{2\pi m_X} \right)^{1/2}, \tag{7.5}$$

which can be compared to our crude estimate in Equation 7.2. We now further allow for a "sticking probability" S, which is the fractional rate at which colliding atoms actually remain stuck to the grain (i.e., $0 < S < 1$). The rate of *mass* gain of the grain is then just

$$\frac{dm_{grain}}{dt} = 4\pi a^2 n_X m_X S \left(\frac{kT_{gas}}{2\pi m_X} \right)^{1/2}. \tag{7.6}$$

Now, it is also true that in general the mass gain rate can be written in terms of a and the grain material density ρ:

$$\frac{d}{dt} m_{grain} = \frac{d}{da} \left(\frac{4}{3}\pi a^3 \rho \right) \frac{da}{dt}. \tag{7.7}$$

$$\frac{dm_{grain}}{dt} = 4\pi a^2 \rho \frac{da}{dt}. \tag{7.8}$$

Equating this with the previous expression for mass growth rate yields

$$\frac{da}{dt} = \frac{n_X S}{\rho} \left(\frac{kT_{gas} m_X}{2\pi} \right)^{1/2}. \tag{7.9}$$

If n_X and T_{gas} are *not* time dependent we can immediately integrate this expression to obtain

$$a(t) = a_0 + \frac{n_X S}{\rho} \left(\frac{k T_{\text{gas}} m_X}{2\pi} \right)^{1/2} t, \tag{7.10}$$

where a_0 is an initial (or "seed") grain radius. Thus, the grain grows linearly with time. Let us now look at the implications for the possible routes for dust grain origins of this simple rate expression. Assuming that $a \gg a_0$ and rearranging Equation 7.10 we obtain

$$t(a) \simeq \frac{a\rho}{n_X S} \left(\frac{2\pi}{k T_{\text{gas}} m_X} \right)^{1/2}. \tag{7.11}$$

Now, a typical interstellar dust grain size is $0.1\,\mu$ m. T_{gas} can be approximately 100 K in the diffuse interstellar medium, and $m_X \sim 10 m_H$ (where m_H is the mass of a hydrogen atom) based on observation. Assuming also a grain density of $\rho = 2.2$ g cm^{-3}, then, for these grains to have actually *formed* in this interstellar environment would have required approximately

$$t \simeq \frac{2 \times 10^{12}}{n_X S} \quad \text{years.} \tag{7.12}$$

The density of hydrogen in the diffuse interstellar medium is typically in the range $n_H \sim 100\text{--}300$ cm^{-3} while, based on element abundances, the density for a condensing atomic species is more like $10^{-4} n_H$. Since $S < 1$ then we arrive at

$$t > \frac{2 \times 10^{15}}{n_H} \quad \text{years,} \tag{7.13}$$

which means that it would take longer than the present age of the Universe to grow dust grains to the observed sizes in this environment! We have therefore demonstrated from first principles that interstellar dust is *not* likely to form in the diffuse interstellar medium. If instead we consider the conditions of dense clouds ($n_H \sim 10^{4-8}$ cm^{-3}, $T \sim 10\text{--}30$ K) then we find

$$t > \frac{6 \times 10^{6-10}}{n_H} \quad \text{years.} \tag{7.14}$$

Thus, in the densest, coldest clouds in the interstellar medium there appears to be a good chance that some dust grains can grow on short enough timescales to be relevant—*if* the depletion of the gas phase is not important. We explore this further below.

7.3.3   Growth Rate with Depletion

Let us consider the further complications to modeling dust growth that we previously ignored. Amongst these are **depletion** (i.e., allowing for the reduction in gas atoms/molecules $n_X(t)$), grain evaporation, **sputtering** (i.e., ejection of atoms by collision of high-energy particles), chemical sputtering, grain–grain collisions and shattering or vaporization, and electrostatic tension and shattering (i.e., charge buildup creating stress forces).

Here we will consider the simplest of these additional factors, namely depletion. As atoms condense onto grains then the number of available gas atoms or molecules (n_X) will of course decrease—in other words the gas phase is depleted.

Given a dust grain number density of n_d and assuming all grains are of radius a, and that this grain number density does *not* change with time, then dn_X atoms are depleted in time dt:

$$\frac{dn_X}{dt} = -4\pi a(t)^2 n_X(t) n_d S \left(\frac{kT_{gas}}{2\pi m_X}\right)^{1/2} \tag{7.15}$$

and as before

$$\frac{dm_{grain}}{dt} = 4\pi a^2 n_X m_X S \left(\frac{kT_{gas}}{2\pi m_X}\right)^{1/2}. \tag{7.16}$$

Thus,

$$\frac{da}{dt} = \frac{n_X(t) S}{\rho} \left(\frac{kT_{gas} m_X}{2\pi}\right)^{1/2} \tag{7.17}$$

still holds. Substituting in $\frac{d}{dt}(dn_X/dt)$ we arrive at

$$\frac{d^2a}{dt^2} = -4\pi a^2 \left(\frac{kT_{gas}}{2\pi m_X}\right)^{1/2} S n_d \frac{da}{dt}. \tag{7.18}$$

As atoms are depleted to zero the grain radius must asymptote to a final radius, a_∞, and $da/dt = 0$.

We can write

$$\frac{d^2a}{dt^2} = \frac{da}{dt} \frac{da/dt}{da}, \tag{7.19}$$

and integrate over da to obtain

$$\frac{da}{dt} = \frac{4\pi}{3} \left(\frac{kT_{\text{gas}}}{2\pi m_X} \right)^{1/2} S n_{\text{d}} (a_\infty^3 - a^3). \tag{7.20}$$

We want to get an expression for a_∞, and can employ the following sleight-of-hand. If $a_\infty \gg a_0$ we can write

$$\frac{da}{dt} \simeq \frac{4\pi}{3} \left(\frac{kT_{\text{gas}}}{2\pi m_X} \right)^{1/2} S n_{\text{d}} a_\infty^3, \tag{7.21}$$

and at $t = 0$ the gas is not yet depleted, so it is still valid to use

$$\frac{da}{dt} = \frac{n_X S}{\rho} \left(\frac{kT_{\text{gas}} m_X}{2\pi} \right)^{1/2}. \tag{7.22}$$

Equating this to the previous expression and re-arranging yields:

$$a_\infty \simeq \left(\frac{3 m_X n_X (t=0)}{4\pi \rho n_{\text{d}}} \right)^{1/3}. \tag{7.23}$$

Thus, grain growth slows with time and grain size asymptotes to a final radius a_∞ if we allow for depletion. This behavior is sketched out in Figure 7.3. Recall that our earlier estimate of grain growth timescales was *conservative*, in the sense that it assumed an infinite supply of undepleted gas (§7.3.2). In this case—where we include gas phase depletion—if we inspect Equation 7.23 we can see that, for a given composition, it is just the ratio of initial gas number density to the dust number density that governs a_∞. In order to attain a final grain size of $\sim 0.1 \mu$ m therefore requires (assuming $m_X \sim 10 m_{\text{H}}$ and $\rho = 2.2$ g cm^{-3}) a gas species (e.g., carbon) to "seed" grain *number* density ratio

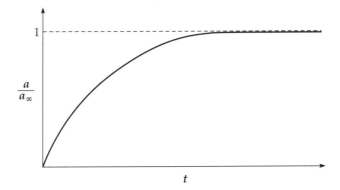

Figure 7.3. The growth behavior of a dust grain when gas depletion is allowed for. The grain size asymptotes to its final value as the gas is depleted.

of $n_X/n_d \sim 10^9$. By mass, dust makes up about 1% of the interstellar medium. If we consider what this means in terms of space number density relative to a given gas species (typically 10^{-4} times the hydrogen number density) we find—and this is a good exercise to try—that $n_x/n_d \sim 1$ in the *densest* interstellar clouds. Thus, the relative density of gas to dust is too low by about a factor of a billion! Including gas depletion in our calculations therefore tells us that, even in the densest interstellar clouds, grain growth is going to be too slow to explain the quantity of observed dust.

7.4 The Circumstellar Medium

If not in interstellar space then where does cosmic dust form? From the previous sections we have seen that high n_X produces bigger grains faster—and the best place to boost n_X is in the immediate environment of massive post main-sequence stars. Most stars of mass $0.5 < M_* < 9M_\odot$ end their fusion driven lifetimes as Asymptotic Giant Branch (AGB) stars. These objects comprise of a core of carbon and oxygen surrounded by a layer of helium undergoing fusion, and outside this a layer of hydrogen undergoing fusion. The outer atmosphere beyond this is enormously extended, and convective, and consists of metal enriched gas that cools dramatically as the atmosphere expands.

Such objects can shed this outer envelope of gas through radiation pressure–driven winds. In cosmic terms this gas is dense and cools quickly as it leaves the stellar surface, reaching temperatures below ~ 1000 K at relatively short distances from the star. Similar conditions can also arise in core collapse supernova ejecta, and in fact may even represent the dominant mechanism for dust production and dispersal (e.g., Sugerman et al. 2006). Supernova dust production is somewhat more complex though, since the energy of the shock wave will also destroy pre-existing dust in the interstellar medium, and so there is a balance between production and destruction that we do not yet fully understand.

7.4.1 Circumstellar Dust Formation

As a workable example to illustrate some of the considerations necessary for calculating dust grain growth in the denser environments around stars or supernova we here consider the circumstellar environment of a massive, post-main sequence, carbon rich star. This is, in many ways, a good deal less physically complex than the environment of a supernova.

It makes physical sense to state that circumstellar dust will begin to form only *below* a **condensation** temperature, T_{cond}, where atoms or molecules can attach to a grain surface without being immediately detached or sublimated.

Let us assume that grain seeds already exist (we will deal with the actual origins of grain seeds below) and can be treated as ideal black bodies (i.e., with no reflectivity). The temperature of a grain in thermal equilibrium with the radiation field of a star at distance r is then of course

$$T = \left(\frac{L_*}{16\pi r^2 \sigma} \right)^{1/4}.$$

(7.24)

Thus, the condensation radius from a star is just

$$r_{cond} = \left(\frac{L_*}{16\pi \sigma T_{cond}^4} \right)^{1/2}.$$

(7.25)

Now, in order to evaluate the growth of a grain at, or beyond, r_{cond} we also need to evaluate the density of a given condensing gas species X: $n_X(r)$.

Consider a spherically symmetric shell of wind material (i.e., gas being blown off the surface of the star). We assume that the shell has mass dM, thickness dr, and a constant velocity equal to that of the stellar wind: $\frac{dr}{dt} = V_{wind}$. We can then write the **mass loss** rate of species X from the star as

$$dM_X = 4\pi r^2 n_X(r) m_X dr. \tag{7.26}$$

If we then divide both sides by dt and rearrange we obtain (denoting $\frac{dM}{dt} = \dot{M}$)

$$n_X(r) = \frac{\dot{M}}{4\pi r^2 m_X V_{wind}}. \tag{7.27}$$

For a typical, carbon-rich, AGB star then $L_* = 10^3 L_\odot$, $V_{wind} = 10$ km s^{-1} and the mass loss is (conservatively) $\dot{M}_{carbon} = 10^{-9} M_\odot$ yr^{-1}. If $T_{cond} = 1000$ K for carbon we can easily obtain the conditions at r_{cond}; since $r_{cond} = 2.45$ AU then $n_{carbon} = 1.862 \times 10^9$ cm^{-3} at this condensation radius (which is more than the factor 10^9 boost we required from our calculation including depletion growth in the interstellar medium). If we refer back to our earlier (very crude) estimate of grain growth timescale ($t \simeq (2 \times 10^{12})/(n_X S)$ years) then we see that in this circumstellar environment the growth timescale should be something like 1000 years.

However, this is not a static environment. The grains will themselves be subject to the photon pressure of the star and will participate in the stellar wind flow. Thus, the grain growth rate will change with time according to the position in the outflowing circumstellar medium. Since $r = V_{wind}t$ to first order we can immediately rewrite the above equation to yield

$$n_X(r) = \frac{\dot{M}}{4\pi m_X V_{wind}^3 t^2}. \tag{7.28}$$

If we keep the problem simple by ignoring the depletion of the gas due to condensation (i.e., assume that changes in n_X are dominated by the dependency on distance from the star) then

$$\frac{da}{dt} = \frac{n_X(t)S}{\rho}\left(\frac{kT_{gas}m_X}{2\pi}\right)^{1/2}, \tag{7.29}$$

as we had before. Thus, substituting for n_X,

$$\frac{da}{dt} = \frac{\dot{M}S}{4\pi m_X V_{wind}^3 \rho}\left(\frac{kT_{gas}m_X}{2\pi}\right)^{1/2}\frac{1}{t^2}, \tag{7.30}$$

where we are also assuming that T_{gas} remains constant—which is likely a somewhat poor assumption. If we now integrate from t_{cond} to ∞ we obtain

$$a_\infty = \frac{\dot{M}S}{4\pi m_X V_{wind}^3 \rho}\left(\frac{kT_{gas}m_X}{2\pi}\right)^{1/2}\frac{1}{t_{cond}}, \tag{7.31}$$

where $t_{cond} = r_{cond}/V_{wind}$—in other words we have switched coordinates to one of time rather than distance from the star. From the previous example we find that $t_{cond} \sim 424$ days. Thus, ignoring depletion due to condensation (which will be secondary compared to the variation and replenishment of n_X due to the stellar wind), then for carbon grain growth around our example of a red giant star (and where we can assume for carbon $\rho = 2.2$ g cm^{-3} and $S = 1$) we find $a_\infty \sim 2 \times 10^{-6}$ cm (0.02 μm), implicitly assuming that $a_\infty \gg a_{cond}$. This is in the correct regime for observed dust, and immediately suggests that the circumstellar environment can work as a dust producer. This is borne out by observations of objects such as AGB stars, where the infrared emission of warm, new dust is seen.

7.5   Nucleation

Until now we have blithely ignored the initial conditions for dust grains—how does the first phase transition from gas to solid actually occur in the environments where dust is formed? To understand this we need to consider the process of **nucleation** (an analogy would be ice crystals here on Earth). This process can be broadly categorized as being either **homogeneous** or **heterogeneous**:

- **Homogeneous nucleation**: In this case a chain of chemical reactions of a *single* species (e.g., carbon) may allow enough energy dissipation that a **phase change** is energetically favorable. In this case a "cluster" of material may form. Interestingly, there is then a critical size above which clusters actually become *more* stable as they grow (the Gibbs free energy decreases). However, homogeneous nucleation may be a very slow process.

- **Heterogeneous nucleation**: This occurs when the nucleation is chemically *different* from the condensing material. For example an ion or molecule such as CO (which is chemically stable at the typical circumstellar wind temperature of $\sim$1000 K) provides an attractive site for carbon condensation.

It would seem that homogeneous nucleation followed by heterogeneous nucleation is a logical pathway for grain formation (e.g., silicon carbide, SiC). And in the relatively thick material in a stellar outflow (or supernova ejecta) we can imagine this condition being met. But in order to understand how this might work in more detail we need to begin to consider the chemistry taking place in either the interstellar, circumstellar, or supernova ejecta environments (§7.7). We must also consider the observed distribution of dust grain sizes and compositions, and what these imply.

7.6 Dust or Molecule?

Although it sounds like a semantic question, it is nonetheless critically important to be aware of the apparent "classes" of grains observed in the interstellar (or indeed circumstellar) medium. The smallest observed objects that might be classified as grains are ≤ 10 Å in size, and consist of fewer than 100 atoms, seen in infrared cirrus clouds in interstellar space. This is therefore approaching the scale of what we might otherwise term large molecules. The measured temperatures of such clouds can be quite high. In some cases, if the clouds are in thermodynamic equilibrium, then they are too hot for silicate or ice grains to survive.

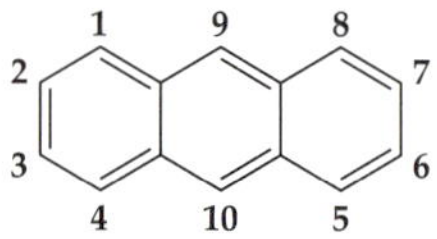

Anthrathene

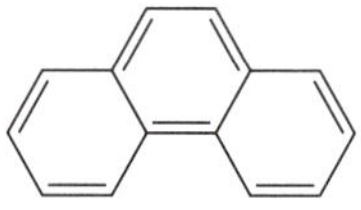

Phenanthrene

Figure 7.4. The molecular structure of the smallest PAH, anthrathene (and its isomer phenanthrene). Three benzene ring structures form a linear chain.

Carbon compounds on the other hand have the necessary resilience. This suggests (and we will explore this more below) that carbon chemistry may play a role in dust nucleation. Since carbon chemistry is also central to the fundamental questions of astrobiology this is clearly an extremely important subject to investigate.

7.6.1 Polycyclic Aromatic Hydrocarbons

If any known carbon structure straddles the regime between dust and molecules it is the family of molecules known as **polycyclic aromatic hydrocarbons** (PAHs). The basic component of a PAH is the **benzene ring** : C_6H_6. Attaching multiple benzene rings together enables the formation of a vast family of compounds. For example, the smallest molecule classified as a PAH is anthracene (and its isomer phenanthrene), with a chemical formula $C_{14}H_{10}$. This consists of a short chain of three benzene rings in the linear arrangement shown in Figure 7.4 (together with the isomer phenanthrene). Progressively larger PAHs include pyrene ($C_{16}H_{10}$), which consists of four benzene rings, coronene ($C_{24}H_{12}$) which consists of seven benzene rings, and so on. In Figure 7.5 a typical large PAH structure is illustrated, together with other carbon structures expected (or observed) in interstellar, and interplanetary space. Amongst these are **fullerenes** (C_{60}) [1] which have been found in meteorites. In general, PAHs are very stable molecules owing to the covalent bond strengths in the benzene ring structures. This helps them resist dissociation by ultraviolet photons in a wide range of environments. Large PAHs can exhibit both chemical (molecular) behavior and physical (grain) behavior. For example, they can collide and aggregate into bigger and bigger clusters, such as graphite-like "platelets," amorphous carbon particles, and even diamond structures (Figure 7.5).

Until quite recently the direct evidence for PAHs in interstellar space was incomplete—while there were many infrared spectral emission features seen in interstellar nebula (such as the cirrus clouds mentioned above) that appeared to match the spectral features produced in laboratory PAHs, there were still discrepancies. Much of the difficulty in uniquely identifying PAHs arises from the fact that large molecules

---

1. Sometimes known as "buckyballs" after the architect Buckminster Fuller who used geodesic domes in his work—with the same structural geometry.

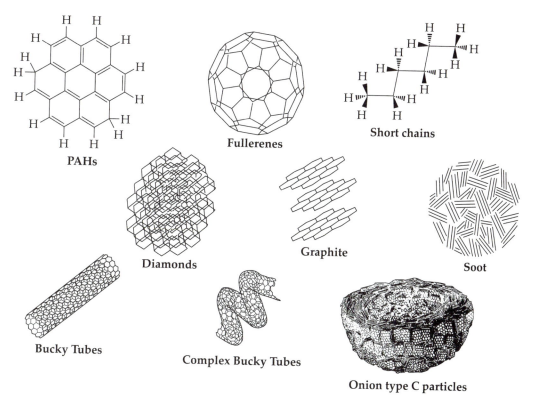

Figure 7.5. Some of the carbon structures observed, and expected, in the interstellar medium (after Ehrenfreund & Charnley 2000). A complex PAH is shown in the upper left. Middle of the upper row is a fullerene ("buckyball") of C_{60}. Upper right is a short chain carbon molecule. Center row left is a diamond structure, center row middle is a graphite structure, and center row right is a schematic of an amorphous carbon "soot" particle, composed of many misaligned graphite type structures. Bottom left is a form of one of the many types of extensions to the basic fullerene structure, in this case a buckytube (sometimes referred to as a carbon nanotube), bottom middle is a more complex version of this, and at the bottom right is a sketch of a so-called onion type-C particle—consisting of multiple "seed" centers for nucleation and many substructures of graphite-type units.

such as these absorb and emit photons in quite broad energy **bands**, typically in the infrared. These bands are due to rotational and vibrational states of the molecules. Thus, unlike individual atoms, or very simple molecules, PAHs do not exhibit discrete emission or absorption wavelengths. Detecting their presence in the interstellar medium, and identifying a particular species, is therefore very dependent on models of their emission bands, and hard to do.

The possible role of PAHs as one stepping stone between simple molecules and the nucleation of larger dust particles looks quite compelling when we study the observed size distribution of interstellar material and place the estimated abundance of PAHs on this plot (Figure 7.6). If PAHs really do act as the building blocks for many of the larger structures then we would expect them to be more numerous, and for their number density to fall along the same curve as the other particulate/solid matter in interstellar space. In fact, it is generally (although not universally) accepted now that PAHs are the most abundant free organic molecules in interstellar space. Precisely how PAHs form in cosmic settings is not fully understood. However, their relatively strong molecular bonds imply both that they are energetically favorable structures to form, and that they are resilient. Indeed, terrestrial PAHs (and structures such as fullerenes) are readily formed almost every time that something carbon based (e.g., wood, gasoline) is incompletely burnt. Much of the soot in a chimney will contain a rich mixture of PAH compounds. It seems quite likely then that PAHs can form in the carbon-rich "sooty" circumstellar (or supernova ejecta) environments we have discussed for dust production. There is also some observational evidence for fullerene ions (infrared bands associated with C_{60}^+) around carbon-rich stars. Some calculations suggest that PAHs may also form, at a slower rate, in dense interstellar clouds, and via energetic processing of ices (see below).

From the perspective of astrobiology PAHs are intriguing molecules. They bear a resemblance to many biologically critical molecules (i.e., ring structures and chains, such as amino acids), and it has been suggested that PAHs could have played a role in the origins of life (the "PAH-world" due to S. N. Platts, see also Ehrenfreund et al. 2006.). This is largely due to their wide range of behaviors, including the ability to build primitive membrane structures, carry information, and even act in a metabolic (energy-carrying) role. PAHs are also found in both meteorites and comets (Chapter 8), suggesting that the early Earth could have been well supplied with them. This relates to one further intriguing characteristic of PAHs. In general, they react with water in the presence of oxygen and ultraviolet light to produce new, complex organic molecules. It is very tempting therefore to imagine a scene on a young terrestrial-type planet where PAHs are exposed to conditions that en-

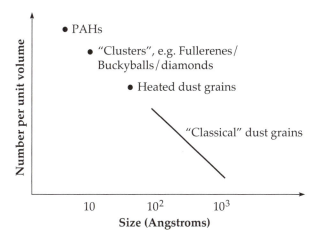

Figure 7.6. A sketch of the estimated relative abundances of grains and large molecular structures in the interstellar medium. Below 10 Å PAHs and some of the simpler fullerene/buckball/diamond arrangements are the principle structures. Between 10 and 100 Å the particle distribution is dominated by what are likely to be heated dust grains—either unable to retain a volatile mantle (see text), or no longer encapsulated by one. Silicon and carbonaceous compositions dominate. At the larger scales—up to at least 1000 Å—are the classical interstellar dust grains, seen in cold environments and likely harboring volatile mantles. The relative abundance of these structures strongly suggests a continuum of sizes, and a relationship between nucleation particles/molecules and the larger dust grains.

courage the production of many critical ingredients in a pre-biotic (see below) milieu.

7.7 Formation of Complex Molecules

There are over 140 unique molecular species currently identified in the interstellar medium, and many more complex species (such as amino acids and PAHs) detected in comets and meteorites (Chapter 8). Table 7.1 lists a selection of these compounds by way of illustrating the amazing diversity seen (see Ehrenfreund & Charnley 2000 for an excellent

Table 7.1. Tabulation of 129 detected molecular species in interstellar space, arranged in columns according to the number of atoms, from 2 to 13. Question marks denote observations yet to be confirmed. Data taken from the National Radio Observatories online resource: http://www.cv.nrao.edu/~awootten /allmols.html

2	3	4	5	6	7
H_2	C_3	c-C_3H	C_5	C_5H	C_6H
AlF	C_2H	l-C_3H	C_4H	l-H_2C_4	CH_2CHCN
AlCl	C_2O	C_3N	C_4Si	C_2H_4	CH_3C_2H
C_2	C_2S	C_3O	l-C_3H_2	CH_3CN	HC_5N
CH	CH_2	C_3S	c-C_3H_2	CH_3NC	$HCOCH_3$
CH^+	HCN	C_2H_2	CH_2CN	CH_3OH	NH_2CH_3
CN	HCO	CH_2D^+?	CH_4	CH_3SH	c-C_2H_4O
CO	HCO^+	HCCN	HC_3N	HC_3NH^+	CH_2CHOH
CO^+	HCS^+	$HCNH^+$	HC_2NC	HC_2CHO	
CP	HOC^+	HNCO	HCOOH	NH_2CHO	
CSi	H_2O	HNCS	H_2CHN	C_5N	
HCl	H_2S	$HOCO^+$	H_2C_2O	HC_4N	
KCl	HNC	H_2CO	H_2NCN		
NH	HNO	H_2CN	HNC_3		
NO	MgCN	H_2CS	SiH_4		
NS	MgNC	H_3O^+	H_2COH^+		
NaCl	N_2H^+	NH_3			
OH	N2O	SiC_3			
PN	NaCN	C_4			
SO	OCS				
SO^+	SO_2				
SiN	c-SiC_2				
SiO	CO_2				
SiS	NH_2				
CS	H_3^+				
HF	SiCN				
SH	AlNC				
FeO(?)	SiNC				

8	9	10	11	12	13
CH_3C_3N	CH_3C_4H	CH_3C_5N?	HC_9N	$CH_3OC_2H_5$	$HC_{11}N$
$HCOOCH_3$	CH_3CH_2CN	$(CH_3)_2CO$			
CH_3COOH?	$(CH_3)_2O$	NH_2CH_2COOH?			
C_7H	CH_3CH_2OH	CH_3CH_2CHO			
H_2C_6	HC_7N				
CH_2OHCHO	C_8H				
CH_2CHCHO					

review of interstellar organic chemistry). The largest confirmed organic molecule is $HC_{11}N$. Organic species are dominant—this is undoubtedly due to the versatility and resilience of carbon-based compounds, as well as the relatively high abundance of C, N, and O from stellar nuleosynthesis. However, in typical environments (either high or low temperatures, low densities, and high UV radiation) these species do not form particularly readily, or persist for very long. The major problems they face are (1) very low probabilities of collision and bonding events (1/100, 000 collisions result in a new, bound molecule) and (2) dissociation by UV/X-ray/cosmic-ray photons or particles: a typical dense stellar environment UV flux implies only a 300-year lifetime for most molecules.

How then do the observed species manage to form and survive? Some do probably form in the densest interstellar clouds (at low temperatures, 10–30 K) through gas phase chemistry. These include CO, N_2, O_2, $C_2 H_2$ (and C_2H_4), HCN and some very simple carbon chains. This is, however, a slow and tricky business. The solution to forming most of the observed species appears to be our friend dust. Dust may form from less fragile molecules (e.g., PAHs, silicates) and therefore be present from an early stage. The surface of dust grains provides a catalyzing and relatively benign environment, by both prolonging any period of atomic/molecular interaction and by shielding against radiation. In fact, in cooler and denser molecular clouds, grains can form **ice "mantles"** that provide an excellent environment for complex chemistry. An efficient route to forming the ice mantles in clouds with sufficient atomic hydrogen is through the hydrogenation of surface atoms—producing CH_4, NH_3, H_2O. In addition, CO can be directly accreted from the gas phase. To briefly summarize this icy grain mantle structure: a typical situation might be a SiO grain core surrounded by an amorphous mantle of solid H_2O, NH_3, CH_3OH, CO, CH_4, and CO_2. PAHs may also be incorporated into these mantles (Figure 7.7). In cold environments, any accreted H, C, N, and O are highly mobile on, and within, this icy surface and react promptly to incorporate themselves. Quantum tunneling can also act as a "search" mechanism for atoms seeking chemical bonds, and in fact may be much more efficient at low temperatures.

The main subsequent processes that drive a more complex chemistry of this starter-mix are (1) UV photolysis (Figure 7.8) and (2) thermally

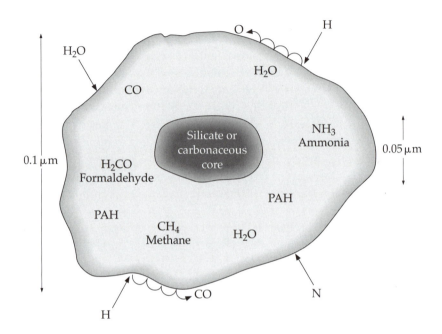

Figure 7.7. A sketch of the structure and composition of an icy grain mantle. Surface reactions produce a series of simple molecules that form the mantle, including water, ammonia, methane, formaldehyde, and carbon monoxide. PAHs may become embedded in this structure, which surrounds a silicate or carbonaceous—non-volatile—core (after Allamandola, Sandford & Valero 1988).

driven polymerization. In the case of (1) the UV photons can arise deep in molecular clouds due to the de-excitation of H_2 molecules, which have been excited by high-energy cosmic rays (e.g., electrons, protons, etc.; see §7.7.1 below). In both cases, a wide range of organic chemistry results, producing molecules such as ethers, alcohols, nitriles, isonitriles, amides, ketones, and polymers such as polyoxymethylene (chains of $-(CH_2-O)_n-$). Eventually these species can be returned to the interstellar medium through evaporation, which in turn results in new gas phase chemistry.

Thus, as we will discuss later on, the chemistry becomes even more rich and interesting when the environment begins to change (e.g., warm up) as consequence of cloud collapse and star formation (Chapter 2). To

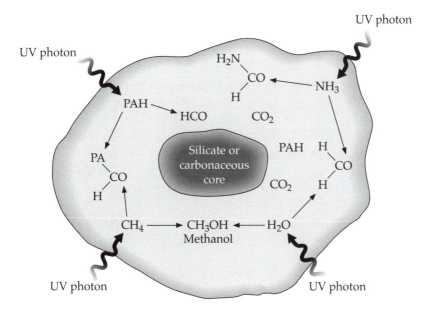

Figure 7.8. An extension to Figure 7.7 illustrating the additional chemistry enabled by energetic processing, in this case by UV photons. An array of more complex molecules can be formed in the ice matrix, including alcohols and PAHs with side groups. If this grain later experiences an increase in local temperature then the volatile mantle, together with these molecules, will evaporate into the gas phase—where new chemical pathways can be followed.

start this discussion, however, we will first look at some examples of fundamental interstellar chemistry within relatively typical, quiescent environments.

7.7.1 Fundamental Processes: H_2 Formation

Our discussion here of the formation of complex molecules and the accompanying chemistry is very limited—this is a potentially vast topic. The aim is to provide a small amount of insight to some of the fundamental situations that must be considered (see also Herbst 1988). As we have mentioned previously, the interstellar medium is approximately 99% gas by mass, and the remainder is dust. Inside a dense molecular cloud (such as that which forms or will form stars, Chapter 2) molecules act as the principal coolants owing to the variety of de-exicitation mechanisms they enable—for example, molecules have rotational and vibrational energy levels. Forming molecules is however a tricky business. Even understanding the formation of molecular hydrogen, H_2, is still a subject of active research—especially with respect to understanding how molecular cooling in the early universe operated in the absence of heavier elements. Indeed, while gas-phase formation of H_2 can, and still does, occur, it is in general far less efficient than H_2 formation on dust grain surfaces, as we will discuss below.

In a very dense molecular cloud ($n(H_2) > 10^4$ cm^{-3}) the dominant ionizing radiation sources are often cosmic rays (high-energy particles). This is because these clouds are otherwise opaque to optical or UV photons. As a side note: it is also true that the free electrons generated by cosmic rays ionizing H_2 can in turn excite H_2 molecules to higher electronic states. Subsequent decay results in the emission of UV photons. Thus, while shielded from external sources of UV photons, the gas in a dense molecular cloud will experience a local flux of cosmic ray-induced UV photons. This may be significant, and will contribute to the internal cloud chemistry. Given a cosmic ray flux, any existing molecular hydrogen may then participate in reactions such as

$$H_2 + \text{cosmic ray} \longrightarrow H_2^+ + e, \qquad (7.32)$$

where the ionization potential of H_2 is 15.4 eV. We denote the *rate* of this reaction by ξ (s^{-1}), which depends on the cosmic ray rate. Once this first step is complete a simple chain of reactions can take place:

$$H_2^+ + H_2 \longrightarrow H_3^+ + H, \qquad (7.33)$$

with a rate coefficient k' (m^3 s^{-1} per molecule). Then:

$$H_3^+ + e \longrightarrow H_2 + H, \qquad (7.34)$$

with a rate coefficient k (m^3 s^{-1} per molecule). Thus, molecular hydrogen can, through cosmic ray interactions, generate further molecular hydrogen and ions. In the steady state (i.e., equlibrium) the rates of these three reactions must obey

$$\xi n(H_2) = k'n(H_2)n(H_2^+) = kn(H_3^+)n(e). \qquad (7.35)$$

Now, since $n(H_2)$ is large compared to the other species this tells us that the second reaction ($H_2^+ + H_2$) can proceed the fastest, and so this implies that $n(H_3^+) > n(H_2^+)$ and $n(e) \approx n(H_3^+)$ if k' and k are comparable. Thus

$$n(e) = \left(\frac{\xi}{k} n(H_2) \right)^{1/2}.$$

(7.36)

If we put in some experimentally reasonable numbers, $\xi = 10^{-17}$ s^{-1} and $k = 10^{-12}$ m^3 s^{-1} per molecule, then we can estimate the **ionization fraction** as

$$\frac{n(e)}{n(H_2)} = \frac{3 \times 10^{-3}}{[n(H_2)]^{1/2}}.$$

(7.37)

In a dense molecular cloud $n(H_2) \sim 10^{10}$ m^{-3}, which implies an ionization fraction of $\sim 3 \times 10^{-8}$ at any given time—in other words a very low level of ionization. This immediately suggests that chemical processes within a dense molecular cloud are going to take a long time to build any substantial quantity of more complex molecules without some encouragement.

Let us now consider some other possible routes. Free hydrogen atoms can also combine via reactions on dust grain surfaces:

$$H + (grain - H) \longrightarrow H_2 + (grain).$$

(7.38)

We can use this most basic example to see how one explores grain catalysis. The above reaction requires that the first "attached" H (in the (grain–H)) remains on the grain surface long enough for the second H to arrive and locate the first by *random walking* (Chapter 9) across the grain surface. An incoming H atom experiences a long-range Van der Waals force from the net electrostatic potential of the grain. This weak atom–surface interaction is called **physisorption**. For the reaction to take place, the rate of arrival of H atoms must *exceed* the rate of evaporation of H atoms from this electrostatic potential. We can write this requirement as

$$n(H)\pi a^2 \bar{v}_H > v e^{-q/(kT_{grain})},$$

(7.39)

where the left-hand side of this expression is the *collision rate*, for a cross-sectional area of interaction πa^2 and an H atom mean velocity $\bar{v}_H$. The right-hand side is the fraction of a Maxwellian distribution of atoms with energy greater than the energy barrier or potential well depth q, which is the product of ν (the frequency of the H atom oscillation within a potential well of depth q) with the classical evaporation rate $e^{-q/(kT_{grain})}$. Rearranging we then obtain an expression for the required grain temperature to enable grain-catalyzed reactions:

$$T_{grain} < \frac{q}{k}\left(\ln \frac{\nu}{n(H)\pi a^2 \bar{v}_H}\right)^{-1}. \tag{7.40}$$

If the appropriate values are used then one finds that $T_{grain} \leq 12\,K$ is the requirement for H atoms to combine on a grain surface.

This does not mean that H atoms cannot combine on hotter grains (for example, T_{grain} is typically observed to be at least 100 K in many dense clouds). The solution is that the H atoms *must* become more strongly bonded to the grain surfaces if they are to avoid quick evaporation. They must therefore form covalent chemical bonds. This atom–surface interaction is termed **chemisorption** onto the grain surfaces.

7.7.2 Routes to Complex Molecules: Ion–Molecule Chemistry

While 100 K temperatures may inhibit certain routes to chemical reactions as described above they are nonetheless not sufficiently high to enable other reactions that require molecules or atoms to have enough kinetic energy to overcome the potential barriers between species. For example, at 100 K, oxygen atoms will not have sufficient energy to overcome the potential barrier with hydrogen atoms.

However, H_3^+ can catalyze this reaction. Since H_3^+ is produced by high-energy photons or particles as described above, the resulting chemistry is very dependent on the specifics of the cloud environment and the sources of such radiation. The reaction proceeds as

$$O + H_3^+ \longrightarrow OH^+ + H_2, \tag{7.41}$$

then

$$OH^+ \xrightarrow{H_2} OH_2^+ \xrightarrow{H_2} OH_3^+, \tag{7.42}$$

which is a hydrogen abstraction reaction. Finally

$$OH_3^+ + e \longrightarrow \begin{cases} OH + H_2 \\ H_2O + H \end{cases}, \qquad (7.43)$$

which can then lead to

$$C^+ + OH \longrightarrow CO^+ + H \qquad (7.44)$$

$$CO^+ + H_2 \longrightarrow HCO^+ + H. \qquad (7.45)$$

With the formation of the HCO group, further reactions can produce molecules such as methanol:

$$COH \xrightarrow{+3H} CH_3OH. \qquad (7.46)$$

All of these can occur within the grain mantle environment with much greater efficiency than in the gas phase.

7.7.3 Influence of the Proto-stellar Environment: Hot Core Chemistry

Chemistry in a molecular cloud does not of course just end with grain-surface interactions. As already mentioned in §7.7, if a region of a cloud is eventually involved in the collapse to a proto-stellar system then subsequent changes in the local environment will have a dramatic impact on the chemistry. For example, as a proto-star heats up its surroundings, the icy mantles of grains can be evaporated, leading to complex gas phase chemistry. Furthermore, UV radiation from the proto-star can ionize surrounding gas and add another driver for chemistry. Such a situation is most likely to occur around *high-mass* proto-stars which, much like their main sequence descendants, experience quicker and more violent evolution. For lower mass stars such as our Sun, the timescale of events is longer, and more distinct phases of physical environment and chemistry are thought to result—but many of the basic phenomena are likely to be the same.

During the early phases, as more of the interstellar cloud material falls into the system, the density increases dramatically. The temperature however is still relatively low, and grains likely experience an enhanced mantle growth consisting of the direct accretion of CO, H and O. Then,

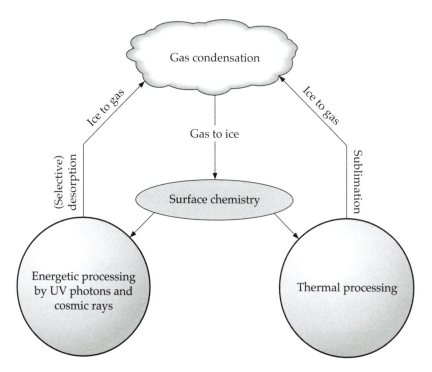

Figure 7.9. A schematic of the cycles of chemistry that take place between dust grain mantles and the gas phase in the interstellar and/or circumstellar medium. Gas condensation initially results in the formation of grain seeds (e.g., SiC) subsequent condensation in cold environments results in the formation of a grain mantle and a range of surface and interior chemistry, enabled by energetic and thermal processing. Sublimation/evaporation and desorption due to UV photons or cosmic rays return molecules to the local gas phase, which can result in a wealth of further chemistry.

as the proto-star heats up the dust grains, their chemically processed mantles will evaporate and release their molecular content into the gas phase - enabling new chemistry to take place. Increased irradiation by UV photons will also stimulate further chemistry in the remaining grain mantles. A schematic of the cycling, and re-cycling, of material and chemistry between solid and gas phases is shown in Figure 7.9.

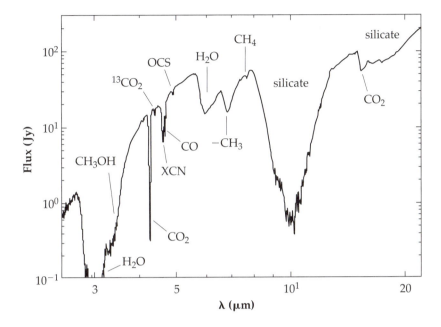

Figure 7.10. An example of a real observation of spectral features due to molecular species around a proto-stellar system (in this case the proto-star W33A, data from Gibb et al. 2000 and Langer et al. 2000). The proto-star's infrared spectrum is dominated by absorption features due to the cold, dense, surrounding material. These features include those due to both silicates (dust core) and a range of organic species, as well as water.

This creates a hot core of gas ($T \geq 100$ K, $n \geq 10^6$ cm^{-3}), rich in basic molecules (alcohols in particular help drive the subsequent chemistry), which can now participate (due to the increased temperature) in the formation of more complex molecules such as ethers, organic acids, and cyanides, to name but a few. During the later phases, proto-stellar winds and outflows can act to redistribute gas and dust, pushing some back out to colder zones where grains can once again build icy mantles, now with the addition of more complex molecules. In Figure 7.10 an example of a spectral observation of a proto-stellar system is shown, many clear features due to a wealth of molecular species are observed.

In summary: the rich chemistry of a proto-stellar and proto-planetary nebula is a product of both earlier interstellar chemistry (driven strongly by grain surface chemistry) and the recycling between hot/warm gas and solid phase chemistry. This further processing is typically enabled by the collapsing proto-stellar system and its thermal and radiation impact. The raw material that winds up on the surface of a forming terrestrial-type world, especially those deposited by late-time impactors, will therefore *already* be a rich chemical mix, including many of the molecules that can be directly incorporated into terrestrial-type life. It should be cautioned however that any material delivered to the surface of a planet via impacts will be subject to significant thermal extremes (including those from the energy of impact)—which may limit some of the molecules that survive (see however Chapter 8).

7.7.4 Prebiotic Molecules

In the context of astrobiology the term **prebiotic** simply refers to chemistry or molecules that are either natural precursors to the production of biologically important molecules, or represent the basic "building blocks" of biological chemistry—without the presence of biological production routes (i.e., **abiotic** production). Compounds such as amino acids (Chapter 5) can therefore be considered as prebiotic molecules if they originated in non-biological systems. The endpoints of the interstellar, or cosmo-chemistry we have summarized in this chapter appear to often consist of just such species.

As we will see in Chapter 8, there are direct observations that confirm that prebiotic chemistry occurs all over the place—and is most certainly not just confined to local planetary surface or subsurface environments. This is quite a radical change in perception compared to only fifty or so years ago. This is not to say that *all* prebiotic molecules on a young terrestrial planet need have arisen elsewhere, but it does seem unlikely that a young world will have to start its organic chemistry from scratch.

References

Allamandola, L. J., Sandford, S. A., & Valero, G. J., (1988). Photochemical and thermal evolution of interstellar/precometary ice analogs, *Icarus*, **76**, 225.

Ehrenfreund, P., & Charnley, S. B. (2000). Organic molecules in the interstellar medium, comets, and meteorites: A voyage from dark clouds to the early Earth, *Annual Review of Astronomy and Astrophysics*, **38**, 427.

Ehrenfreund, P., Rasmussen, S., Cleaves, J., & Chen, L. (2006). Experimentally tracing the key steps in the origin of life: The aromatic world, *Astrobiology*, **6**, 490.

Gibb, E. et al. (2000). An inventory of interstellar ices toward the embedded protostar W33A, *Astrophysical Journal*, **536**, 347.

Herbst, E. (1998). Interstellar molecular formation processes, *Reviews in Modern Astronomy*, **1**, 114.

Langer, W. D., et al. (2000). Chemical evolution of protostellar matter, In *Protostars and Planets IV*, Eds. Mannings, V., et al., Univ. Arizona Press, 29.

Sugerman, B. E. K., et al. (2006). 'Massive-star supernovae as major dust factories, *Science,* **313**, 196.

Problems

7.1 This question requires some research. Describe the scientific aims of the NASA STARDUST mission, and the technology used to accomplish them. To date what has the analysis of the STARDUST captured particles revealed about the dust population in our solar system?

7.2 (a) For dust formation, the final spherical grain radius, allowing for depletion, is given in terms of the gas atom mass, m_X, the gas number density, n_X, and the grain number density, n_d:

$$a_\infty \simeq \left(\frac{3 m_X n_X (t=0)}{4 \pi \rho n_d} \right)^{1/3} .$$

The *observed* number-per-mass distribution of interstellar dust ($n(m)$) obeys the relation

$$n(m) dm \propto m^{-k} dm,$$

and measurements suggest $k = 11/6$. Assuming $n(a) da = n(m) dm$, and that grains of different sizes form from independent seed populations, compare the measured distribution with that predicted by the above equation. What *physical* mechanism might be responsible for any difference (think of what may happen in violent collisions)?

(b) At a fixed distance from a star the solar radiation exerts a force on the dust grains proportional to their geometric cross-section. Derive the *proportionality* relation between the *acceleration* experienced by a grain and its mass m. Demonstrate that the rate of *mass outflow* at a given radius is dominated by small grains.

(c) In the Solar System the *size* distribution of objects in the asteroid belt is $n(a) \propto a^{-3.4}$. How can this be so similar to the dust grain size distribution? Can the physical mechanism you gave in (a) be applied to larger bodies? Given what you know about planet growth, what stage of evolution might set the final $n(m)$?

7.3   Dust grains can experience evaporation (e.g., the sublimation back to a gas phase of material), sputtering (the ejection of atoms due to high-energy particle interaction), and even shattering (the fragmentation of a grain into smaller parts due to electrostatic forces or shock waves). Evaporation and sputtering return atoms to the surrounding gas phase. Discuss how this would alter our arguments about gas phase depletion. Under what circumstances could grain growth reach an equilibrium state?

7.4   Described the nature of PAHs and their potential role in the formation of macroscopic structures in interstellar space.

In the paper "Experimentally tracing the key steps in the origin of life: The aromatic world" (Ehrenfruend et al. 2006, see references for this chapter), a hypothesis is made in which PAHs form an integral part of a rudimentary form of life. Describe (and if possible, critique) these arguments, in particular those relating to proto-cells.

7.5   In some so-called reflection nebulae there is evidence for continuum (blackbody) emission from dust at temperatures of 1000 K. This temperature is well in excess of what would be expected to be the thermal equilibrium temperature between the dust and its stellar illumination. The answer appears to be that only a small fraction of the dust grains are this hot. They absorb high-energy photons and remain hot only for a short time, cooling quickly to the equilibrium temperature of the rest of the dust cloud.

Assume grains can be treated as pure blackbodies. The equilibrium temperature of a dust cloud is 100 K. 60% of the total observed intrinsic luminosity of the cloud at any given time is due to hot (1000 K) grains. Compute the relative numbers of hot and cool (equilibrium) dust grains in this cloud at any given instant.

7.6 Describe the role of grain mantle chemistry in the formation of complex molecules. Include discussion of the physical processes that help drive this and further chemistry.

  Using Table 7.1, what percentage of known interstellar molecular species do *not* involve carbon? Why is carbon chemistry so prevelant?

7.7 Consider the following reaction chain:

$$H_2 + \text{cosmic ray} \longrightarrow H_2^+ + e$$

$$H_2^+ + H_2 \longrightarrow H_3^+ + H$$

$$H_3^+ + CO \longrightarrow HCO^+ + H_2$$

If the abundance of CO is $n(CO) = 10^{-4} n(H_2)$ and $n(H_2) = 10^{11}$ m^{-3} (a constant), the reaction rate $\xi = 10^{-17}$ s^{-1} and the rate coefficients for the 2nd and 3rd reactions are $k_2 = 10^{-15}$ m^3 s^{-1}, $k_3 = 10^{-15}$ m^3 s^{-1}, then determine the abundances of H_2^+ and H_3^+.

CHAPTER **8**

Comets, Meteorites, and Proto-planetary Disk Structure

8.1 Introduction

We have a unique opportunity to probe our own proto-planetary disk through the study of its "fossil" remains—namely the planets, asteroids, meteoroids, and comets. Indeed, the study of the smallest chunks of material in our solar system has been going on since well before there was a concept of a proto-planetary disk. Much of our present understanding of planet formation has its direct origins in the study of these small and large objects. Meteors have been observed throught human history, and Meteorites—the name given to smaller remnant solids that survive passage through the Earth's atmosphere and reach the surface—have been physically collected. Comets have also been witnessed by our human ancestors and documented by astronomers from the ancient Chinese dynasties through the Middle Ages and up to modern times. Even the crater structures left from more substantial impact events on Earth have been scrutinized since humans were around to explore them. With the advent of telescopic instruments, and more recently planetary exploration, we have also learnt that almost every body in the solar system with an observable surface shows evidence for past impacts. Some are ancient, and some are very recent. All provide not only a geological "clock," helping us estimate the age of the surface (either locally or globally), but also a probe of the surface composition, and even of the nature of the impactors themselves. Combined with actual encounters with comets and asteroids (including the direct investigation of their external and internal composition), and the wealth of collected "fallen" material

333

on the surface of the Earth, we have a remarkable dataset with which to test models of both the proto-planetary disk structure and chemistry and planet formation itself.

Detailed astronomical observations of proto-planetary disks and young stars provide the further possibility of witnessing analogs to our own solar system, and mapping out the full diversity of chemistry and dynamics that is likely. These data, together with that of formed exoplanetary systems, also provide a means to evaluate whether or not all systems form via the same mechanisms, or through some combination thereof. These investigations also have far-reaching implications for the quest to place life in full context. The environmental conditions of the earliest stable bodies in a nascent stellar system—namely the young planets and planetesimals—may have much to tell us about the emergence of the complex phenomenon of life (Chapters 1, 5 and 10). A deeper understanding of the physical and chemical conditions of proto-planetary disks and their youthful planetary bodies is crucial. The combination of a hands-on study (often literally) of the remains of this period in our own solar system with the study of distant systems places a unique probe at our disposal.

However, the questions and topics raised by both the study of comets and meteorites—from impact events to organic chemistry and volatile delivery—and the study of the proto-planetary disk are numerous and complex. In this chapter we will restrict ourselves to rather briefly reviewing some of the simpler observations of meteorites, comets, impact events, and proto-planetary disks as well as considering a few of the implications for life.

8.2   Comets

Comets (from the ancient Greek for "long-hair") have without a doubt been observed throughout human history, although clear records span only the past couple of thousand years. Comets represent one end of the compositional variation in small bodies in the solar system—with a high internal volatile content. They are generally classified as being either **long period** comets (orbital periods greater than 200 years) or **short period** comets (orbital periods less than 200 years). All cometary orbits are highly eccentric, and it is only during their passage through perihe-

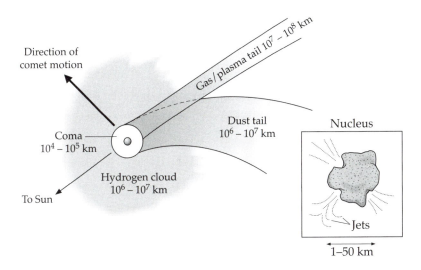

Figure 8.1. Schematic of generic cometary structure during active episodes. The central nucleus is a solid body typically 1–50 km across. Around perihelion the solar radiation creates outgassing jets and sublimation of volatiles (e.g., water, carbon dioxide). This gas, together with entrained dust, forms a **coma** around the nucleus. Beyond the coma a larger halo, or cloud, of hydrogen exists—unseen at optical frequencies. Radiation pressure pushes the particles of the coma away from the comet. As their distance from the Sun increases their tangential velocity decreases due to angular momentum conservation and they spread out in a curved region behind the direction of motion of the nucleus. This dusty **"tail"** can extend to 10^6–10^7 km. Comets can also exhibit a bluish plasma, or ion tail of charged species that follow interplanetary magnetic field lines in an anti-solar direction, pushed by the solar wind. These tails can reach 10^7–10^8 km beyond the nucleus before dissipating or becoming too faint to detect. Comets are therefore messy beasts, spewing dust and particles as they pass through the inner solar system.

lion that they are readily seen. This is a result of the solar heating they experience, and the consequent sublimation of volatiles that, together with entrained dust, "blows" from them in extended structures, including "tails" that emit and reflect radiation (Figure 8.1). While a typical cometary body is some 1–50 km in size, the temporary structures caused by solar heating can span scales of up to 10^7–10^8 km.

Approximately 1000 comets have been cataloged to date. Long period comets have semi-major orbital axes of $1 - 5 \times 10^4$ AU and are thought to originate directly from the postulated Oort cloud of some $\sim 10^{12}$ such objects left from the formation of the solar system (Chapter 1). This class of comets also has orbits that are randomly oriented with respect to the plane of the inner solar system. By contrast, short period comets have pro-grade orbits close to the plane of the inner solar system. It is possible that a few of the short period comets originate from the Oort cloud population, but dynamical interactions with the inner planets have perturbed their orbits over time until they are reduced to less than 200 year periods. The majority of short period comets however are likely to come from a distinct population, specifically the Kuiper belt (Chapter 1)—which already lies in the same plane as the inner system.

The precise origin and formation history of all comets is however currently unclear. To summarize, in our own solar system models range from formation scenarios at the orbit of the outer planets (e.g., Neptune, Uranus) to far out in the Oort cloud. Specific models include:

- Interstellar origin: interstellar dust grains agglomerate to form comet nuceli in cold regions of nebulae far from proto-star.
- Complete chemical equilibrium model: presolar material is altered and chemically equilibrated as it falls into the proto-planetary system from interstellar space (undergoes both sublimation and co-condensing of water and organic species).
- Intermediate model: presolar material is partially chemically and physically processed.

The general assumption is that a comet in fact contains a mixture of true interstellar material and material formed during the proto-stellar cloud collapse, including that from the outskirts of the proto-stellar environment. This mixture will vary significantly between comets, depending on their precise formation pathways and histories. While we still do not know in any detail the generic makeup of comets, from a few examples we do know that they are generally amorphous and composed of a mix of water ice ($\sim$30%), silicates ($\sim$26%), large organic molecules (known as "CHON" particles for the principal elements involved) ($\sim$9–

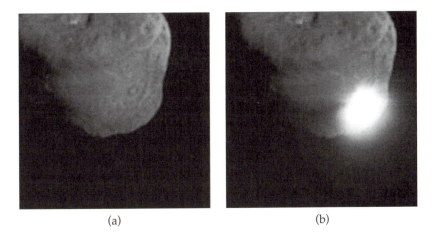

(a) (b)

Figure 8.2. Images taken by the Deep Impact flyby spacecraft showing comet Tempel 1 immediately before (left image) and 2.52 seconds after impact by a 370-kg impactor craft (right image). The bright plume of material excavated by the impact is clearly visible—saturating the camera sensitivity. Also visible is the *shadow* of the plume to the immediate left of the brightest feature. The final crater was estimated to be some 200 meters across. The nucleus seen here is approximately 8 by 5 km in size. From A'Hearn, SCIENCE **310** : 258(2005). Reprinted with permission from AAAS.

23%) and assorted other miscellaneous materials (e.g., see Ehrenfreund & Charnley 2000). The Deep Impact space mission to comet Tempel 1 (involving a deliberate collision designed to excavate subsurface cometary material, e.g., A'Hearn et al. 2005, and see Figure 8.2) revealed not only a clear difference in chemical composition between the comet surface (which is water poor, or "dry") and its immediate interior (which is water rich, or "wet") but also a wealth of complexity in composition. This includes the observation of clay particles and olivine (both potentially formed in the presence of liquid water!), as well as some heavy metals and a diversity of high-temperature carbon compounds. Earlier data from comets have also revealed an enormous array of organic compounds, including heteropolymers, alcohols, aldehydes, ketones, acids, PAHs, aliphatic hydrocarbons, alkynes, dienes, large nitriles, imino and amino end groups, nitrogen heterocycles, pyrrole, pyrroline, purine, adenine, and so on.

8.2.1 Deuterium-to-Hydrogen Ratios: D/H

The environments necessary for some of the observed chemistry to have taken place also provide clues to the physical location of comet formation. As discussed elsewhere (Chapters 1, 9) there is an observed enhancement, relative to the Earth, of the isotopic ratio of deuterium to hydrogen in some comets and in some molecular species. The bulk D/H ratio on Earth is some $\sim 1.5 \times 10^{-4}$, whereas the water D/H ratio observed in the comets Halley, Hale-Bopp, and Hyakutake is some $\sim 3.1 \times 10^{-4}$, and possibly higher. Cold temperatures favor deuterium chemistry due to subtle differences in the electronic energy levels of a deuterium atom versus a hydrogen atom. Thus, the observed D/H ratios in comets compared to that on Earth suggest that at least some of the formation processes leading to comets were definitely in very cold environments, such as those in the outer solar system and beyond. Figure 8.3 summarizes some of the known variations in D/H ratios within the solar system (based on Robert 2001). While the Sun and the gas giants (Jupiter and Saturn) share very similar D/H ratios, which are considered to reflect the global ratios of the original circumstellar disk, other bodies exhibit values intermediate to these and those in the true instellar medium. In carbonaceous chondrite meteorites (§8.3 below) a range of D/H ratios is seen. While organic molecules in these meteorites tend to have an enhanced deuterium content, water contained in clays within these objects tends to have a D/H ratio matching that seen on Earth. Intriguingly, based on measurements of Martian meteorites (see below) and the Martian atmosphere, Mars has a D/H ratio more in keeping with that estimated for comets. This suggests, but by no means proves, that the water content of Mars may have had a greater contribution from cometary material than the Earth, which may have had most of its water supplied by late-time meteorite and asteroid impacts (see below). In all cases, enhanced D/H ratios compared to those of the circumstellar disk suggest that low-temperature environments—at significant distances from the proto-sun—played an important role in the processing of both cometary and meteoritic materials.

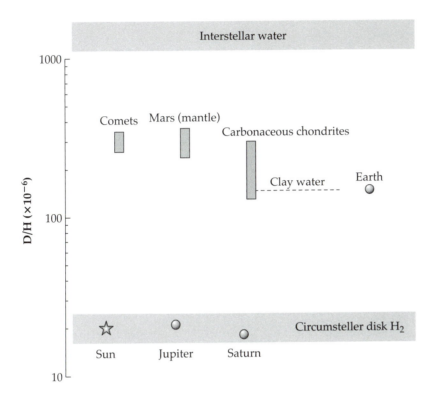

Figure 8.3. An illustration of the variation in deuterium to hydrogen (D/H) ratios seen in the solar system (based on Robert 2001). While interstellar water exhibits high observed D/H ratios (uppermost bar) the ratio deduced for the proto-stellar or circumstellar disk of the solar system is significantly lower (by two orders of magnitude). Solid bodies in the solar system typically exhibit intermediate D/H values. Observation of sublimated water from comets indicates a ratio a factor $\sim 2-3$ times higher than seen on Earth. This ratio is however more in agreement with that seen in the atmosphere of Mars and in Mars derived meteorites here on Earth. Carbonaceous chondrite meteorites exhibit a wide spread in D/H, however most of this is due to the difference between the deuterium replacing hydrogen in organic compounds and that in the water-bearing clays seen in these objects. The D/H ratio in the water-bearing clays is typically identical to that seen on Earth, with a small scatter.

8.3   Meteorites

As unfair as it might seem, it is nonetheless appropriate to term meteorites as leftovers, the unused bits and pieces from the formation of planets, including "primitive" planetesimal material, and material produced by large impacts—literally chippings from internally differentiated planetesimals or asteroids. This demeaning label does however belie their extraordinary importance as direct probes of the formation history of planets and the chemistry of the proto-planetary disk.

Meteorite is the term given to a small body (a **meteoroid**) if it survives passage through the Earth's atmosphere and arrives intact on the surface. Such objects come in a bewildering array of types. Most small bodies never make it to the ground in any recoverable form (indeed, they are termed **bolides** if they explode during transit through the atmosphere), but those that do are classified according to their general appearance. Because of the historical context of their discovery and examination, this classification is not necessarily an accurate way of evaluating the origin and history of a particular meteroid, but it serves as a primary discriminator.

More than about 30,000 pieces of meteorites are currently in known collections. Listed here, in no particular order, are some of the principal meteorite classes (see for example Lissauer & de Pater 2004):

- **Achondrites** Material originating from **differentiated** parent bodies—those that experienced total or partial melting during which heavier elements (e.g., iron) would have sunk to the center, leaving lighter elements (e.g., silicon, carbon) in the outer surface/crust (Chapter 3). These include meteorites identified on the basis of composition and noble gas content to originate from the Moon and from Mars—the products of major impact events on those bodies (approximately 60 such samples have been collected).

- **Stony Irons:** Also originating from differentiated parent bodies. Containing similar amounts of macroscopic metallic and rocky components, further subdivided into classes:
  - **Mesosiderites:** composed of broken planetesimal/planetary mantle rock that has been reassembled with lighter silicate rock (these are **polymict breccias**).

- **Pallasites**: formed at the mantle–core boundary in planetesimals/proto-planets.

- **Irons:**: Iron–Nickel alloy, representing planetesimal/planetary outer core material.

- **Tektites and Impactites:** Natural silica glasses, formed from molten material ejected from the Earth by impacts which quickly re-entered the Earth's atmosphere and in turn impacted on the surface.

- **Chondrites:** So-called **primitive meteorites**, unlike any terrestrial rock. These represent the majority of recovered meteoritic material on the Earth. Chondrites have never melted and are composed of material directly agglomerated from the proto-stellar cloud and disk. Some show modification by processes involving *liquid* water as well as thermal processes indicating a range of environmental histories. The term **chondrite** arises from the presence in most such objects of small **chondrules** (0.1–2 mm sized spherical silicate intrusions, Figure 8.4) that are assumed to have formed directly from dust that was subject to intense heating (at least 1600 K) in the proto-planetary disk (e.g., perhaps during a proto-star's flaring events). The typical elemental composition of chondrites is very similar to that of the Sun. These primitive meteorites also often contain **refractory** (a term referring to resistance to high temperature) inclusions rich in calcium and aluminum (known as **CAI**s). Such high-temperature structures may have formed close to the proto-star before being transported outwards and incorporated into structures in a much cooler environment. Both the chondrules and CAIs are embedded in a dark, fine grain **matrix**, with a range from 0% to as much as 80% of the mass of the meteorite being in these inclusions. In short: chondrites represent a grab-bag of material, some likely originating from true interstellar space, and some from later condensation and processing of material in the proto-planetary disk.

  Nine classes of chondrites have been established, grouped into three sets:

  - **Carbonaceous Chondrites**: The most volatile-rich subset of chondrites. High carbon content, containing organic molecules, including amino acids—some could have possibly formed from cometary material and show signs of never having been heated to

Figure 8.4. Image of a cut section of the carbonaceous chondrite known as the Allende meteorite. This meteorite fell in 1969 in Mexico, scattering several tons of material over more than a 300 square kilometer area. The mottled appearance is due to the millimeter sized chondrules embedded in a black matrix material. The lighter features (indicated by the pointer) are CAIs. Reproduced by permission of David Kring, all rights reserved.

more than about 500 °C. See discussion below of the Murchison Meteorite.

- **Ordinary Chondrites**: The most common type of primitive meteorite.
- **Enstatite Chondrites**: Chondrites dominated by the mineral enstatite ($MgSiO_3$).

The amazing diversity seen in meteoritic types is an indicator of the extraordinary level of activity, and the occasional violence of events, in the forming solar system. The specific physical conditions necessary to produce the range of meteoritic material seen, and the detailed isotopic, chemical, and mineralogical components of meteorites, have provided major clues aiding the reconstruction of the history of planet formation in our own solar system. This information is also critical for the interpretation of extra-solar proto-planetary systems, as well as formed exoplanets.

8.3.1 Radionuclides: Dating and Heating

Radionuclide dating of meteoritic material has played a central role in establishing the age of the solar system and the bodies within it. For example; radioactive decay chains of uranium ($^{238}_{92}$U and $^{235}_{92}$U) result in stable lead isotopes ($^{206}_{82}$Pb and $^{207}_{82}$Pb). The abundance ratio of these Pb isotopes in CAIs in chondrites has been used to date their formation at $4.563 \pm 0.004 \times 10^9$ years ago. From detailed studies of specific meteorites, the orbital configuration of their parent bodies can be estimated (e.g., Trieloff & Palme 2006, Lauretta et al. 2005). The majority of all known meteorites appear to have originated from bodies in the asteroid belt between Mars and Jupiter (apart from the Martian and Lunar meteorites). Asteroids are the likely descendants of the population of planetesimals (i.e., of size less than 1000 km in diameter) in the proto-planetary disk (Chapter 3). We can combine this information with further analyses of elemental composition and the observation that the majority of meteorites are derived from *undifferentiated* material (i.e., material from parent bodies that never underwent significant heating due to large coalescence events and radiogenic heating). It then appears that a total of about only 100 unique planetesimal bodies are needed to produce the observed population of meteorites (e.g., Lauretta et al. 2005).

The minority population of achondrites clearly originates from bodies that underwent at least partial differentiation, and some ordinary chondrites show evidence for higher temperature origins. In large planetary embryos, the major source of heating in the late stages of formation came from collisions (e.g., Chapters 3, 9). In smaller planetesimals it was primarily the radioactive decay of ^{26}Al with some contribution by ^{60}Fe that was responsible (Chapter 3). By studying the way in which meteoritic material cools it is possible to deduce that while the outer layers of a typical large planetesimal (e.g., an object 100 km across) could radiate thermal energy away quite rapidly, the deeper cores probably retained thermal energy over periods of at least 100 Myr—allowing the buildup of temperatures as high as 1200 K.

The detailed interpretation of chondrite composition and structure therefore provides a wealth of information on the *extremes* of thermal, chemical, and radiation environments in the proto-planetary disk. This is well beyond the scope of our discussion here, but we provide some useful references as starting points (e.g., Lissauer & de Pater 2004, Trieloff & Palme 2006). Below in §8.5 we present a summary of the current understanding of proto-planetary disk structure and chemistry; much of this has its origins in the study of meteorites. We should emphasize the remarkable uniformity of isotopic composition in meteorites. Given the potential range of origin points for meteoroids this indicates that the material of the proto-planetary disk must have experienced extensive mixing at some stage.

8.4   Late-Time Impactors and Early Earth Chemistry

In addition to the rich chemistry seen in comets, the carbonaceous chondrites also exhibit a wide range of organic content. This includes amines, amides, alcohols, aldehydes, ketones, aliphatic and aromatic hydrocarbons, PAHs, sulfonic and phosphonic acids, amino acids, hydroxycarboxylic acids, carboxylic acids, purines, pyrimidines, and kerogen-like material (see Ehrenfreund & Charnley 2000 for a summary).

The best known proto-type for this characteristic is the **Murchison meteorite**, which landed in Murchison, Australia in 1969. About two hundred pounds of material was deposited over a 5 mile area, much of which was in fragments large enough to be collected. The meteorite has been estimated to be approximately 4.5 billion years old, and contains a complex "tar-like" goop with over 16 identified amino acids and grains of silicon carbide. Furthermore, in the Murchison organic mix there is an enrichment, compared to terrestrial levels, of deuterium, especially in the amino and hydroxycarboxylic acids (but not in the water-bearing clays). Enhanced deuterium to hydrogen (D/H) ratios can be obtained via low-temperature, dust grain catalyzed interstellar chemistry (§8.2.1). This suggests that some fraction of these organic acids were genuinely formed in interstellar space, and therefore pre-date the solar system. Other compounds in the meteorite were likely formed during "re-heating" episodes as material was accreted onto the proto-planetary disk. The characteristics of the Murchison meteorite and other carbonaceous chondrites, combined with the deuterium ratios seen in water (§8.2 above), provide startling evidence that the early Earth could have obtained an extensive mix of volatile and organic material from impactors. Clearly much of this could be dissociated during a violent impact event (see below), however much could also survive—either being dispersed into the atmosphere or remaining incorporated in solid objects. There are many potential implications, but these can be complex to understand. By way of a relatively simple example we pause here to consider a possible connection to an issue of terrestrial biology first raised in Chapter 5.

8.4.1   Chirality

As we have briefly discussed in Chapter 5, amino acids exhibit left- and right-handed (structural) forms (Figure 5.9) or *enantiomers*, which are non-superposable mirror images of each other while sharing the same chemical composition. This is known as **chirality**. In terms of chemistry, left- and right-handed molecules "do not mix." This is to say that if a pro-tein was constructed from a mix of left and right enantiomers it would not fold in the same (and biologically critical) way and would there-fore not carry out its "correct" structural duties. In typical terrestrial conditions, if one leaves a collection of purely left-handed enantiomers (e.g., amino acids) in a test tube for long enough it will gradually con-vert to an equal mix of left- and right-handed molecules—a process called **racemization**. In fact this will happen at any finite temperature, although much more slowly in colder environments.

Life on Earth operates almost exclusively with left-handed amino acids and right-handed sugars. Why this should be so is arguably one of the biggest, and most fascinating, unanswered questions in biol-ogy. It certainly seems entirely reasonable that life operates with one handedness in order for the complex and interdependent chemistry of *populations* of organisms to work. However, there does not appear to be *any* simple chemical or energetic argument to favor left- over right-handed biotic chemistry. Over the years various claims have been made to suggest that there *is* in fact a universal bias towards left handedness. These include arguments for a deep, quantum-level, energetic "symmetry breaking" whereby left-handed chemistry results in a minutely lower final energy level and is therefore favored. While appealing in some ways there has been no verification of such ideas. The more mainstream schools of thought fall into three principal categories:

- Terrestrial conditions on the pre-biotic Earth created local "patches" of left and right chemistry—for example stemming from calcite surface chemistry. Whichever region first produced life would them simply come to dominate.

- In a similar vein, life could have originated in both left- and right-handed forms and then by simple random chance, and natural selection, one or the other would have grown to dominate. Since the chemistry of life and all organisms is so interconnected this would be a runaway process whereby a small excess of left-handed resources (e.g., amino acids) would make it increasingly favorable for life operating with that handedness.

- The late-time impactor chemical "wash" that the Earth received might have been rich in left-handed enantiomers due to some astrophysical phenomena in the interstellar/proto-planetary chemistry, and therefore it would be inevitable that life would wind up left handed.

None of these solutions is completely satisfying. They still don't fully answer, for example, the question of why life on Earth doesn't exist of both left- and right-handed organisms—even if the right handed are a tiny, isolated, minority. The seemingly obvious answer to that question is that the Earth's resources just don't allow for two chemically incompatible biota occupying the same niches (or even different parts of the planet). But this has many potential loopholes—for example, if dead organisms will, through natural racemization, return to an equal mix of enantiomers then they are not sequestering resources much beyond their lifetimes.

Intriguingly, measurements of the enantiomeric content of some carbonaceous chondrite meteorite amino acids have suggested a small, ~10% *excess* of left-handed amino acids (although these results are not without question). If true this suggests an extraterrestrial bias, and supports the third category in the above list. Given the estimated age of the amino acids in, for example, the Murchison meteorite (based on the D/H measurements), considerable racemization would be expected, so any excess today suggests a high left-handed purity originally.

What astrophysical phenomena could produce a excess of left- (or right-) handed enantiomers in interstellar chemistry? One possibility is that circularly polarized ultraviolet light, which has been shown to "flip" chiral molecules between different handedness at a small, but not insignificant rate, could irradiate material in interstellar space. We know that some astrophysical sources can produce strongly circularly polarized light - for example, reflection nebulae in star forming regions (e.g., Bailey et al. 1998).

If then our solar system formed from a particular cloud of molecular gas and dust which had been irradiated with one sense of circularly polarized light it might then contain an excess of left-handed molecules which ultimately led to a bias in the pre-biotic and early biotic chemistry on Earth. This has several broader implications. For example, we might then reasonably expect that life (extinct or extant) on Mars would share our left-handedness. If it *didn't* that would point towards a different origin for homochiralty.

8.4.2   Impact Events on the Earth

Extensive impact cratering is evident throughout the solar system, from our own Moon to almost all objects with solid surfaces. Indeed, the level of observed impact cratering is a useful tool in dating surfaces (see §8.4.3 below). Youthful surfaces that have been, or are still in the process of being, renewed and altered by local geophysics should clearly exhibit less cratering. Here on Earth it can be surprising to learn that there are over 170 confirmed impact structures on the present-day dry land mass of the planet (and an essentially unknown number in oceanic regions), ranging from tens of meters in diameter to hundreds of kilometers. In Figure 8.5 a few examples are given, ranging from ~1 km to ~300 km in diameter and ages of ~50, 000 years to ~2 Gyr.

(a) Barringer Crater, USA
Diameter: 1.2 km
Age: 49,000 years

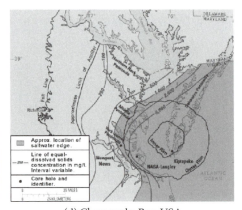

(b) Wolf Creek, Australia
Diameter: 0.9 km
Age: 300,000 years

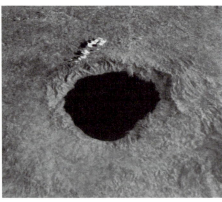

(c) Bosumtwi, Ghana
Diameter: 10.5 km
Age: 1.3 million years

(d) Chesapeake Bay, USA
Diameter: 90 km
Age: 35.5 million years

Figure 8.5. Six examples of terrestrial impact craters (see also the online Earth Impact Database: http://www.unb.ca/passc/ImpactDatabase/). Over 170 impact structures have been confirmed on the surface of he Earth, ranging from as small as 15 meters in diameter to an estimated 300 kilometers. Many are directly identified from their visual appearance, others (especially those associated with sea/land boundaries) require extensive geological detective work to confirm their nature. Image credits: (a) USGS, (b) Photo by Dainis Dravins, Lund Observatory, Sweden, (c) NASA/Landsat, (d) USGS, (e) NASA, (f) NASA.

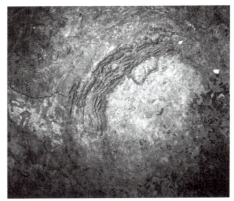

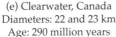

(e) Clearwater, Canada
Diameters: 22 and 23 km
Age: 290 million years

(f) Vredefort, South Africa
Diameter: 300 km
Age: 2.02 billion years

Figure 8.5 *(continued)*.

The present-day rate of impact events can be approximated by a consideration of the known distribution of so-called **Near Earth Objects**, or NEOs (and ultimately calibrated by observation of other bodies, see below). These are asteroids and comets with perihelion distances of less than 1.3 AU. In the case of comets, they are further constrained to have orbital periods of less than 200 years (i.e., short-period comets). Their discovery and orbital determination are an ongoing activity. At present more than 4000 NEOs are known, including over 700 objects with diameters larger than 1 km.[1] We can consider the present population of NEOs as representative of the objects that are most likely to impact the Earth.

The distribution of NEO sizes is given as (approximately, Collins et al. 2005)

$$N(> L) \approx 1148 L_{\text{km}}^{-2.354}, \tag{8.1}$$

_____

1. See for example the NASA/JPL NEO search programs, http://neo.jpl.nasa.gov /neo/.

where N is the number of objects with diameter greater than L in km. Present estimates of the probability of a single object in the NEO population colliding with the Earth are $\sim 1.6 \times 10^{-9}$/year. To first order, the impact energy is just the kinetic energy of the object: $E = (mv^2)/2 = (\pi/12)\rho L^3 v^2$, where ρ is the material density and v is the relative velocity of impact. Thus, multiplying the number of objects with a given impact energy by the probability of collision yields the present **recurrence interval** T_{RE} (i.e., the time between impacts) for an impact of initial energy E_{mt} in *megatons* (where $1\,\text{Mt} = 4.18 \times 10^{22}$ ergs):

$$T_{\text{RE}} \approx 109 E_{mt}^{0.78}, \tag{8.2}$$

where T_{RE} is given in years. For a rocky asteroid NEO a typical density is some 2000–3000 kg m^{-3}, and a typical impact velocity is some 12-20 km s^{-1}. Thus, a 1 km diameter asteroid may have a pre-impact energy of some $2 - 8 \times 10^4$ megatons, and would be expected approximately every 230,000 to 730,000 years. Translating the properties of an impacting body into a quantitative description of the resulting cratering effect on a planet is a highly complex problem, involving a thorough understanding of both the atmospheric passage of the body and the mechanics of the impact itself. Nonetheless, based on experiment and computer modeling it is possible to establish a series of scaling laws and generic features that allow an estimation of the impact characteristics (e.g., see Collins et al. 2005). In Figure 8.6 the approximate impact rates and corresponding crater sizes as a function of original object size are given for the Earth. Both cometary (water ice dominated) and asteroidal bodies are considered. While cometary impacts are expected to occur less frequently, the higher typical impact velocities (~ 50 km s^{-1} compared to ~ 17 km s^{-1}) suggest greater impact energies. The most noticeable difference between cometary and asteroid derived impactors arises at the lower mass end of things, where the ablation and disruption in the Earth's atmosphere remove cometary bodies somewhat more efficiently than rocky objects.

8.4.3   Impacts and Extinction Events

Impacts capable of forming craters release a tremendous amount of energy (see above). Consequences include both "instantaneous" effects, such as fireballs, shock waves, and tsunami, and longer lasting effects such as the lofting of material high into the atmosphere (see Chapter 9)

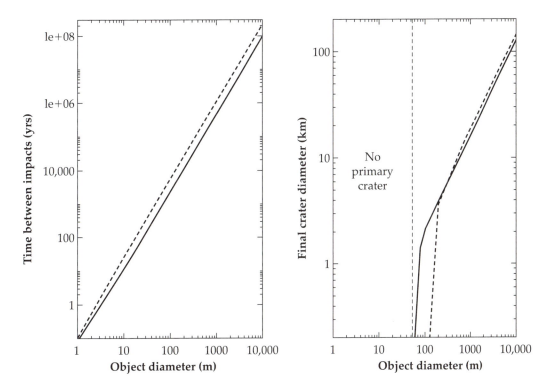

Figure 8.6. The approximate characteristics of Earth impact events. (a) Left-hand panel: The mean time between impacts by bodies is plotted versus original body diameter, based on NEO characteristics. Dashed curve corresponds to icy bodies with impact velocities of 51 km s^{-1} (typical for cometary objects). The solid curve corresponds to dense rocky bodies with impact velocities of 17 km s^{-1} (typical for asteroidal objects). (b) Right-hand panel: The estimated typical final crater diameter is plotted versus original body diameter for both comet-derived (dashed line) and dense rocky (solid line) bodies assuming the same velocities as in (a). The impact surface is assumed to be sedimentary rock and the impact angle is assumed to be 90°. Shallower angles reduce the final crater size. The vertical dashed line indicates the impactor diameter below which primary crater formation is unlikely due to atmospheric disintegration. (Numbers computed from the online impact calculator http://www.lpl.arizona.edu/impacteffects/, by H. J. Melosh, and G. Collins, see also Collins, Melosh & Marcus 2005).

and the widespread ignition of organic material in a terrestrial type (oxygen rich) atmosphere. In very massive impacts there is even the possible triggering of major volcanic activity. Effects such as these seem likely to have a relationship with the ability of pre-existing life to survive.

Table 8.1. The six largest known mass extinctions of species on
Earth. Estimated species loss is highly approximate
and includes different estimation techniques and does
not always differentiate between marine and land
organisms.

Period	Time (Myr ago)	Estimated species loss
End Precambrian	~650	70%
End Ordovician	~450	50%
Late Devonian	~360	70% (over ~20 Myr)
End Permian	~250	96% (marine) 70% (land)
End Triassic	~200	20%
End Cretaceous	~65	50%

On the Earth the fossil record of life indicates that there have been many episodes of apparent **extinction**, from individual species to entire swaths of organisms (Chapter 5). The largest known of such events[2] are summarized in Table 8.1 (e.g., Raup & Sepkoski 1982).

The possible causes of these extinction events are highly uncertain, and it seems likely that no single phenomenon is responsible for all. Nonetheless, the periodic impact on Earth of large bodies may well play a role in some cases. The most investigated example of such a potential connection is the End Cretaceous, or **Cretaceous–Tertiary** (known as **K–T**) boundary event that marks the end of the age of giant lizards, or dinosaurs. Two principal facts provide good evidence for the K–T boundary coinciding with a major asteroid impact (Alvarez et al. 1980). First, an enhanced **iridium** abundance exists in the sedimentary rock dated to this boundary. Iridium is a terrestrially rare member of the platinum-like metal family and would normally be expected to be slowly deposited on the surface of the Earth by micrometeorites. Instead, in rock of a clay-like origin at about 65 Myr ago there is a clear

---

2. There has been discussion that we are currently experiencing a mass extinction event, the **Holocene** event. A combination of the loss of species following the last ice age (~10,000 years ago) and the accelerating loss of species due to human activity since then suggests a comparable event to those listed in Table 8.1.

enhancement by more than a factor of ten in the iridium abundance. Such a jump suggests the rapid deposition of non-terrestrial material on a planet-wide scale, such as might be caused by a major impact. Second, following the iridium measurement, a major impact site, of a corresponding age, was identified beneath the shore of the Yucatan Penninsula (Chicxulub, Mexico) with a diameter of 150–300 km. This suggests an original impactor of approximately 10 km diameter—such a collision is expected to occur every 68 to 151 million years according to Equation 9.2.

An impact of this size would have a profound effect on the Earth, both on a short timescale and on a longer timescale owing to the effect on global climate (Chapter 9). There are many theories that attempt to explain the mechanical details of the subsequent, or contemporaneous, K–T extinction event. It may well be that this impact was solely responsible; it may have also coincided with a longer term change in the global population of organisms. In either case, it does provide good evidence that at least some terrestrial extinction events may have a connection to major impacts.

In terms of our seeking the potential for life on exoplanets then impact events must be considered to play some role. However, here on Earth we also see that following extinction events there is new experimentation by life, and a quite rapid re-filling of ecological niches—to the extent that it may even play a beneficial role in terms of the ultimate, long-term survival of life on a world (Chapter 5). Interesting questions are also raised when we consider the possibility of past life on worlds such as Mars. There is some evidence that the Martian climate was more temperate in the past, possibly to the extent that surface life (in the terrestrial mold) could have existed. However, the cratering rate seen on Mars (see below) is higher than that on Earth due to the increased number of asteroidal bodies crossing Mars's orbit. Major impacts could then have played an equally important role in the evolution of any hypothetical Martian biosphere, perhaps more so if the Martian climate was more fragile than that of the Earth (Chapter 9).

8.4.4 Impact Events on Other Bodies

On the Earth we are familiar with the general concept of soil—it is a highly complex material resulting from processes of erosion—often driven by water—and the incorporation of biological matter to form a matrix of particles. On other worlds, in particular those without significant surface solvents (i.e., liquid water) the equivalent of terrestrial soil is generally termed **regolith**. The regolith of a body is just the outermost "rocky" layer of material. In the case of (for example) the Moon there has been no erosion in the terrestrial sense; however, a thick layer of regolith exists, from several meters to over ten meters in the lunar maria and highlands respectively. The lunar regolith is a consequence of impact events over the ~ 4 Gyr lifetime of the moon. Large impacts result in deep and extensive fracturing and rearrangement of the lunar crust, smaller impacts churn over the outer few meters, and the smallest—sub-mm and micrometer—impacts effect only the outermost centimeter or so. Over billions of years these impacts have ground down the original lunar rock to a mix of fine dust, small rocks, and boulders. Thus, even in the absence of terrestrial-type erosion, bodies in the solar system are actively **weathered** by impacts over long timescales. The presence of an atmosphere serves to modify the rate and characteristics of this impact weathering. Put simply, smaller bodies stand less chance of directly impacting the surface if they have to first get through an atmosphere, due to drag and thermal or shock disruption (as seen in the bright trails of meteors). Not surprisingly the presence of a significant planetary atmosphere greatly modifies the cratering history of an object. As seen in Figure 8.6, on the Earth it effectively disrupts smaller objects that would otherwise produce extensive surface cratering. Atmospheric modification of the distribution of crater sizes is seen on Mars, and in fact it has even been considered as a means by which the atmospheric history of Mars might be probed, by serving as a proxy for the measurement of atmospheric density (Chappelow & Sharpton 2005). In Chapter 9 we also discuss some of the possible implications of atmospheric disruption of impactors for the habitability of a planet.

If the surface of a body (from asteroids to planets) is initially pristine following its formation (and solidification after any melting), or after any geological process that refreshes it (e.g., plate tectonics), then the actual number of observable craters can provide a first-order "clock" to evaluate its age. This further requires a model describing the variation of impact rates as a function of both time and location within the Solar System. The degree of cratering actually observed in the solar system ranges from zero on geologically active Io around Jupiter to **saturation cratering** on parts of the Moon and other bodies such as Mercury. If cratering is saturated then the entire surface is cratered, such that any new impact events will simply erase or modify existing craters.

Radioisotope dating on rocks retrieved from the moon during the Apollo missions[3] has been correlated with the observed surface density distribution of craters from the regions surrounding these samples. On the basis of such studies it can be established that the cratering rate was significantly higher between 3.5 and 4.5 Gyr ago—lending support to the notion of the so-called **late-heavy bombardment** (e.g., Chapters 3 and 9). From about 3–3.5 Gyr ago until now, the lunar cratering rate appears to have been significantly lower and essentially constant with time. As a rule of thumb it appears that within the orbit of the Earth (i.e., <1 AU) cratering rates are lower (c.f. Mercury, approximately 50% the rate of Earth impacts) and at larger radii the rates increase up to the orbit of Jupiter (e.g., Callisto has a rate some 200% of that of the Moon) and then appear to drop again to about 50% of the Earth rate for the moons of Saturn and Uranus. Using the lunar calibration, and these observed general rates, then we can immediately state with some confidence that the surfaces of some objects must be subject to significant renewal due to geological processes—e.g., the absence of craters on Io, the low crater numbers on the icy moon Europa, and so on.

8.5   Chemistry of the Proto-planetary Disk

As we have described above, the study of proto-planetary disks has, until recently, been based almost solely on the fossil remnants of our own disk—in the form of meteorites and comets, and even the giant planet atmospheres (c.f. the minimum mass solar nebula concept, Chapter 3). At the time of writing, the *direct observation* of true proto-planetary disks in other systems is now providing an enormous amount of new information. Here we give a very brief overview of the current status of our general model for proto-planetary disks.

---

3. A total of 382 kg of lunar material was returned to the Earth by the Apollo astronauts.

8.5.1 Basic Disk Structure and Chemistry

Compared with our earlier, theoretical consideration of the Minimum Mass Solar Nebula (MMSN, Chapter 3), observed disks can be very extended—up to a hundred or few hundred AU in radius (recall that this would readily contain the Kuiper belt region of our own solar system at some 50 AU), although many as yet unobserved disks might be smaller. In terms of surface density and midplane temperature, measurements tend to bracket the functional forms discussed in Chapter 3. Namely: surface density follows $\sigma(r) \propto r^{-p}$, where p ranges from 0 to 1, and midplane temperature $T(r) \propto r^{-q}$, where q ranges from 0.5 to 0.75 (compared to 2/3 from our very simplistic physical argument in Chapter 3). The actual physical value for disk surface densities ranges from some 0.1 to 10 g cm^{-2} at 100 AU and, sampling at 1 AU from the protostar, the disk temperatures are some 100–200 K, dropping to several 10's of Kelvin at ~ 100 AU (see Bergin et al. 2007, Kitamura et al. 2002). The total mass of proto-planetary disks is seen to range from 0.001 to $0.1 M_\odot$.

Although we have discussed the *radial* variation of disk surface density (σ, Chapter 3) we have not quantified the *vertical* distribution of gas. If we recall Equation 3.19 and make the substitution $\rho = \mu P / kT$ (assuming an ideal gas) we then obtain

$$\frac{dP}{P} = -\frac{\mu g_z}{kT} dz. \tag{8.3}$$

We can then define a vertical **pressure scale height** $H_z = kT/\mu g_z$ (c.f. Chapter 6) and integrate to obtain

$$P_z = P_{z0} e^{-\frac{z}{H_z}}. \tag{8.4}$$

Thus, gas pressure drops off with vertical height in the disk. Since $g_z = GM_* z/r^3$ then clearly H_z *increases* with distance from the central object and thus the pressure also drops off more slowly with height at larger radii. For an isothermal disk, the gas density (ρ) is found to actually drop off with vertical height z like a Gaussian distribution whose width increases with r. As a consequence, proto-planetary disks exhibit a **flared** structure—shown in Figure 8.7.

Complexity in the thermal structure of the disk comes largely from the fact that it is irradiated by both the nascent proto-star and the sur-

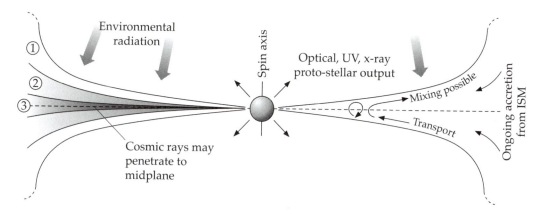

Figure 8.7. Schematic illustration of the structure of a proto-planetary disk. The disk exhibits a flared structure with radius owing to the requirements of vertical hydrostatic equilibrium. The disk at and beyond ~ 100 AU can be divided into approximate zones according to temperature and density. Region (1) is a photon-dominated layer, (2) is a warm molecular layer, and (3) is the midplane "freeze-out" layer. The surface layer (1) is irradiated by UV photons from both the proto-star and the external interstellar radiation field. In the warm molecular layer (2) the temperature is typically several 10's of K and the density is $> 10^6$ particles per cm^3. This layer is therefore shielded from most of the UV photons and molecules can exist without being dissociated. Water can remain on grain mantles in this environment (Chapter 7) and extensive chemistry can occur. At the midplane, temperatures are typically ~ 20 K, which is lower than the freeze-out temperature of carbon monoxide, which is highly abundant. Significant gas-phase depletion is therefore expected in this layer. Within 100 AU the midplane temperature rises and the **snow line** for volatiles occurs at different locations (e.g., 3–4 AU for water in a solar-system analog). Thus different sublimation zones and different chemical zones are expected. Very close to the proto-star high-energy particles and X-ray photons can play a role in heating solid materials in the disk.

rounding environment (which may be, for example, a cluster of other forming stars, see Chapter 2). The radiation from the central proto-star is absorbed by dust grains at the surface of the disk. These are heated up and re-radiate the energy as thermal, or infrared photons, which then heat the interior of the disk to the extent that they can penetrate the optically thick material. Thus, contrary to what we might expect from gas in simple hydrostatic equilibrium, with no additional energy input, the inner and midplane disk temperatures are typically *less* than that at the disk surface due to being shielded. However, this is probably not true at smaller radii (less than a few AU from the proto-star). Here mass accretion onto the disk heats it up as kinetic energy is transferred, and so the disk temperature at the midplane can be larger (e.g., 1000 K). This central disk and the proto-star may be accreting as much as $\sim 5 \times 10^{-6} M_\odot$ per year—in accord with the discussion of "inside out" star formation in Chapter 2.

At large radii, beyond ~ 100 AU, and well away from the denser inner disk, the chemical structure of the disk can be broadly divided into *three layers* (Figure 8.7). In the surface, **photon-dominated region**, the irradiation of the disk by UV photons from the proto-star ionizes and photo-dissociates molecules and atoms. At intermediate heights in the disk a **warm molecular layer** exists. Here the temperature is a few 10's of K, and the density is high enough to maintain molecular species, even if some UV photons manage to penetrate this far down. It can be cold enough for water ice to be frozen onto dust grains, thereby trapping oxygen, but warm enough for carbon monoxide–rich gas (carbon monoxide is an abundant and volatile component of the interstellar medium). Thus, all manner of rich and varied carbon chemistry can occur in this layer. Deeper down, at the **midplane freeze-out layer**, the temperature is typically less than the freezing point in vacuum of carbon monoxide (20 K) and all manner of molecular species solidify onto dust grains, thereby depleting the gas (c.f. Chapter 7). Observation of protoplanetary disks at X-ray, UV, optical, infrared, sub-mm, and radio wavelengths yields a wealth of complex and fascinating data that help inform this picture of structure. In the infrared (for example 1–200 μm in wavelength) observations have demonstrated the presence of amorphous silicates, water ices, and CO_2 and CO ices, as well as good evidence for the family of PAH compounds discussed in Chapter 7. Gas phase molecules can been seen through a multitude of emission pathways (e.g., vibrational–rotational emission lines). At short wavelengths (e.g., X-ray) an extensive array of higher energy processes close in to the proto-stellar object are probed. At longer wavelengths (e.g., sub-mm and mm) the dust emission is mainly optically thin, and at temperatures of 20–30 K (corresponding to ~ 100 AU radii) both sub-mm and mm emissions are dominant and allow a probe of the dust disk properties.

Detailed observations of the effective temperature and luminosity of the central proto-star, as well as measurements of the luminosity of the central accreting disk structure (yielding an estimate of the mass accretion rate), provide a "clock" with which to gauge the evolution of the disk (e.g., Kitamura et al. 2002).

Within about 100 AU the midplane layer warms up enough for many of these icy dust grain mantles to sublimate. This is exactly the argument we introduced in Chapter 3 for a "snow-line" in the proto-planetary disk. It is also similar to the situation described for hot-core chemistry in the wider proto-stellar environment described in Chapter 7. In fact, not only will water have a snow-line, but of course so too will other volatile ices—at differing radii. Thus, as the disk evolves and material enters different zones, an extraordinary variety of sublimation and gas phase chemistry can occur. Smaller molecules can sublimate, become incorporated into larger molecules in the gas phase, and then re-deposit onto dust grain mantles.

8.5.2 Radiation and Chemistry

It is important to stress the key role that radiation plays in the protoplanetary disk chemistry, in addition to simply imparting thermal energy. In the less dense parts of the disk, cosmic rays can play a role in driving reactions of ions and molecules and the production of free-radicals. UV flux from T-Tauri stars can be intense, and is likely produced as accreted material is shock heated on the stellar surface. Because it originates at the center of the system, geometry dictates that it will irradiate the flared part of the disk and contribute significantly to dissociation of molecules. Proto-stars also exhibit intense X-ray emission, again due to violent events at the stellar surface. X-ray photons can ionize atoms and molecules, and create high-energy photo-electrons that heat the disk gas and collide with atoms and molecules creating excitations. In the latter situation, the excited atoms or molecules will then emit UV photons. Because the responsible X-rays can penetrate deep into the disk this results in a UV photon flux in layers otherwise shielded from stellar UV radiation.

8.5.3 Disk Chemistry and Planet Formation

With this very brief overview of disk chemistry in hand we can begin to glimpse a complete physical description of a protoplanetary disk and the stages leading to planet formation. Earlier, in Chapter 3, we knowingly sidestepped many things, including the question of exactly how microscopic dust turns into planetesimals. This is a tough and currently unanswered question. In part, this is due to the intricate way in which we think protoplanetary disks evolve. There are many connected, but different, mechanisms and phenomena at play—from dynamics to molecular chemistry. While tiny (micron scale) dust grains are colliding, coagulating, fragmenting, and settling, the disk itself is dynamically evolving - accreting material, transferring angular momentum outwards, and mass inwards. Simultaneously, the chemistry of the disk is evolving—driven by the radiation environment and the thermal structures in the disk. As the chemistry progresses, more complex molecules are produced. At the cooler disk midplane they are then deposited or frozen out onto dust grain mantles, and thereby preserved when the gaseous disk is eventually dispersed as the proto-star heats up. All things are connected though. For example, as grains coagulate the total cross section for interaction with atoms and molecules is decreased, and hence the disk chemistry will alter. Furthermore, with fewer small grains more UV photons will penetrate into the disk—again, the disk chemistry will alter.

It does seem, then, that before we arrive at a complete picture of the formation of planets, and in particular the surface chemistry that terrestrial worlds may be initially equipped with, we must understand the protoplanetary disk and dust chemistry. Intriguingly we have one set of "fossils" to work with (planets, meteorites and comets of our own solar system), and a burgeoning supply of data on the early stages of *other* protoplanetary systems—eventually these two ends must meet.

References

A'Hearn, M. F., et al. (2005). Deep Impact: Excavating Comet Tempel 1, *Science,* **310**, 258.

Alvarez, L. W., Alvarez, W., Asaro, F., & Michel, H. V. (1980). Extraterrestrial cause for the Cretaceous tertiary extinction, *Science,* **208**, 1095.

Bailey, J., et al. (1998). Circular Polarization in star-formation regions: Implications for biomolecular homochirality, *Science,* **281**, 672.

Bergin, E. A., Aikawa, Y., Blake, G. A., & van Dishoeck, E. F. (2007). The chemical evolution of protoplanetary disks, In *Protostars and Planets V,* eds. B. Reipurth, D. Jewitt, and K. Keil, University of Arizona Press, 751–766.

Chappelow, J. E., & Sharpton, V. L. (2005). Influences of atmospheric variations on Mars's record of small craters, *Icarus,* **178**, 40.

Collins, G. S, Melosh, H. J., & Marcus, R. A. (2005). 'Earth Impact Effects Program: A Web-based computer program for calculating the regional environmental consequences of a meteoroid impact on Earth, *Meteoritics & Planetary Science,* **40**, 817.

Davies, A. (2005). Cosmochemistry: A breath of solar air, *Nature,* **434**, 577.

Ehrenfreund, P., & Charnley, S. B. (2000). Organic molecules in the interstellar medium, comets, and meteorites: A voyage from dark clouds to the early Earth, *Annual Review of Astronomy and Astrophysics,* **38**, 427–483.

Kitamura, Y., et al. (2002). Investigation of the physical properties of protoplanetary disks around T Tauri stars by a 1 arcsecond imaging survey: Evolution and diversity of the disks in their accretion stage, *Astrophysical Journal,* **581**, 357–380.

Lauretta, D., Leshin, L. A., & McSween, Jr., H. Y. (Eds.) (2005). *Meteorites and the Early Solar System II,* University of Arizona Press

Lissauer, J. J., & de Pater, I. (2004). *Planetary Sciences,* Cambridge University Press.

Raup, D. & Sepkoski, J. (1982). Mass extinctions in the marine fossil record, *Science,* **215**, 1501–1503.

Robert, F. (2001). The origin of water on Earth, *Science,* **293**, 1056–1058.

Sakamoto, N. et al. (2007) Remnants of the early Solar System water enriched in heavy oxygen isotopes, *Science,* **317**, 231.

Trieloff, M., & Palme, H. (2006). The origin of solids in the early Solar System, In *Planet Formation: Theory, Observations, and Experiments,* eds. Klahr, H., & Brandner, W., Cambridge University Press, 64.

Problems

8.1 In addition to studying the D/H ratios in the solar system as a means to learn about the temperature environment and transport of material in the proto-planetary disk (as well as the origin of terrestrial water) other isotopic ratios are important. The possible enrichment of water (relative to the modern Earth) in the proto-stellar nebula with the isotopes ^{17}O and ^{18}O is a good example. Study of these oxygen isotopes is also used to investigate the origin of different bodies in the solar system. By using Davies (2005) and Sakamoto et al. (2007) as starting references, describe some of the uses of oxygen isotopic measurements in the study of the early solar system. What are some of the physical mechanisms that may be at play in setting oxygen isotopic ratios?

8.2 (a) The short-period comet 55P/Temple-Tuttle (about 1.5 km radius) has an orbital period of approximately 33 years and an orbital eccentricity of 0.982. Assuming that the comet is in thermal equilibrium with the solar insolation at any point in its orbit, and has a surface albedo of 0.1, compute the *difference* in its temperature between aphelion and perihelion.

   (b) Water sublimates rapidly in a vacuum at temperatures above about 170 K. The rate at which water is lost from an ice surface can be stated in terms of a *surface lowering* rate (i.e., cm s^{-1}) s:

$$s = \frac{P_{vap}}{\rho} \left(\frac{m}{2\pi RT} \right)^{1/2}.$$

   The gas constant $R = 8.31451 \times 10^7$ erg mol^{-1} K^{-1} at low temperatures, the density of porous water ice is $\rho = 0.92$ g cm^{-3}, and the molecular weight of water is $m = 18.01508$. The water vapor pressure over an ice surface is

$$\ln P_{vap} = \frac{-4.77 \times 10^{11}}{RT} + 28.9.$$

   Compute the depth of pure water ice that Temple-Tuttle could lose if it spends approximately one year around perihelion. Given the actual size of Temple-Tuttle and the fact that it still exists can you comment on its likely composition and the probable physical distribution and surface characteristics of any water it contains?

8.3 Describe the characteristics of meteorites belonging to the class of chondrites. Discuss how the composition and internal structure of chondrites can be related to the environments in the proto-planetary disk. What sets carbonaceous chondrites apart?

8.4 One of the radionuclide dating schemes used for meteoritic samples is that involving the decay of $^{238}_{92}$U to $^{206}_{82}$Pb.
   The age of a given sample is generally estimated using

$$t = \frac{1}{\lambda} \ln \left(1 + \frac{N_D}{N_P} \right)$$

   where λ is the decay constant, and N_D and N_P are the number of atoms of the daughter (decay product) and parent (original) isotopes in the sample. The decay constant λ is related to the isotopic half-life since $N_P(t) = N_P(0) \exp^{-\lambda t}$ and the half-life is therefore $t_{1/2} = \ln 2/\lambda$.

The decay of ^{238}U to ^{206}Pb actually takes a total of 14 decay steps, with intermediate radioactive isotopes at each step. The first step $^{238}_{92}$U $\longrightarrow ^{234}_{90}$Th is the slowest, with $t_{1/2} = 4.46 \times 10^9$ years. All other steps are more than 20,000 times faster.

Show that, over 4 Gyr, this still implies that the ^{238}U to ^{234}Th decay effectively controls dating made using ^{238}U.

For a chondrite sample that is 4.5 Gyr old what is the expected ratio between ^{238}U and ^{206}Pb atoms (assuming there was no initial ^{206}Pb content)?

8.5 Describe the nature of bio-chirality. Discuss the possible reasons for homochirality on the Earth. In "A search for chiral signatures on Mars," Sparks et al. (Sparks, W. B., Hough, J. H., & Bergeron, L. E. (2005), *Astrobiology*, **5**, 737) attempted to search for polarization signals in telescopic data of Mars. Review this paper, describing their methodology. Do you think this experiment was likely to yield a positive result even if surface life existed on Mars? What makes this a tricky experiment? You may wish to search for information on polarization signals seen in terrestrial situations.

8.6 You will need to access 2 web sites for this problem:
http://www.unb.ca/passc/ImpactDatabase/
and
http://www.lpl.arizona.edu/impacteffects/
    The first is a catalog of known asteroid/comet impact sites on Earth, together with crater sizes (diameters) and approximate ages. The second is an impact calculator that allows you to assess the physical effects of the impact of a certain sized object on the Earth.

(a) Using the impact database determine a mean impact rate (impacts per Myr) and a mean impact crater size (km) for the periods (i) present day to 100 million years ago, (ii) 100 million years to 2 billion years ago. For ages given as < inequalities just assume the upper limit that's listed. Impact rates should be converted to global rates assuming a dry land area that is 30% of the total surface area of the Earth.

(b) Assume initially that the rates calculated in (a) are not biased by geological weathering/removal of craters. Using the impact calculator in the second web site and assuming that the impactors over the Earth's history have been 50% icy comets and 50% iron meteorites by number, use the mean quantities from (a) to determine the total mass delivered to Earth by such objects over the past 2 billion years. You may assume typical impact characteristics (as listed on web site) and sedimentary rock impact sites.

(c) The Earth's outer crust has mass $\sim 4 \times 10^{22}$ kg. Extrapolating the answer in (b) to 4 billion years (assuming no change in impact rates) evaluate the apparent fraction of the crust composed on impacted material. The true answer (not the one you will get) is that about 1×10^{22} kg of the Earth's crust comes from impactors following the primary formation of the planet. By referring to your answers in part (a) can you give *three* explanations for why your estimated total impactor mass differs from the true value? You may also want to play around with the impact simulator to consider what happens to some comets or meteors upon entering the atmosphere.

8.7 Describe what we know about the typical structure of a proto-planetary/circumstellar disk. In particular, discuss the zoning of volatiles and the impact of high energy radiation.

CHAPTER 9

Habitable Zones

9.1 Introduction

Terrestrial life seems to require a mixture of order and chaos. The biochemical processes driving living organisms are not in thermodynamic equilibrium with their surroundings, yet organisms maintain a temporary steady state. Additionally, the external factors driving the evolution of an organism cannot vary too dramatically with respect to biochemical responses or thermal limitations. Neither can they be entirely absent, or natural selection will not equip an organism with the ability to withstand change, or even survive the propagation of tiny chemical mishaps resulting from quantum mechanical uncertainty. For at least these reasons it seems likely that planetary bodies offer the phenomenon of life a principal foothold in the Universe. In many cases, planets should be capable of sustaining environments where the full array of temperate molecular chemistry can occur, and where stability exists over periods commensurate with the needs of biological adaptation.

These are, however, somewhat qualitative characteristics. We would like to utilize some more exacting parameters in quantifying the suitability of a given world as a harbor for life. Arguably, the two most compelling specific requirements for life are the presence of liquid water and the presence of energy to drive metabolism. Invoking these requirements enables a more quantitative approach, but also immediately introduces enormous complexity into the question of the suitability of a given planet as a host for either nascent or established life.

365

In this chapter we will investigate some basic questions about the suitability of a planet for life, based on these very rudimentary—but demanding—criteria. We will also extend this discussion to consider far-reaching constraints on the suitability (or habitability) of an environment—such as the location of a system within its parent galaxy, or indeed the particular characteristics of our parent universe. However, the notion of a so-called **habitable zone** remains a rather loose thing. Habitable, in this sense, is shorthand for "capable of sustaining active life." Thus, a habitable zone may range from a lush, tropical planet, to a small aquifer of salty water lurking a kilometer down in the porous rock of an otherwise inhospitable world. The original concept grew out of historical considerations of temperature and climate, and it is there that we begin this discussion after considering the nature of water itself.

9.2   Water

The dipolar nature of the water molecule and its specific geometry (see Chapter 1) result in a particularly complex phase diagram (the physical configuration as a function of pressure and temperature). A sketch of this is shown in Figure 9.1.

There are several key features of the phases of water. The first is that water exhibits a primary **triple point**. At a pressure of approximately 661 Pa (about 0.006 atmospheres) and a temperature of 273.16 K *all* three major phases of water—ice, liquid, and vapor—can co-exist.[1] The second is that, below this pressure, there is no liquid phase for water—it condenses or sublimates directly between solid and vapor phases. Third, at higher temperatures the pressure controls which phase water exists at - increasing pressure will take water from being a vapor, to being a liquid, to being a solid. The only place where this doesn't quite hold is at temperatures close to 273 K. As we can see in Figure 9.1, over almost two orders of magnitude in pressure, the phase of water can stay more or less the same between the triple point and the point at which it must become a solid. The conditions on the Earth (shown in

---

1. Contrary to popular belief, this is not the only triple point that water exhibits. Several others exist between two different phases of ice and liquid water, and between three different phases of ice.

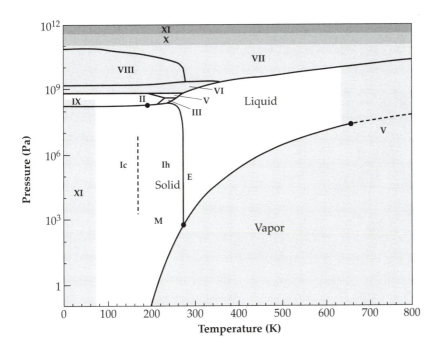

Figure 9.1. The phase diagram of pure water. The approximate surface conditions for the Earth, Mars, and Venus are indicated with labels E, M, and V respectively. The diagram is extended to higher pressures than typically shown to illustrate that above a few 10^9 pascals (Pa) of pressure (note that 1 atmosphere equals $\sim 1.013 \times 10^5$ Pa) it becomes increasingly hard for liquid water to exist. It is also worth noting that in the solid phase (ice) water has fourteen crystalline forms (not all seen here), with nine of these occurring at pressures greater than about 10^8 Pa. At the triple point all three primary phases can coexist. If Mars for example were slightly warmer (or if there are local "hot-spots") then liquid water, along with ice and vapor, could be present. At pressures below the triple point ice sublimates directly into vapor (figure based on http://www.lsbu.ac.uk/water/phase.html).

Figure 9.1) are such that small variations in local temperature can readily "switch" water between liquid and solid phases. Fourth, a **critical point** exists at approximately 646 K and a pressure of $\sim 22 \times 10^6$ Pa (about 200 atmospheres). At pressures and temperatures higher than this there is no physical distinction between a gas and a liquid. In this situation water behaves as a so-called **supercritical** liquid (e.g., as can be the case around hydrothermal vents, Chapter 5) which has chemically and physically *different* behavior than liquid water at our norm of 1 atmosphere. In this state, water molecules exist in small, liquid-like clusters that in turn behave as particles in a gas. Viscosity and compressibility change in this case, and water's behavior as a solvent also alters—switching from being a polar solvent to a non-polar solvent. Finally, it is intriguing to note that there are *fourteen* known phases of solid water—from ice-I to ice-XIV. These are due to different crystalline arrangements, from cubic to hexagonal (the norm on Earth), and tetragonal to rhombohedral, to name but a few. The different solid phases of water may become relevant when the interior structures of water-rich planets and moons are modeled (see Chapter 3). Most of these fascinating properties of water are ignored in our following treatment of habitability, but it is important to be aware of them.

The presence of highly soluble compounds can also dramatically alter the phase diagram of water. For example, ammonia (NH_3) dissolves readily in water. At a concentration of about 32% by weight it lowers the freezing point of water to a minimum of about 177 K at one atmosphere pressure. Although such a high ammonia concentration may create other difficulties for organisms, it is important to note that extreme environments may exist (e.g., the interior of Titan, see Chapter 10) where water molecules can nonetheless be mobile.

9.2.1 The Origin of Water on Terrestrial Worlds

A vexing, and still not fully resolved question is that of the origin of water on the Earth. By extension, this may provide some insight to the origin of water on planets elsewhere. Covering 70% of the planetary surface to a mean depth of about 4 km and with a total mass of $1-2 \times 10^{24}$ g, the present-day Earth harbors a significant water component. However, as we saw in Chapter 3, the Earth formed well within the circumstellar snow-line, and so there would have been little in the way of solid water to accumulate from the local proto-planetary disk.

Since we will equate liquid surface water with one of the principal tenets of habitability, it is important to understand where terrestrial type planets get their water from in the first place. The first serious considerations of this problem revolved around the theory that both the oceans and the atmosphere of the Earth originated from the volcanic outgassing of volatiles (i.e., water, carbon dioxide, etc.) from the interior of the planet (see Kasting 1993 for a review). This is really the secondary atmosphere; the true primordial atmosphere was likely accreted gas from the original circumstellar nebula, dominated by hydrogen and helium as for the giant planets (Chapter 3). However, the temperature of the early Earth, its relatively small gravity well, and the low mass of hydrogen and helium molecules would have resulted in the loss of this primordial gas (Chapter 6). While it is likely that outgassing contributed at some level to the later volatile content of the planetary surface it is now generally thought that the majority of volatiles were delivered directly to the surface during the later stages of planet formation.

By the time the forming Earth was about a third of its present mass, the enhanced infall velocity (i.e., that due to gravitational acceleration rather than just planetesimal/embryo relative velocities) would have been sufficient to ensure the partial or total vaporization of impacting bodies. From our discussions of planet formation and proto-planetary systems (Chapters 3, 8) we know that many of these bodies (planetesimals or planetary embryos) would have had a substantial volatile component—originating from beyond the snow-line in the circumstellar disk. Thus, the surface and atmospheric volatile content of the Earth could have been dominated by this vaporized deposition. The formation of the Moon (see §9.5.1) presents a potential difficulty for this picture of the water supply for the Earth. The impact of the lunar progenitor would have almost certainly re-vaporized any formed surface oceans on the proto-Earth, as well as likely stripped away much of the atmosphere. A possible solution to the problem lies in the late-time heavy bombardment (Chapters 3, 8) over the hundreds of millions of years spanning this period, combined with the later contribution of cometary bodies from the Uranus–Neptune region. Indeed, it has been suggested that such cometary material could have been a major source of the Earth's water (e.g., Chyba 1987). The tricky part to this is that we know from observations of comets such as Halley, Hale-Bopp, and Hyakutake that their deuterium-to-hydrogen ratio is approximately twice that of the Earth's oceans (Chapter 8). This can be understood in terms of the low-temperature formation of these objects and the subtle differences in chemical reaction rates between hydrogen and its heavier isotope. This strongly indicates that only about 10% of the Earth's water could be due to such distantly formed material. The majority of the Earth's water is therefore likely to have originated from bodies formed and processed much closer to the snow-line itself.

The final amount of water accumulated by an inner, rocky world may then be largely determined by the stochastic processes of late-stage planet formation. In this case the fact that Earth has substantial surface water may be sheer luck. Some numerical simulations of planet formation and volatile delivery (e.g., Raymond et al. 2004) indeed suggest that not all inner terrestrial planets end up water rich. However, those that do might also be much more water rich than the Earth.

9.2.2   *Early Oceans and a Cool Earth*

Some evidence suggests that the Earth has harbored significant reservoirs of *liquid* water since about 4.4 Gyr ago. This is a quite remarkable finding, and if borne out by future results also indicates that the surface temperature of the Earth 4.4 Gyr ago was not so very different from that today (e.g., see Valley et al. 2002).

As described previously, the early Earth prior to about 4.5 Gyr was both actively accreting and experiencing extensive heating from radioactive decay. The planet was also possibly "re-melted" during collision with a Mars-sized body at about 4.5 Gyr ago (see Chapter 3 and §9.5.1 on lunar origin below). It is not hard to imagine therefore that the Earth just after 4.5 Gyr ago was a hot place. However, the study of ancient **zircon crystals** indicates that by some 4.4 Gyrs ago the surface may have cooled dramatically. Zircon (zirconium silicate, $ZrSiO_4$) is a common trace mineral in rocks such as granite. Zircon crystals, or **zircons**, are typically small (a few hundred micrometers in size), extremely resistant to weathering, and survive temperatures that typically alter other rocks. This resilience has resulted in their identification (using radioactive dating based on traces of uranium and thorium isotopes in the crystals) as the oldest unaltered minerals on the planet, dating to some 4.4 Gyr ago. As discussed in Chapter 5, this predates the oldest rock formations known.

What makes zircons particularly interesting for the study of the early Earth is that the isotopic content of the oxygen in them can be used as a thermometer. Put simply, the lighter isotopes of oxygen are less likely to be chemically bound at a given temperature. Thus, measuring the relative abundance of a heavy isotope such as ^{18}O in a zircon can, by proxy, indicate the ambient temperature experienced by the ingredients prior to the formation of the crystal. Measurements made of ancient zircons have shown that the ^{18}O content varies little from 4.4 to 2.6 Gyr ago. Since we have excellent evidence for worldwide oceans 2–3 Gyr ago (from sedimentary deposits) then an obvious explanation is that the Earth was already cool enough for liquid water to exist 4.4 Gyr ago (Valley et al. 2002). It is important to note however that "cool" in this context means that surface temperatures can be constrained to be less than about 200 °C.

9.3   The Classical Circumstellar Habitable Zone (CHZ)

Owing to the central role that liquid water plays in life on Earth it seems only natural to require that a planetary body capable of harboring life has a surface environment that can provide it. As we have already seen in discussing the extremes of life on Earth, this is certainly a desirable circumstance, although it may neither be entirely necessary or sufficient to ensure the presence of life on, or in, a world. Liquid water can certainly exist beneath the surface of a world, where environments perfectly suited to life (for example prokaryotes) occur. Determining the surface environment of a planet (e.g., temperature, composition) is however a much more tractable problem than some of these alternatives. As we saw in Chapter 4, astronomical techniques allow us (in principle) to probe the external characteristics of a planet, including its atmospheric composition and response to seasonal variations (Chapter 6). For these reasons the classical approach has been to equate liquid surface water with suitability for life—or **habitability**. We will illustrate this more explicitly in the following sections.

If we start by simply assuming an ambient pressure of approximately 1 atmosphere then the presence of liquid water requires that a planetary surface must have a temperature $T_{surface}$ between 273 and 373 K.

As with our previous discussion of planetary emission and reflection, we can approximate a planet as a blackbody with a Bond albedo A, so that the distance of a planet from a star at which an equilibrium emission temperature T_p is attained is:

$$a = \left(\frac{(1-A)L_*}{16\pi \sigma T_p^4} \right)^{1/2} . \qquad (9.1)$$

Putting in values appropriate to the Earth–Sun system (i.e., $L_* = L_\odot$ and $A = 0.3$) we find that the temperature range 273–373 K corresponds to a distance range, or **circumstellar habitable zone (CHZ)**, of $a = 0.87-0.47$ AU. Since the Earth is outside of this orbital range, yet exists with a habitable surface temperature range, we are clearly missing something in our simple model.

What we are missing first and foremost is, of course, the *atmosphere* of the planet. An atmosphere is a major complication that dramatically modifies the energy processing and redistribution of a world. It also provides the pressure conditions that water requires to exist in the liquid phase. The detailed temperature structure and composition of a planetary atmosphere can be complex. In Chapter 6 we discussed some aspects of this, in particular the *vertical* pressure and temperature structure. Here we will look at some further aspects of planetary atmospheres that are of direct relevance to questions of habitability.

9.3.1 The Greenhouse Effect

In order to begin to understand the impact an atmosphere has on the thermal conditions of a planet we will consider the example of the Earth as a test case. Despite our earlier approximations, the Earth is of course *not* an ideal blackbody. It has significant variation in surface albedo, an absorbing and reflecting atmosphere, and a complex hydrological cycle—including extensive condensate structures (clouds) in the atmosphere (Chapter 6). However, solar radiation that does make it to the Earth's surface will heat the land or oceans much as for a blackbody. The surface then re-radiates energy approximately according to its blackbody emission temperature. We can therefore immediately appeal to Wien's Law to estimate the peak wavelength of this emitted radiation:

$$\lambda_{max} T = 0.29 \text{ cm K.} \qquad (9.2)$$

For a mean surface temperature measured for the Earth of approximately 287 K, the emitted spectrum peaks at ~ 0.001 cm or 100,000 Å, which is in the *infrared* part of the electromagnetic spectrum. Atmospheric gases such as CO_2, H_2O and CH_4 preferentially absorb/scatter radiation very strongly in the infrared (see Chapter 6). As a consequence, infrared photons will be scattered in essentially random directions at every level of the atmosphere. The net effect is that much of the emitted

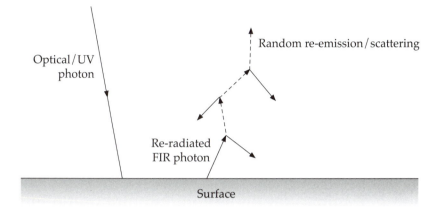

Figure 9.2. A sketch of the basic radiation processes taking place at the Earth's surface and within its atmosphere. Solar photons penetrate to the ground (ignoring cloud albedo effects), and some fraction are absorbed. The finite temperature of the Earth's surface results in blackbody emission peaked in the infrared. However, infrared photons are strongly scattered/absorbed in the atmosphere, resulting in a net downward flux that heats the planetary surface above the externally observed emission temperature. This is known as the greenhouse effect.

infrared flux from the surface will simply find its way back down, as well as heating the atmosphere itself. This creates a difference between the externally observed effective temperature (T_p) of the planet and its surface, or ground, temperature T_g. This mechanism is sketched out in Figure 9.2.

In order to have a slightly more quantitative grasp of this effect it is useful to understand the concepts of **optical depth** and **radiative transfer**. We will not devote a great deal of space to this, but once again, sketch out the essence of the idea.

9.3.2   Optical Depth and the Greenhouse

The concept of **optical depth** is fundamentally important in many areas of physics and astronomy and provides a useful means for describing the transfer of radiation through a medium.

There are different approaches to defining optical depth, and so we will take this opportunity to do it by also introducing the concept of **random walks**. Consider a photon traversing a medium (e.g., a gas) that

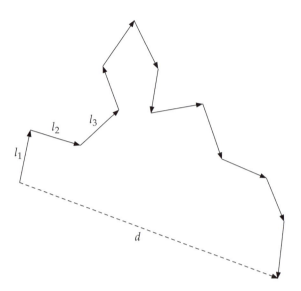

Figure 9.3. Illustration of the random walk taken by a photon in a scattering medium. Each step is of average distance l (the mean free path), but the net distance traveled after N steps is d.

scatters the photon (i.e., alters its direction but not energy). The photon can then undergo a random walk (shown in Figure 9.3). After N random "steps" the photon will have traveled a net distance d.

This can be simply described in terms of vectors:

$$\vec{d} = \vec{l}_1 + \vec{l}_2 + \cdots + \vec{l}_N. \tag{9.3}$$

Let us assume that the *mean* step length is l. l is often then termed the **mean free path** of the photon. This is entirely analogous to our discussion of planetesimal interactions in Chapter 3, so that $l = 1/n\sigma$ where n is the density of scattering molecules/particles in this case and σ is a cross section of interaction between the photon and the molecules/particles. We can now write

$$\vec{d} \cdot \vec{d} = d^2 = Nl^2 + l^2[\cos\theta_{12} + \cos\theta_{13} + \cdots], \tag{9.4}$$

where θ is the angle between two vectors. If each step is random then for large N all $\cos \theta$ terms must sum to zero, in which case

$$d = l\sqrt{N}. \tag{9.5}$$

We can then define an **optical depth** τ given by $\tau^2 = N$ for a medium, such that when $\tau \sim 1$ the characteristic (i.e., typical) physical distance travelled inside this medium is equal to the mean free path, that is, $d = l$.

If $\tau \gg 1$ then the medium is effectively opaque to photons; if $\tau < 1$ then, on average, some—but not all—photons will traverse the medium without scattering (i.e., the radiation is **attenuated**). In general τ is a complex function of the medium composition and the photon wavelength.

The intensity of the radiation will therefore be reduced in passing through the medium according to

$$I_{\text{final}} = I_{\text{initial}} e^{-\tau}, \tag{9.6}$$

Thus, for $\tau = 1$, radiation will in fact be attenuated by a factor e^{-1}.

A treatment of the infrared radiative transfer in an atmosphere such as the Earth's allows us to compute the actual flux difference between the surface of the planet and the top of its atmosphere, by using these concepts. For a relatively simple model (which ignores *convection*-driven energy transfer in the atmosphere) one arrives at (e.g., see Lissauer & de Pater 2004)

$$T_{\text{g}}^4 = T_{\text{p}}^4 (1 + \frac{3}{4}\tau_{\text{g}}), \tag{9.7}$$

where τ_{g} is the infrared optical depth to the ground from the top of the atmosphere. T_{p} is the effective, or top of the atmosphere, emission temperature of the planet. From direct measurement the Earth has an optical depth of $\tau_{\text{g}} \sim 0.83$ in the infrared, and thus:

$$T_{\text{g}} = T_{\text{p}} (1.62)^{1/4}. \tag{9.8}$$

From Equation 6.2 we can estimate $T_{\text{p}} \approx 255$ K for the Earth, and therefore $T_{\text{g}} \approx 287$ K—which is remarkably close to the observed mean surface temperature of the Earth! Here then is the **greenhouse effect**.

The infrared opacity of the Earth's atmosphere results in a 30–40 K increase in T_g from what it would otherwise be. Reapplying Equation 9.1 yields a new orbital range for "habitable" Earth-type planets around an $L_\odot$ star of $a \approx 0.55$–1.1 AU, and we are saved! For comparison, the thick atmosphere of Venus has $\tau_g \sim 60$, so very little solar radiation makes it all the way to the ground, and similarly, the far infrared emission from the surface is very much "trapped" by the atmosphere. Thus, the Venusian T_g of some 750 K (compared to its $T_p \sim 264$ K) is a direct result of its strong greenhouse effect. Mars on the other hand, with $\tau_g \sim 0.2$, has a very weak greenhouse effect, raising the surface temperature by only about 10 K to reach its measured mean of ~ 223 K.

As a point of interest, water ice transmits visible light to a depth of several centimeters, but is mostly opaque to infrared radiation. Thus a version of the greenhouse effect—the **solid-state greenhouse effect**—can occur, so that the subsurface of an ice crust may be significantly warmer than the outer, ice albedo would suggest. This may be relevant in situations that occur on icy bodies, such as the moon Europa (see Chapter 10).

The atmospheric content of greenhouse gases is, at least here on Earth, in dynamic equilibrium. It is also true that over the course of the Earth's history the amount of solar radiation received has varied according to the main sequence age of the Sun (e.g., Chapters 1 and 6). Thus, while we have defined an instantaneous circumstellar habitable zone, it is really a much more complex function of time. We will explore this a little more below.

9.3.3   The Carbon–Silicate Cycle

Carbon dioxide is the second most important greenhouse gas in the modern Earth's atmosphere (after H_2O which is responsible for about 60% of the terrestrial greenhouse effect), with a 26% contribution to the trapping of infrared radiation. (Other gases such as CH_4, N_2O, and O_3 make up most of the rest of the effect, with human-made compounds such as chloro-fluorocarbons adding in some 5–6%.) However, the atmospheric content of CO_2 is a function of one of the most important *long-term* feedback cycles on the planet. CO_2 abundance is naturally regulated by the **carbon–silicate** cycle, often termed just the **carbon–cycle** (Figure 9.4). This is a cycle (with a characteristic timescale of over 10^6

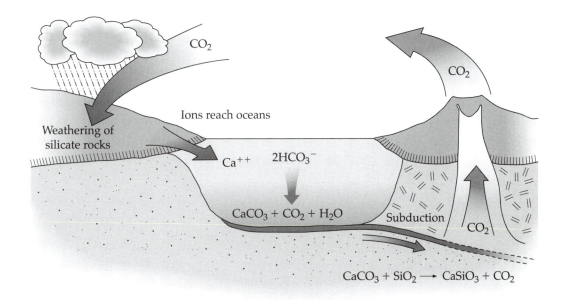

Figure 9.4. The carbon–silicate cycle on Earth. Atmospheric carbon dioxide dissolves in water vapor to produce carbonic acid which in turns weathers silicate rocks via reactions such as that of wollastonite summarized in Equation 9.9. The products of weathering (e.g., Ca^{++} and $2HCO_3^-$) are washed from continental land masses into the planet's oceans. Carbon is incorporated into calcium carbonate (on the modern Earth organisms play a major role in this on short timescales; however, the process will occur abiotically as well), which is ultimately incorporated into seafloor sediments. Plate subduction at the oceanic/continental lithosphere interfaces results in this carbon being driven into environments of high pressure and temperature where carbonate metamorphism (Equation 9.11) reverses the chemical processes of weathering and produces CO_2. Volcanic activity then returns the carbon dioxide to the atmosphere—and the cycle is complete. Equilibrium states appear to exist for a range of planetary temperatures, water contents, and tectonic activity such that the atmospheric carbon dioxide content is stabilized through negative feedback on $\sim 10^6$ year timescales.

years) that is intimately linked to the volcanic and tectonic activity of the planet, as well as its hydrological processes (Walker et al. 1981).

In very simple terms; atmospheric CO_2 dissolves in water to produce carbonic acid (a relatively weak acid); this in turn **weathers** (dissolves) rocks and minerals on the Earth's surface—e.g., during rainfall - locking up some of the carbon and oxygen in mineral form (see below). CO_2 is of course also removed from the atmosphere by the action of photosynthesis in organisms. However, even in the absence of organisms, CO_2 will be removed from the atmosphere through weathering. An example of this process is the breakdown of **wollastonite** ($CaSiO_3$):

$$CaSiO_3 + 2CO_2 + H_2O \longrightarrow Ca^{++} + 2HCO_3^- + SiO_2, \qquad (9.9)$$

and,

$$Ca^{++} + 2HCO_3^- \longrightarrow CaCO_3 + CO_2 + H_2O. \qquad (9.10)$$

Thus, the atomic components of atmospheric CO_2 are incorporated into silicon oxide (SiO_2, quartz) and calcium carbonate ($CaCO_3$, limestone). Again, organisms can also be responsible for producing limestone (e.g., marine life); however, they are *not* necessary for this process to occur, although they will clearly contribute greatly here on Earth.

What happens to this material? Most will eventually wash into the oceans on Earth and become incorporated into marine sediments. Over geological timescales a tectonically active world (such as the Earth) will eventually subduct this oceanic material under the continental crust and into the upper mantle of the planet (Chapter 5). The rate at which these combined processes occur on the present-day Earth is such that in about 400 million years all the carbon in the atmosphere and oceans would be removed. However, this is not a one-way process. Once subducted, the carbonate sediments are subject to high temperatures and pressures and **carbonate metamorphism** occurs such that the previous reactions are reversed:

$$CaCO_3 + SiO_2 \longrightarrow CaSiO_3 + CO_2. \qquad (9.11)$$

Thus, volcanism will eventually *return* this CO_2 back to the atmosphere (Figure 9.4). If present, organisms will also release incorporated CO_2 by decay (facilitated by bacteria), respiration, and combustion (given sufficient atmospheric oxygen). However, the feedback of CO_2 released by organisms operates on much shorter timescales.

Thus there is (with or without life) a fundamental carbon–silicate cycle on the Earth. Although this is a very simplified version of things, it is enough to let us see that there is potential for the carbon–silicate cycle to **self-regulate**. Consider the situation if the planet were to cool (perhaps due to some external effect). A cooler planet means more ice cover, and less effective weathering due to a decrease in liquid water precipitation and lower reaction rates due to lower temperatures. In this case the CO_2 removal rate would decrease, atmospheric CO_2 would increase, and the greenhouse effect would be enhanced. Thus $T_{surface}$ would be forced to increase, mitigating the cooling. How could the planet cool in the first place? The simplest would be if the stellar input were decreased, which could conceivably happen during the passage of the solar system through a denser region of interstellar gas or dust. Alternatively, the dynamical nature of the planet's climate (see below) could just result in it drifting towards a different equilibrium.

Conversely, if the planet were hotter, increased weathering would remove *more* CO_2 from the atmosphere and therefore lower the greenhouse effect. If tectonic and volcanic activity remains constant over long timescales, then again this would act to "regulate" the surface temperature.[2] Any such self-regulation implicitly requires active tectonics and the presence of liquid surface water. On a world such as Mars, with little or no present activity and no apparent liquid surface water, there would be little or no regulation through the carbon–silicate cycle. This is an interesting observation; there is sometimes talk of "terra-forming" Mars (making it habitable), but without somehow restarting Martian tectonics and volcanism, the long-term stability of such an environment is questionable. It is generally assumed that Mars has ceased major geological activity due to its lower mass ($0.1M_{\oplus}$), which results in lower radioactive heating rates (Chapters 3 and 8) and faster radiative heat loss. The atmospheric loss of water (Chapter 6) could also contribute since water may be an important tectonic lubricant.

The carbon–silicate **negative feedback** loop therefore appears to help keep things in check on the Earth by acting to keep global temperatures within the liquid water range. However, this raises the question of boundary conditions. Specifically, to what extent can this self-regulation

---

2. There are some (controversial) investigations of past tectonic activity that suggest that plate subduction might have halted at certain times, such as 1.5 Gyr ago, with far-reaching effects on climate.

maintain a temperate environment when external factors vary? To put this another way—the carbon–silicate cycle has the potential to extend the circumstellar habitable zone as determined by Equation 9.1, but by how much? There is some ambiguity in the criteria that one uses to determine the inner edge (lower orbital radius) limit, but the two that seem most applicable are (a) the surface temperature at which water vapor enters the stratosphere (see Chapter 6) and (b) the surface temperature at which the oceans evaporate altogether (Kasting et al. 1993). Of these, (a) is the more conservative (in the sense of a lower temperature). Once water vapor makes it into the stratosphere it is readily dissociated by UV photons. The light hydrogen atoms can then rapidly escape to space and thus the planet begins to "dry out." For the Earth this would occur if the stellar flux was about 10% larger than it is at present.

Where the strict outer edge of the circumstellar habitable zone occurs is also subject to some ambiguity about which physical criteria are the most important. The regulating power of the carbon–silicate cycle relies on the atmospheric CO_2 being in the gas phase. However, if temperatures get sufficiently low, then at certain altitudes the *condensation* of CO_2 can occur. This signals a profound change in the way a planet's environment behaves (e.g., Kasting et al. 1993). As CO_2 condenses it forms clouds, and these clouds increase the planetary albedo. In addition, as condensation occurs, then heat is released, and this influences one of the ways that energy is transported in the atmosphere (convection) and actually results in a *diminished* greenhouse effect. In combination then, the onset of CO_2 condensation appears to initiate a positive feedback loop— more radiation is reflected and less is trapped, temperatures drop even further, and more CO_2 condenses out. The continued supply of gaseous CO_2 by volcanoes simply fuels this process.

In all calculations of circumstellar habitable zones it is also important to consider that the precise conditions of a terrestrial-type planet are not just given by an "instantaneous" model. In the simplest terms, since a regulating mechanism like the carbon–silicate cycle operates on timescales of 10^6 years and longer, we must allow for at least this amount of time to see if a planet at a given orbital radius can settle to a habitable state. Many factors contribute to the surface environment, including atmospheric composition (see §9.3 below), and many of these are "non-reversible" phenomena (i.e., one cannot simply follow a cycle of processes and return to the same starting point). It is therefore critically important to consider the variation with time of planetary conditions and external forcings.

9.4   Habitability through Time

9.4.1   The Faint Young Sun Paradox and the Continuously Habitable Zone

As we have discussed before, the output of a Main Sequence star increases with time according to (for a star of the mass and composition of the Sun)

$$L(t) \approx \left[1 + \frac{2}{5} \left(1 - \frac{t}{t_\odot} \right) \right]^{-1} L_\odot, \tag{9.12}$$

where $t_\odot$ and $L_\odot$ correspond to the present age and luminosity of the Sun. The reason for this gradual evolution is that the central composition of the star changes with time—as hydrogen is processed into heavier elements. The central gas pressure therefore changes and the star must adjust to maintain hydrostatic equilibrium (see Chapters 1 and 2). This seems to present a difficulty for estimating the conditions on the early Earth. There is good evidence (§9.2.2, Wilde et al. 2001) indicating that 4 Gyr ago (give or take 0.5 Gyr) there were planet-wide bodies of liquid water. Whether or not these were oceans in the modern sense depends in part on the amount of dry land (continental crust, Chapter 5) and the net water volume. However, it seems likely that liquid water was ubiquitous, and that much of the Earth was temperate—as it is today. The problem with this observation is that, according to Equation 9.12, the Sun was about 30% *fainter* 3–4 Gyr ago than it is today. If all other things are kept at their present values, then the mean surface temperature of the planet would have been as much as 20 K *less* than it is today. Thus, instead of a balmy 287 K average surface temperature, the Earth could have had a frigid 267 K surface. Here then is the **Faint Young Sun paradox**—if the young Sun was fainter, how did the Earth manage to maintain surface conditions suitable for liquid water (and de facto life)?

This problem is still a matter of considerable debate. However, there are a variety of potential solutions. Foremost amongst these is evidence that the atmosphere of the Earth was profoundly different 3–4 Gyr ago, and thus the impact of the greenhouse effect was much greater. Two principal possibilities are that (1) the early Earth had a much higher atmospheric CO_2 abundance—possibly as much as 80% by mass—and (2) in addition, the atmospheric content of CH_4 could have been substantially larger. In the case of enhanced CO_2, the carbon–silicate cycle would again act to stabilize any environment around the point where liquid water was abundant. Enhanced CH_4 is a natural consequence of an early Earth dominated by methanogenic organisms (Chapter 5), suited to warm, oxygen free atmospheres (e.g., Kasting 2004).

Since the luminosity of a Main Sequence star increases with time, and the atmospheric content of a planet may vary with time, it is clear that the concept of a circumstellar habitable zone needs to be somewhat modified. In fact this gives rise to the definition of a **continuously habitable zone**. This corresponds to a range of orbital radii within which a planet would remain habitable over the entire Main Sequence lifetime of the star.

9.4.2   Modeling Habitability

It should now be apparent that to accurately predict either the circumstellar or continous habitable zone for a terrestrial type planet it is really necessary to construct a full model, or simulation, of the planetary atmosphere as a function of time.

To do this one would like to take a model with the sophistication of a full global climate model (such as is applied to the Earth, §9.4.5) and incorporate factors such as varying stellar input, atmospheric composition, volcanic outgassing, and so on. This is an *enormous* computational challenge, not least because of the uncertainties in how to even begin to model some of these characteristics for the *present-day* Earth (see below). What can be done in lieu of such sophistication is to construct relatively rudimentary models. These drastically simplify most aspects of a planetary atmosphere (such as radiative transport, circulation, and so on), but nonetheless provide some insight as to what factors have the greatest potential influence on a planet's surface conditions. In effect these models amount to a large set of coupled differential equations, which are then solved numerically—typically with time as the fundamental variable (see Hart 1979, Kasting et al. 1993).

Here we present a brief description of some of the phenomena that need to be incorporated into a model, in no particular order of importance.

- **Degassing**: The loss of planetary atmosphere due to diffusion. For the Earth the diffusion timescale is some 800 million years

- **Condensation of H_2O**: The variation of the boiling/vapor point (b.p.) of water with pressure P:

$$\text{b.p} = 373.15 \frac{(5.78 - 0.15 \log P)}{(5.78 - 1.15 \log P)} \text{ K,} \qquad (9.13)$$

  where P is measured in atmospheres.

- **Photodissociation and escape of hydrogen**: For example, H_2O is dissociated by UV photons in the atmosphere; hydrogen atoms can then escape from the atmosphere on a relatively short timescale.

- **Oxidation of surface minerals**: For example, iron compounds remove oxygen from the atmosphere:

$$2\text{FeO} + \text{O} \longrightarrow \text{Fe}_2\text{O}_3. \qquad (9.14)$$

Such a process has removed the present O_2 abundance in the Earth's atmosphere several times already over Earth's history.

- **Weathering**: The rate of CO_2 removal/return from/to the atmosphere as a function of planet age, T_{surface}, and composition. This is described by the carbon–silicate cycle (§9.3.3). The relationship between the amount of CO_2 in the atmosphere can be quantified through the *rate* of global weathering (W) relative to the present-day weathering rate on the Earth (W_0) via

$$\frac{W}{W_0} = \left(\frac{P}{P_0}\right)^{0.3} e^{(\Delta T/13.7)}, \qquad (9.15)$$

where P_0 is the partial pressure of CO_2 at the present time and global temperature (e.g., $T_0 = 287$ K), and P is the partial pressure of CO_2 when the global temperature deviates from the present value by $\Delta T = T - T_0$ (Walker et al. 1981). In any equilibrium it can be assumed that the rate of *release* of CO_2 by volcanoes (through carbonate metamorphism) (V) must be approximately proportional to W, thus $V/V_0 \approx W/W_0$.

- **Solubility of gases**: For example, N_2, O_2, H_2, H_2O, CO_2, CO, CH_4, NH_3, and Ar. (Why is argon in this list? Argon is 0.93% of the Earth's atmosphere today and actually increases with time since it comes from the radioactive decay of K_{40}. While it is an inert noble gas it is quite soluble in water and therefore contributes to the global cycle of gas exchange with surface water).

- **Solar luminosity evolution**: As described above (Equation 9.12).

- **Albedo**: The albedo must change with time, and be a function of terrestrial and atmospheric composition, for example:

$$A = f_{cloud} \cdot A_{cloud} + f_{ice} \cdot A_{ice} + f_{ocean} \cdot A_{ocean} + f_{rock} \cdot A_{rock}, \quad (9.16)$$

where f is the fraction of the planet surface covered/obscured by a given material or structure.

- **Greenhouse effect**: As described above, a given CO_2, H_2O, or CH_4 atmospheric content produces a given $\Delta T_{greenhouse}$ addition to the global mean surface temperature.

- **Initial conditions**: A set of initial conditions must be supplied to the model, for example the initial temperature, stellar luminosity, orbital configuration, planet size and so on. For the Earth it has often been claimed that the initial atmospheric conditions some 4.5 billion years ago that result in a best fit to current conditions might require an atmosphere of 85% CO_2, 10% CH_4 (and other reduced carbon compounds), 5% N_2, and trace amounts of other gases.

With such a model at our command we can ask fundamental questions about the extent of habitable zones. In Figures 9.5 and 9.6 two examples are shown. In Figure 9.5 the habitable zone for a $1M_\odot$ main sequence star (see Chapter 1) is plotted against time. In this plot the possibility for an initially "cold" planet (i.e., one that begins outside of the CHZ) warming up and becoming habitable is allowed. The inner limit assumes that stratospheric water loss is important (see above) and

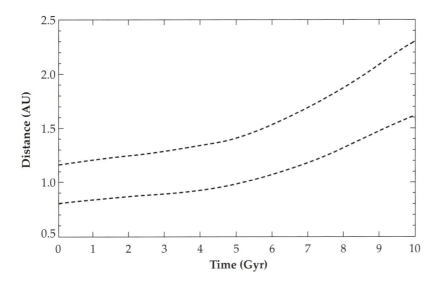

Figure 9.5. An example of the predicted evolution of the circumstellar habitable zone (CHZ) around a $1M_\odot$ star for an Earth-type planet, following Kasting et al. 1993. The upper curve corresponds to the outer limit of the CHZ assuming (conservatively) that the onset of CO_2 condensation in the atmosphere immediately initiates a positive feedback of increased albedo and decreased greenhouse effect. The lower curve corresponds to the inner limit of the CHZ assuming (again, conservatively) that once water vapor reaches the stratosphere, dissociation and rapid hydrogen escape to space result in the large-scale loss of water by the planet. Different assumptions (e.g., runaway greenhouse and maximum greenhouse effects) can broaden the CHZ by some 40–50% at any given time. It is also assumed in this plot that a planet can "cold-start," in other words, if it is initially beyond the outer CHZ limit it can still warm up as the stellar luminosity increases. The alternative is that a planet initially beyond the outer CHZ is never capable of "switching on" the necessary mechanisms for climate maintenance.

the outer limit assumes the initial condensation of CO_2 as critical. These are the most conservative limits, others (such as runaway greenhouse effects) tend to broaden the CHZ by 40–50%. A similar plot is shown in Figure 9.6 but now illustrating the difference in CHZs for stellar parents of masses ranging from 0.5 to $1.5M_\odot$. In this plot the different rates of stellar evolution (Chapter 1) are incorporated. It is interesting to note that at stellar masses just a little above $1M_\odot$ the increasingly rapid evolution of

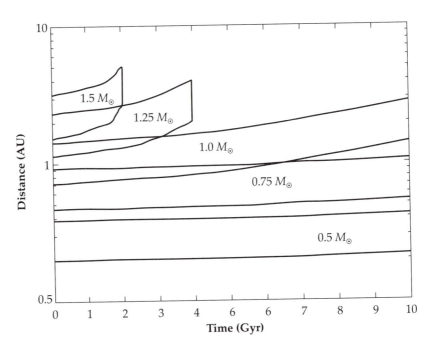

Figure 9.6. A further example of the predicted evolution of the CHZ is shown for a range of stellar parent masses. As in Figure 9.5, cold starts are allowed. The slower luminosity evolution with lower stellar mass is clearly seen in the almost unvarying CHZ for a $0.5M_\odot$ stellar parent, compared to the rapid evolution for a $1.5M_\odot$ star. In the latter case, by about 2 Gyr the star finishes its main sequence life, at which point the CHZ is much more uncertain.

stars results in a greatly shortened span of time where *any* CHZ can exist in a system. In all cases the CHZ moves outwards with time. Finally, in Figure 9.7 the width of the *zero-age* main sequence (Chapter 1) habitable zone is plotted against the mass of the stellar parent—together with the distribution of planets in our solar system. Also indicated is the approximate orbital range for the formation of terrestrial-type planets (Chapter 3) based on a minimum mass solar nebula scaling for the circumstellar disk. As would be expected, the CHZ moves inwards for lower mass, lower luminosity stars.

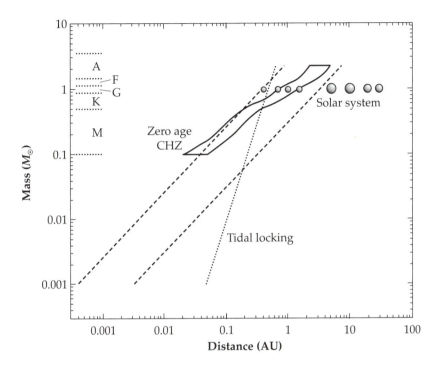

Figure 9.7. Plot showing the zero-age main sequence circumstellar habitable zone (closed region marked as CHZ) for an Earth-type planet (i.e., with a commensurate atmosphere and feedback processes) with different mass stellar parents—ranging from $0.1 M_\odot$ M-dwarfs up to $\sim 3 M_\odot$ A stars. Dashed lines indicate the approximate orbital range for terrestrial planet formation based on a standard minimum mass solar nebula scaled to the stellar parent mass. The dotted line indicates the radius within with a planet of rotation period 13.5 hours and specific dissipation function $Q \approx 100$ would be tidally locked into synchronous orbit after 4.5 Gyr. The location of the major planets in the solar system is also shown as filled circles. (based on Figure 16 in Kasting et al. 1993)

This creates another potential complication for habitability calculations. Within a certain distance of a parent star a planet will, over time, become tidally locked into a **synchronous orbit**. In other words, tidal dissipation of energy (see §9.5.1 below) will eventually result in the spin period of a planet equalling its orbital period (in the absence of any external forcings or resonances). The orbital radius at which this will occur after a certain time is given as (Peale 1977)

$$a_{\text{lock}} = 0.027 \left(\frac{P_0 t}{Q} \right)^{1/6} M_*^{1/3}, \tag{9.17}$$

where P_0 is the initial rotation period of the planet and $1/Q$ is the specific dissipation function of the solid body plus ocean.[3] Q ranges between about 13 for the present-day Earth (due to rapid dissipation in shallow seas) to a likely 100 for the early Earth, closer to a pure solid body value. The stellar mass is M_*. In Figure 9.7 $P_0 = 13.5$ hours, $t = 4.5$ Gyr, and $Q = 100$ are assumed. In reality a planet may never actually reach tidal lock due to external perturbations—such as the influence of giant planets in the system—but it is still a distinct possibility.

In such a situation a planet will have a permanent dayside and night-side (and eventually a circularized orbit). This can have profound effects on the redistribution of stellar energy on a world and its global climate (e.g., atmospheric circulation, §9.4.5 below). It is not obvious that a stable, temperate climate can result even if a planet is within the CHZ in this case. Thus, based on Figure 9.7, for stars such as M-dwarfs (between ~ 0.1 and $0.5 M_\odot$) it is unclear whether the simple definition of the CHZ is going to work. The lower temperature of an M-dwarf star (some 3700 K compared to 5700 K for the Sun) also results in a stellar spectrum that peaks in the red and near infrared (Equation 9.2). Phototrophic life on Earth has evolved to harvest photons predominantly from the visible part of the stellar spectrum (Chapter 6). Phototrophic life on a planet in the CHZ of an M-dwarf star would, at minimum, have evolved a different set of pigments for photon harvesting, and most likely different biochemical mechanisms for photosynthesis since the dearth of visible and UV photons would alter the available chemical pathways.

Another possible problem in defining the CHZ around lower mass stars is the question of their increased coronal activity compared to a solar mass star. Many (perhaps 40%) M-dwarfs are known to exhibit powerful **flare** activity. This is similar to the solar flares we see on the

---

3. Q is defined as $Q = 2\pi E_0 / \Delta E$ where E_0 is the peak energy stored in a cycle of a system and ΔE is the energy loss. Thus $1/Q$ is a measure of the energy lost, or dissipated, due to characteristics such as friction.

Sun (Chapter 1), but are much more frequent, and can include X-ray photon emission a factor of $\sim 10^4$ times more intense. This might present a problem both for the atmosphere of a planet in the CHZ of an M-dwarf, and for surface life. However, as we see on the Earth, what at first appear to be adverse conditions may not really be so.

Of course life itself would add another significant factor to the above models. Photosynthetic organisms and respiring organisms can produce carbon dioxide and methane, or oxygen (e.g., cyanobacteria). They can also absorb oxygen and carbon dioxide. Organisms can furthermore dramatically modify the surface albedo of a planet (e.g., forestation, marine algae blooms), and even influence the particulate matter in the atmosphere, with consequences for cloud formation and the atmospheric transmission of radiation. This is a bit of a thorny problem. It doesn't seem unreasonable to suppose that one could create a model world which would not be habitable *unless* life was in fact abundant, and busily modifying the climate (c.f. the early Earth problem described above). Conversely, it seems reasonable to suppose that an otherwise habitable world (according to our T_{surface} definition of habitability) might eventually be rendered uninhabitable by the presence of life, if some positive feedback processes were in effect. In fact, based on what we know about life on Earth, and in particular the history of life on Earth, it seems likely that abundant life is tightly entwined with the atmospheric composition and surface conditions of the planet through a variety of feedback systems.

What then is a planet modeler to do? Well, clearly one can start with a model for a given planet with no life and ask the question of whether that planet ever reaches the criterion for habitability, if it does then life could be "added" (in a crude way as described above, and as in Chapter 6) to see what the effects would be. Indeed, this is the prescription for using such models of the CHZ as a means to assess the potential for habitable worlds in exoplanetary systems.

9.4.3   Impact Events

Among the criteria for habitability we invoked in the introduction to this chapter was that an environment be stable over a timescale commensurate with biochemical processes, and with the evolutionary response time, or adaptation time, of organisms. There are many external events that can perturb a planet, from supernova explosions (§9.5.2 below) to passage through interstellar clouds that might reduce stellar insolation, to the impact events of asteroids or comets. We will take a brief segue here to explore some aspects of the latter.

Apart from the immediate, literally explosive effect of an asteroid or comet impact (see Chapter 8) there is a potential influence on the planet-wide environment over an extended period through the effect on the planetary atmosphere. The atmospheric disintegration of an asteroid or comet, or its impact on the surface, can introduce a **blocking medium** (e.g., dust) into the atmosphere that can increase the albedo of the planet. There are clearly other possible influences on the atmosphere, for example, a major impact will ignite large-scale surface combustion of any suitable material (e.g., plant life) in an oxygen-rich environment. This will release carbon dioxide and particulates that will also affect the energy transfer of the atmosphere. A very major impact also has the potential to trigger volcanic activity. Indeed, there has been some discussion that major impacts set waves in motion in the Earth's crust that propagate around the planet and come to a focus at the anti-impact location (i.e., 180 degrees away on the sphere). Such a phenomenon could trigger so-called "super volcanoes" (such as Yellowstone) that have a far-reaching effect on climate. Here though we will just consider the first, somewhat simpler case of opaque material injected into the atmosphere.

As discussed in Chapters 5 and 8, major impact events over the course of the Earth's history have almost certainly played a role in determining the evolutionary pathways taken by life. The mass extinction of major phyla (e.g., the **Cretaceous–Tertiary**, or **K–T** boundary 65 million years ago between the age of the reptiles/dinosaurs and the age of the mammals) appears connected to specific impact events (Chapter 8), and on a much more fundamental level may be connect to the ancient "deep life" (Chapter 5) of the planet.

9.4.4 Injection of Atmospheric Dust

Let us suppose that a comet suffers a high-altitude explosion or disintegration in the Earth's atmosphere (Chapter 8) and injects its dust content into the atmosphere. We could equally treat dust due to an explosive surface event, but that complicates the relationship between impactor properties (mass, composition, velocity) and dust quantity.

Prior to this event, the surface temperature of the Earth is related to the stellar flux incident at the top of the atmosphere:

$$T_{\text{surface}} \propto f_*^{1/4}, \tag{9.18}$$

where f_* is the stellar flux. If the cometary dust increases the optical depth of the atmosphere then f_* will be diminished by a corresponding factor $e^{-\tau}$ at the surface of the Earth (c.f. Equation 9.6). In other words, the dust-veiled surface temperature T_D will be

$$T_D = T_{\text{surface}} e^{-\tau/4}. \tag{9.19}$$

Now, we're going to perform a little lateral thinking and actually refer to geological and paleontological records to utilize this equation to explore the relationship between comet size and induced temperature drop. To first order let us assume $T_{\text{surface}} \approx 300$ K. Studies of the growth rate of trees over the past few tens of thousands of years, together with measurements of isotopic abundances, point towards occasional global temperature drops of $\Delta T \sim 3$ K, which might conceivably be associated with cometary impacts introducing blocking material into the Earth's atmosphere (e.g., Rigby et al. 2004; all of this is of course highly uncertain, but it serves as a means to play around with this problem). If we assume that this is indeed a correct interpretation then we can immediately see that we require $e^{-\tau/4} = 0.99$ and therefore $\tau \approx 0.04$ of additional optical depth in the atmosphere to cause a 3 K temperature drop on short timescales.

Since this is all in the spirit of illustration let us further assume that the mass of dust put into the atmosphere is half of the cometary mass $M_{\text{dust}} = 0.5 M_{\text{comet}}$ (i.e., the comet is a 50/50 mix of ice and solids). We can further write that the *number* of dust grains N_d of mass m_{grain} is just

$$N_d = \frac{0.5 M_{\text{comet}}}{m_{\text{grain}}} = \frac{0.5 R_c^3 \rho_c}{a^3 \rho_g}, \tag{9.20}$$

where R_c is the original comet radius (assuming it to be spherical, naturally), ρ_c and ρ_g are the mean densities of the comet and dust grains respectively and a is the dust grain radius.

If the comet is a mix of water ice ($\rho = 1\,\mathrm{g\,cm^{-3}}$) and carbon dust ($\rho \sim 2$ $\mathrm{g\,cm^{-3}}$) then $\rho_c/\rho_g \simeq 0.5$, and so $N_d \approx 0.25 R_c^3/a^3$.

We can now estimate the optical depth due to the dust by assuming that all the dust is distributed throughout the atmosphere and recalling that from the definition of optical depth $\tau =$ column density $\times$ cross sectional area $= dn\sigma$. Thus, globally averaged,

$$\tau \approx \frac{N_d a^2}{4 R_\oplus^2}, \tag{9.21}$$

where the factors of π cancel out and $R_\oplus$ is the radius of the Earth. Substituting the previous expression for N_d and rearranging we then arrive at

$$R_c \approx (16 R_\oplus^2 a\tau)^{1/3}. \tag{9.22}$$

Using the value of $\tau = 0.04$ and a dust grain radius $a = 10^{-6}$ m for a largish interstellar/cometary grain we arrive at an initial comet radius $R_c \approx 300$ m.

Now, as artificial as much of this working is, it does indicate that a relatively *small* comet of only 300 m radius could potentially lower the *global* temperature by 3 K! Models that attempt to assess habitability should therefore try to incorporate the effect of impact events, and in particular the longevity of their effects. It is not clear how long the dust of the above example could persist in the atmosphere, nor is it precisely clear how the planet's climate would respond to this perturbation. Our simple blackbody energy model is highly simplistic. Nonetheless, based upon current impact rates in our solar system, the Earth can expect to be hit by a 300 m comet approximately every 68,000 years. Extrapolating this back over the past 4 billion years then we find a *lower limit* (since impacts rates were significantly higher earlier in the solar systems history) of some 60,000 such events over Earth's history. Clearly then the classical circumstellar habitable zone could be rapidly nudged back and forth quite frequently in geological terms—depending on the impact details.

The potential variability of planetary surface conditions over these timescales has some important consequences not only for our consideration of the habitability of the Earth, but also of exoplanetary systems. Different orbital configurations, ages, and metal abundances will result in a wide range of impact histories. These will be very difficult things for us to ascertain remotely (Chapter 4). We should be aware, however, that it is entirely possible that any terrestrial-type world we eventually manage to obtain spectroscopic data from may not only be relatively younger or older than the Earth, but may also be in the throes of recovery from a major (or even relatively minor) impact event.

9.4.5   Climate: Features and Models

In our discussion thus far we have often referred to the "climate" of the Earth, or of terrestrial-type planets. However, we have not been very specific about this term, or about how one might model it in detail. Here we attempt to give a very short overview of some salient aspects and how these fit into efforts to determine both habitability and the observable characteristics of a terrestrial-type world. The climate system of the Earth (and therefore by extension, any similar planet) is defined as the totality of the atmosphere, hydrosphere, biosphere, and geosphere and their interactions (e.g., see McGuffie & Henderson-Sellers (2005) for an introduction). This is a little different from our usual, casual use of the term climate. In fact, the climate of a planet is really an extensive set of interrelated, **physically coupled** (i.e., a causal physical link exists) phenomena. An illustration of the network of processes and how they relate to each other is given in Figure 9.8.

The simplest first step to modeling climate is very much along the lines of our previous efforts to incorporate the greenhouse effect into an estimate of a planet's surface temperature, and to deal with the questions posed by Daisyworld in Chapter 6. These all make use of what would be termed low-order **energy balance models** or EBMs. These are zero, or one-dimensional (if latitudinal variation is included) models dealing with the energy inputs and outputs in an assumed equilibrium state. Models such as these can provide some understanding of, for example, the basic characteristics of planetary temperature according to atmospheric composition (see section above), albedo variation (Chapter 6), and varying stellar input (see section above).

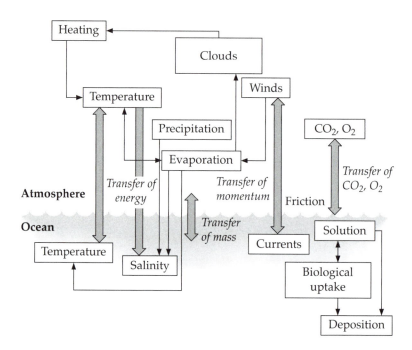

Figure 9.8. A schematic of the principal physical coupling mechanisms in a terrestrial-type planetary climate. The processes operating between the atmosphere and the oceans are of major interest and include (as indicated by the heavier connecting lines) energy exchange (temperature), ocean salinity, mass transfer (evaporation, precipitation), winds, and currents (see §9.3.3), and the exchange of soluble gases (CO_2, O_2). Adapted from McGuffie & Henderson-Sellers (2005).

The atmosphere of a planet (and oceans if present) is however not static, and plays a fundamental role in the redistribution of thermal energy. In Figure 9.9 a schematic of the basic, large-scale, atmospheric flow phenomena is presented for the Earth. These are quite generic features of a terrestrial planetary atmosphere and so would be expected in many circumstances. The **Hadley cell**, for example, is a closed circulation loop—like a donut encircling the planet. Around the terrestrial equator, warm, moist air is lifted aloft in low pressure areas to the **tropopause** (Chapter 6) and carried north or south. This occurs at the equator on the Earth because this is where, on average, the most solar radiation is

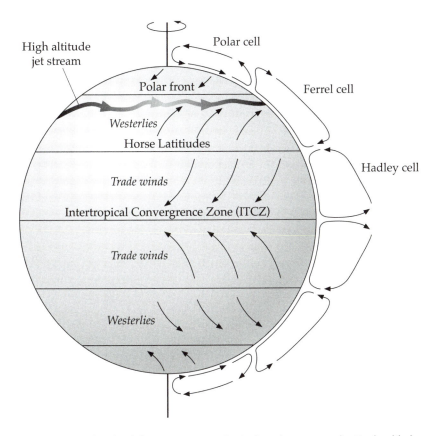

Figure 9.9. A sketch of the major atmospheric flow features on the Earth—likely typical of any terrestrial-type planet with a similar spin-axis orientation relative to the parent star. At the equator (in a region known as the intertropical convergence zone), hot moist air rises to the troposphere and is then carried north and south to cooler (lower pressure) latitudes. This air eventually cools and descends, returning to the equator along the surface. This circulation pattern is known as a Hadley cell. The rotation of the Earth results in a pattern of high and low altitude East–West and West–East air flows, giving rise to phenomena such as the "trade winds" and "westerlies," as well as the high-altitude jet streams in both hemispheres. Two other major circulation cells exist on the Earth, in the polar regions (polar cells) and between these and the Hadley cells (Ferrel cells). Together these circulation patterns act to redistribute thermal energy, water (vapor), and gases around the planet. The latitudinal extent of the major cells is a function of stellar insolation, atmospheric density, gravity, and many other factors.

received. At about 30 degrees latitude (which is specific to this world), this air drops in a cooler high pressure area. Some of this descending atmosphere then moves back to lower latitudes close to the planetary surface, thereby closing the loop of the Hadley cell. This flow creates several other important features of circulation. Both the high-altitude flow away from the equator and the low-altitude return flow (south and north of the equator) are subject to the **Coriolis effect** (sometimes called a "force"), which is a consequence of the rotation of the Earth and the conservation of angular momentum. Since the Earth rotates towards the east this results in the high-altitude flow (away from the equator) turning towards the east in both hemispheres, and moving faster as it reaches higher latitudes due to angular momentum conservation—creating the west-to-east **jet streams**. The low-altitude returning flow on the other hand slows relative to a non-rotating observer and gives rise to winds towards the west and the equator—known as **trade winds**. At both the equatorial region and at the downward edge of the Hadley cells at $\sim 30°$, surface winds can be minimal—giving rise to the wonderfully named equatorial **doldrums** and the sub-tropical **horse latitudes**. Further features include the **Polar cell**—which is not unlike the Hadley cell (see Figure 9.9)—the more complex **Ferrel cell** between the Hadley and Polar cells, and other circulation patterns that are a consequence of terrestrial geography and ocean arrangement. These major circulation systems on Earth act to transport energy away from the equator (thereby *lowering* the equatorial surface temperature) and deposit energy at the poles (thereby *raising* the polar temperature).

Thus, in order to determine even the globally averaged surface conditions of a planet with some real certainty we would need a model capable of reproducing at least some of these major atmospheric characteristics. Ideally we would also include the possibility of a hydrosphere (oceans), atmospheric chemistry, and so on. For the Earth, full three-dimensional models that attempt to simulate the coupled systems shown in Figure 9.8 have been constructed (usually termed **general circulation models** or **GCMs**). They rank amongst the most complex and intensive computer simulations undertaken by humans. They are also, by necessity, highly "Earth-centric," and tuned to very specific things, such as day and year length. To ask questions about the potential habitability of other terrestrial-type planets, where we may have little prior knowledge other than orbital parameters and stellar parent luminosity, it still makes sense to use simpler models.

9.5   Additional Factors

In the above discussions of habitable zones we have barely scratched the surface of considering the possible influences on habitability. For example, as shown in Chapter 4, the majority of currently detected giant exoplanet systems exhibit eccentric orbits. Any terrestrial-type worlds in these systems likely also exhibit eccentric orbits. The effect of strong eccentricity on the climate and habitability of terrestrial-type worlds is not fully understood—much depends on the nature of the planetary atmospheres and how the planet responds to strong variations in stellar insolation. As with all issues revolving around climate this can become a computationally intensive problem. On the Earth, the oceans play a major role in "damping" changes, by virtue of their heat capacity and circulation patterns. Modeling oceans and atmospheres together is a great challenge.

In fact, the variation with time of external irradiation may be one of the major factors to be considered for habitability. As we have suggested before, while the presence of liquid water may be one necessary component for a habitable zone, the *stability* of a system with respect to both short timescale biochemical processes, and longer timescale evolutionary processes, may profoundly effect the ability of living organisms to flourish. Here we consider a few of the other factors that come into play in our own solar system and are likely to be relevant to exoplanet systems as well.

9.5.1   *Lunar Tidal Influence*

The Earth–Moon system offers various insights to characteristics that may influence the habitability of terrestrial-type worlds. While it would be unwise to consider the Earth–Moon system as the norm, it is also not clear that it is particularly abnormal—in the absence of a knowledge of other terrestrial-type systems we should simply proceed cautiously with what we have!

The best current model for the origin of the Moon is that the early Earth (approximately 4.5 Gyr ago, based on the age of the oldest lunar rocks) suffered a collision with a body of a mass similar to that of Mars during late-stage planet formation (Chapter 3). The collision produced a plume of material in Earth orbit that coalesced into the Moon over a period of as little as a month. This material was dominated by differentiated mantle, and thus the Moon formed as a relatively iron-poor object—as observed. Extensive numerical simulations support this picture and in addition provide a good explanation for the unusually high (relative to other solar system objects) angular momentum of the Earth–Moon system. In fact, the observed angular momentum of the Earth–Moon system can be used to demonstrate the likely mass of the original impactor. The angular momentum delivered by an impactor is given by

$$L \approx 1.3 L_{\oplus - M} \, b \left(\frac{M_{\text{total}}}{M_\oplus} \right)^{5/3} \left(\frac{\gamma}{0.1} \right) \left(\frac{v_{\text{impact}}}{v_{\text{escape}}} \right), \qquad (9.23)$$

(following Canup 2004) where the mass of the impactor is given by γM_{total} (γ is the fraction of the total mass of the system that is in the impactor) and b is the impact parameter normalized to the sum of the impactor and target (Earth) radii, given as $\sin \theta$ where θ is the angle between the surface normal and the trajectory of impact (i.e., $\theta = 90°$ corresponds to a grazing impact, $\theta = 0°$ corresponds to a head-on collision). v_{impact} and v_{escape} correspond to the relative velocity of impact, and the escape velocity from the surface of the Earth respectively. Since the Moon is about $M_\oplus/80$ in mass we can assume $M_{\text{total}} \approx M_\oplus$. If we also assume that $v_{\text{impact}} \approx v_{\text{escape}}$ then we find that a grazing impact requires a *minimum* impactor mass of about $0.08 M_\oplus$ to match the observed angular momentum of the Earth–Moon system $L_{\oplus - M}$. Thus, an impactor of the order of $0.1 M_\oplus$ is required—about the size of Mars.

The close proximity of the Moon (0.00256 AU at the present) means that gravitational tidal effects are important. Tidal effects are the result of *gradients* in the gravitational fields. Consider the scheme illustrated in Figure 9.10. The force due to a moon on a small mass m contained within the planet is just $F_m = \frac{GMm}{r^2}$, the variation in this force is therefore just

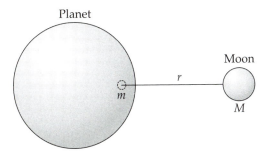

Figure 9.10. A schematic for the simplified arguments shown in Equation 9.24 for understanding the nature of gravitational tidal forces. For any small part (*m*) of a planet, a moon (*M*) will exert a gravitational force.

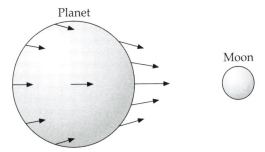

Figure 9.11. Ignoring self-gravity or the effect of rotation, a moon will exert a differential force throughout the planet.

$$dF_m = \frac{dF_m}{dr}dr = \frac{2GMm}{r^3}dr. \tag{9.24}$$

This is simply illustrated in Figure 9.11 for a *static* situation, ignoring the *self-gravity* of the planet.

If we include both the self-gravity of the planet and rotation (centrifugal forces) the net forces on the planet are shown schematically in Figure 9.12.

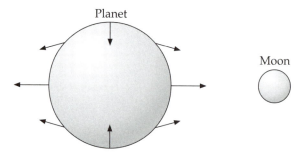

Figure 9.12. If self-gravity and rotation (centrifugal forces) are included, then in the simplest situation the net result of a moon's tidal force is to "stretch" the planet out as an ellipsoid—with a near and far tidal "bulge" (hence two lunar tides per 24-hour period in the case of the Earth).

The exact solution to this problem is mathematically quite cumbersome and we refer the reader to the literature (e.g., Murray & Dermott 1999). The net result is that forces act to stretch and compress *both* the planet and the moon (which is clearly also subject to the forces as described in Equation 9.24 above). These forces are responsible (together with tidal forces due to the Sun) for the oceanic tides on Earth and also contribute to the low level tectonic activity of the planet (e.g., Tolstoy et al. 2002). Put simply, a "bulge" is created (if material can move sufficiently) on the near and far sides of the planet (thereby explaining the existence of two high tides per day at any location on the oceans). As with any real dynamical system there is not only a transfer of energy (from gravitational potential to the motion of oceans and rocky material) but *dissipation* of energy. Friction results in a *finite response time* of the tidal bulge raised on the planet. If the planet rotation/spin period is *less* than that of the moon orbital period (assuming a pro-grade moon orbit) then the tidal bulge will lag behind the point at which it would otherwise be expected relative to the moon position. Figure 9.13 illustrates this situation. Put another way, by the time the tidal bulge reaches its maximum extent the planet has spun past the moon location above it. The actual lag time is complicated to calculate from first principles, since phenomena such as resonances come into play. However, it can

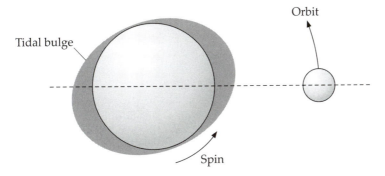

Figure 9.13. An illustration of the nature of the tidal bulge lag in the Earth–Moon system. The Earth's oceans are shown hugely exaggerated as the darkly shaded ellipse. The Moon moves in a pro-grade direction (i.e., in the same sense as the Earth's rotation) and raises the near and far side tidal bulges in the oceans. However, frictional forces result in the tidal bulges taking a finite amount of time to form at any given point. Thus, by the time the bulges have risen to their local maximum (to a stationary observer on the Earth) that part of the Earth has spun *past* the Moon, since the planet's spin period is shorter than the Moon's orbital period. This results in a gravitational torque between the Earth and the Moon.

be observationally determined and, for the equatorial oceans, the time difference between the water reaching maximum height at a location and the Moon being at the zenith of that location is about 12 minutes (or about 3 degrees on the sky).[4]

As a result, the planet exerts a $+ve$ gravitational torque on the moon (which in turn exerts a $-ve$ torque on the planet). The consequences of this can be understood by considering the total angular momentum of the planet–moon (Earth–Moon) system $L = I\omega$, where I is the moment of inertia (the rotational analog of moons) of the system. For a planet–moon system L is the sum of the planet rotational angular momentum,

---

4. The Sun also generates tides on the Earth, with approximately 45% of the force of the lunar tides, and must of course be taken into account to fully describe ocean tides and the long-term dynamical evolution of the system (see Equation 9.17).

the moon orbital angular momentum, and the lunar rotational angular momentum.

Note on derivation of moment of inertia: It may not be immediately obvious how to calculate the moment of inertia of a solid body. Here we give a useful example of how to tackle this problem. For an arbitrary object, the moment of inertia about an axis is defined as

$$I = \int_0^M r^2 dm. \tag{9.25}$$

Thus, for a point mass at a distance r from the axis, $I = mr^2$. Consider a unit thickness, uniform density (ρ) disk, rotating about a central axis. In this case $dm = \rho dV = \rho 2\pi r dr$ and so the moment of inertia of the disk is

$$I = 2\pi\rho \int_0^R r^3 dr = 2\pi\rho \frac{R^4}{4}. \tag{9.26}$$

Substituting $\rho = M/\pi R^2$ we obtain $I = \frac{1}{2}MR^2$ for the disk.

Now, to calculate the moment of inertia of a sphere we can simply treat it as a collection of disks along an axis z.

For a thin disk $dI = \frac{1}{2}y^2 dm = \frac{1}{2}y^2 \rho dV = \frac{1}{2}y^2 \rho \pi y^2 dz$. Then we can integrate for the whole sphere:

$$I = \frac{1}{2}\rho\pi \int_{-R}^R y^4 dz = \frac{1}{2}\rho\pi \int_{-R}^R (R^2 - z^2)^2 dz = \frac{8}{15}\rho\pi R^5. \tag{9.27}$$

Finally, substituting $\rho = 3M/4\pi R^3$ we obtain the moment of inertia of a uniform sphere: $I = \frac{2}{5}MR^2$.

The total angular momentum of the Earth–Moon system is then

$$L = \frac{2}{5}M_\oplus R_\oplus^2 \omega_\oplus + M_{\text{Moon}}\omega a^2 + \Lambda, \tag{9.28}$$

where the terms on the right-hand side of this expression correspond to Earth rotation, lunar orbit, and lunar rotation. The exertion of torques results in the transfer of energy from the Earth to the Moon, as well as the dissipation of energy through tides. We can see how this works by considering Equation 9.28 above. Let us assume that the lunar rotation remains constant ($d\Lambda/dt = 0$), which is actually a fairly good approximation as the Moon approaches **synchronous** orbit—it is tidally locked to the Earth's rotation (see below). From Kepler's laws we know that $\omega = (GM/r^3)^{1/2}$ and so making a substitution for the lunar orbit and differentiating with respect to time,

$$\frac{dL}{dt} = \frac{2}{5} M_\oplus R_\oplus^2 \frac{d\omega_\oplus}{dt} + \frac{1}{2} M_{\text{Moon}} (GM_\oplus a)^{1/2} \frac{da}{dt} = 0, \qquad (9.29)$$

and rearranging we obtain

$$\frac{da}{dt} = \frac{-4R_\oplus^2}{5M_{\text{Moon}}} \frac{d\omega_\oplus}{dt} \left(\frac{M_\oplus a}{G}\right)^{1/2}. \qquad (9.30)$$

Now, torque is $\tau = Id\omega/dt$ in general. So Equation 9.30 tells us that if the torque opposes rotation, i.e., $d\omega_\oplus/dt < 0$—as in the case of the Earth–Moon system then this implies that $da/dt > 0$.

Thus, in the pro-grade Earth–Moon system there is a transferral of energy from the Earth to the Moon (via the torque exerted by the lagging tidal bulge on the Earth) that results in the lunar orbital radius *increasing* with time. In the case of a retrograde moon orbit the moon would experience orbital decay and eventually spiral inwards to the planet (c.f. the Neptune–Triton system).

There are therefore several principal consequences of the Earth–Moon tidal interaction. First, the Earth's rotation rate decreases at a rate of about 0.0016 seconds per century, which will continue until the planet's rotation is synchronous with the lunar orbital period (just as the Moon is in synchronous orbit). Eventually (assuming no external sources of torque) the Earth–Moon system will reach its minimum energy configuration with both bodies in synchronous orbit and a 47×24-hour day length for the Earth. This also indicates that in the past, when the Moon was closer to the Earth, the tides experienced must have been significantly greater. Ancient sea–land ecosystems may have exploited this.

The second consequence is that the direction of the Earth's spin axis is stabilized, which we discuss below. A third consequence, which is still open to debate, is that the same tides that we witness on the surface of the planet must also affect the molten interior. It has sometimes been argued that lunar tides help maintain the internal dynamo of the Earth (Chapter 1) and contribute to the strength of its magnetic field—which in turn affects the degree to which the planet is shielded from particle radiation (Chapter 6). However, the internal heat of the Earth—which is thought to drive convective motions in the liquid parts of the core— appears to be enough to power the observed magnetic field. It may be therefore that the lunar tides have a secondary effect on the long-term state of the Earth's magnetic field. Finally, as mentioned above, it has been confirmed that the rise and fall of the oceans does influence the tectonic activity of regions on the seafloor (Tolstoy et al. 2002). This may be due to the "breathing" of water in and out of tectonic fault areas (Chapter 5), and possibly even influences the flow of nutrients between the subseafloor environment and the lower oceans.

9.5.2   Planet Obliquity

In the absence of any stabilizing forces the **obliquity**, or tilt, of a planet's spin axis will tend to vary and its orientation will precess over time due to interaction with the Sun and other planets (e.g., Jupiter and Saturn). Currently the Earth's obliquity is 23.5° relative to the plane of its orbit and the spin axis precesses around a complete circle every 25,800 years. Mars has a present obliquity of $\sim 25°$, while Uranus has a remarkable obliquity of $\sim 98°$—considered to be evidence of a large collision. As we know on Earth, the fact that the planet's spin axis is not perfectly normal to the plane of the ecliptic gives rise to the distinct seasonal variations between northern and southern hemispheres due to the resultant periodic changes in solar illumination. It also profoundly effects the redistribution of energy by the atmosphere and oceans.

For the Earth, the lunar tidal interaction prevents the spin-axis pre-cession rate from entering into resonances due to the orbital precessions of other planets (e.g., Jupiter and Saturn). This can be thought of in terms of a gyroscopic system subject to an external torque—the torque tends to act to keep the spin axis normal to the mutual plane. As a consequence the Earth's obliquity moves back and forth only some $\sim 2.5°$ over a period of about 41,000 years. This small variation in obliquity, together with changes in the eccentricity and precession of Earth's orbit has been used to help explain the coming and going of ice ages (lowering of planet temperature and increased polar glaciation) through **Milankovitch theory**. This theory states that the variations in stellar input at different latitudes due to these orbital and orientation changes are (at least in part) responsible for changes in global climate. An example of the various periodic variations that have to be incorporated in this theory is shown in Figure 9.14. As an aside: we have described in Chapters 1 and 4 how the majority of currently known exoplanet systems exhibit significantly larger orbital eccentricities than our own. If this property extends to terrestrial-type exoplanets then the variations in stellar input they experience will be highly significant and may dominate their climates.

By comparison to the Earth, the Martian obliquity is presently in a **chaotic** state (Laskar & Robutel 1993). As a consequence it shifts between $15°$ and $35°$ over a few 10^5 years and between $\sim 0°$ and $60°$ over a few 10^7 years. If the Moon did not exist in its present configuration and if the Earth spun slower than about a 12-hour period then it too would experience chaotic obliquity changes from $\sim 0°$ to $85°$ over periods of $\sim 10^7$ years.

Such a dramatic shift in spin axis over relatively brief periods must have a profound impact on the Martian climate—or indeed on that of any planet experiencing large obliquity changes. [5] Although we do not know the precise consequences for habitability, it seems reasonable to speculate that organisms on such a world would be subject to strong environmental shifts over timescales commensurate with their evolution. Models of Martian climate suggest that the major circulation patterns (i.e., the Hadley cells) can undergo radical change, thereby rearranging the transport of energy around the planet.

Finally, the approximate likelihood of the Moon forming at a time when the Earth's obliquity was $<25°$ and locking in this value is $\sim 1/12$. Together with the dependency of the chaotic obliquity state on the orbital configuration of the system *as a whole* this suggests that while the Moon appears to benefit climate stability on the Earth, it will be

---

5. The criteria dictating whether or not a planet's obliquity enters a chaotic regime are somewhat complex, and depend on both the spin rate (faster bodies tend to be more stable) and the configuration of perturbing bodies such as giant planets. See for example Laskar & Robutel 1993 and Waltham 2004.

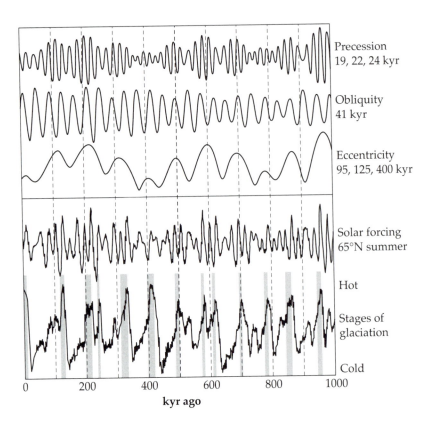

Figure 9.14. An illustration of the orbital and spin orientation variations of the Earth that may be responsible in part for driving periods of glaciation throughout the planet's history, according to Milankovitch theory. The upper three plots show the variation in spin axis precession, obliquity, and orbital eccentricity as a function of time, calculated from orbital dynamics. The fourth plot down shows the effective solar flux variation at a terrestrial latitude of $+65°$ during northern summer as a result of the orbital and spin variations. The lowermost plot shows the actual estimated variation of glaciation on the Earth at the same times. Interglacial periods (warm eras) are shaded rectangles. What is apparent is that while there is a generally positive correlation between warmer global climate and the peaks in stellar insolation, it is not a particularly simple relationship. Since global climate is a highly nonlinear system it is perhaps not surprising that there is no simple, point-by-point correspondence between the lowermost curve and estimated stellar insolation. Adapted from figure by Robert A. Rhode, Global Warming Art.

hard to extrapolate this argument in any general way to the habitability of extrasolar terrestrial-type planets. A more tractable problem is the issue of how climate actually responds to different obliquities.

9.5.3   The Influence of Giant Planets on Habitability

The presence of giant planets such as Jupiter (which is almost 0.1% the mass of the Sun) in a planetary system plays a significant role in shaping the environment of inner, terrestrial-type worlds. We have seen one in the previous section—the influence of Jupiter is felt even in the obliquity stability of worlds such as Mars or the Earth.

In the case of our own solar system, Jupiter also seems to play a "guardian" role by capturing, or ejecting, objects such as comets and asteroids on interior crossing orbits that might otherwise raise the risk of impact on the Earth.

In Chapter 3 we discussed how the orbital migration of giant planets might "sweep" a system of forming, inner, rocky worlds. As noted then, while such evolution may well lower the chances of long-lived terrestrial worlds it almost certainly does not eliminate them, and in simulations at least 30% of such systems appear to still harbor terrestrial-type planets. Furthermore, there are indications that the influence of giant planets (through so-called sweeping resonances, e.g., Nagasawa et al. 2005) can actually *enable* the final assembly of inner rocky worlds, by encouraging material to congregate at certain orbital positions.

9.6   The Galactic Habitable Zone

It should be clear by now that questions of circumstellar habitability can become highly complex as one considers more of the details— many of which do indeed profoundly affect the outcome for planetary climate and surface conditions. This prompts us to step back and ask whether we can pose rather simpler, more far reaching questions about the general likelihood of habitable worlds in the context of our entire galaxy during its history, which is estimated to span approximately the past 10 billion years. Rather than focussing on questions of water and energy availability, we can consider even more basic issues such as the likelihood of planetary bodies, and the hostility of their surrounding interstellar environment.

Consider the basic structure of the Milky Way Galaxy shown schematically in Figure 9.15 (also in Chapter 2). The stellar content of the galaxy is a function of location. In the halo and nuclear bulge most of the stellar population consists old, low metallicity stars. The thin disk contains the most active ongoing star formation and higher metallicity stars—such as our own Sun. This arrangement of stars suggests two principal criteria that we can adopt as being important for potential habitability in the galactic context:

- Sufficient heavy elements must exist in a region to enable the formation of terrestrial type planets. In other words, carbon, silicon, and other heavy elements must be available during planet formation.
- A sufficiently "calm" interstellar environment must exist to allow life to form. In other words, cataclysmic events such as supernovae must be infrequent enough to not sterilize or destroy all planets in their surrounding environment.

Together these criteria can be used to define a **Galactic Habitable Zone** (GHZ) (see for example Gonzalez et al. 2001, Lineweaver et al. 2004). In our own galaxy this amounts to a ring or annulus within the galactic thin disk, since the distribution of heavy elements and of massive stars (which result in supernova) averaged around the disk exhibit a radial dependency. Clearly the composition and structure of an entire galaxy are intimately linked to its formation and subsequent evolution—which is still an actively researched topic. We do however have solid observational measurements of many of these characteristics and can therefore, at least empirically, tackle the GHZ question.

Let us then consider the two main GHZ criteria in a little more detail.

9.6.1 Heavy Elements and the Goldilocks Zone

The criterion for sufficient heavy elements to enable the formation of terrestrial type planets can certainly be mapped out for our galaxy based on direct metal (in the astronomical sense) abundance measurements of stars. Since the stellar metallicity of main sequence stars is expected to reflect the metallicity of the original proto-stellar/proto-planetary nebula and disk, it serves as a direct proxy for the metals available in planet formation. In fact, in the galactic disk, the metal abundances appear to

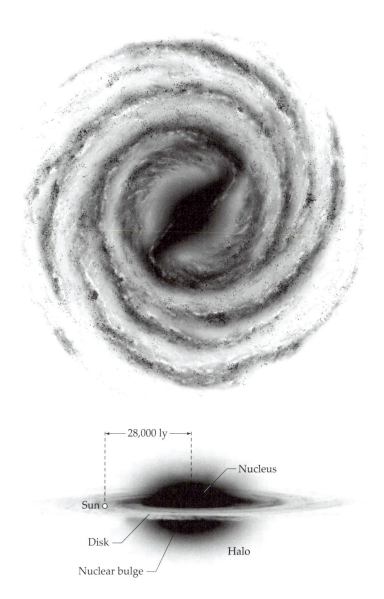

Figure 9.15. A representation of the visible (stellar and interstellar medium) structure of the Milky Way Galaxy. Uppermost is the disk of the Galaxy seen from above, showing the central concentration of stars and the spiral arms. The lower picture shows a high inclination view. The nuclear bulge of densely packed, but older and lower metallicity stars is surrounded by the disk where current star formation is occurring. The Sun is located some 28,000 light years (some 8.6 kpc) from the Galactic nucleus. Adapted from NASA/JPL/R. Hurt.

be directly proportional to the number of stars per unit volume, and so they drop off with radial distance from the center of the Galaxy. Current observations of exoplanet systems indicate that a higher metal abundance of stellar parents is associated with an increased likelihood of gas giant planetary offspring (see Chapter 4). Some 5% of stars with solar metal abundance appear to host giant planets, increasing to about 25% for stars with 3 times solar metal abundance. For stars with only a third of the solar metal abundance the number with planets drops to just a few percent. This suggests that a similar functional form could exist for planet likelihood as a function of galactic radius.

However, an increased likelihood of giant planets might also mean an increased likelihood of some of them orbiting close to the parent star - the population of hot Jupiters (Chapter 4). As we have seen, such planets must have undergone orbital migration, and as a consequence will have a high likelihood of disrupting terrestrial-type planet formation and even scattering formed planets out of systems. While this may not eradicate inner terrestrial-type planets (see Chapter 3) it may dramatically lower their numbers in such systems—even if they too are more populous with increased stellar metallicity. This suggests that there may be a **Goldilocks Zone**, where the metal abundance in the proto-stellar/proto-planetary nebula/disk is "just right," to allow for the efficient formation of terrestrial-type worlds, but only moderate giant planet formation and migration. It is important to emphasize that this is *highly speculative*. The purpose of making the discussion here is to illustrate how one can extend what we have learnt in earlier chapters to these more far-reaching arguments.

How do we place a quantitative estimate on something like the hypothetical Goldilocks zone? (Here we will largely follow the work of Lineweaver et al. 2004.) Current observations of exoplanets provide the number of stars (N_P) hosting "hot" (presumably migrated) giant planets as a function of the parent star's metallicity ([Fe/H]—in the usual astronomical nomenclature of a logarithmic ratio of iron to hydrogen, relative to the solar values). To keep things simple let us assume that the presence of a hot Jupiter implies an absence of terrestrial-type worlds. This is a gross assumption, but it keeps things simple to start with (see exercises at the end of this chapter for a suggested modification to this).

We can then write the relative probability of the *destruction* of a terrestrial planet as a function of metallicity:

$$P_D([Fe/H]) = \frac{N_P([Fe/H])}{N_*([Fe/H])}, \tag{9.31}$$

where N_* is the total number of stars searched for the presence of "hot" giant planets. For the sake of simplicity we further assume (arbitrarily) that the probability of initially producing a terrestrial planet (P_P) varies linearly with [Fe/H] and cuts off at very low metallicity—at one tenth solar abundance—and becomes unity at the highest observed metallicity for an exoplanet system. This currently has absolutely no basis in observation, and may be entirely wrong! (See exercises at the end of this chapter). Nonetheless, it simplifies the mathematics and lets us see how one can go about making a calculation like this for the GHZ. Thus

$$P_P([Fe/H]) \propto [Fe/H]. \tag{9.32}$$

The probability that a stellar system harbors a *surviving* terrestrial-type planet is then just

$$P_{TP}([Fe/H]) = P_P([Fe/H]) \cdot [1 - P_D([Fe/H])]. \tag{9.33}$$

Interestingly, this probability as calculated seems to peak at slightly higher metallicities than solar. In order to apply this to determining the potential for terrestrial-type planets in our Galaxy we need to map out the expected metallicity as a function of galactic position. The metallicity of a given region is of course directly related to the length of time that stars have been present, burning hydrogen to form heavier elements and dispersing them into interstellar space, as well as the local density of stars. The **star formation rate** as a function of time and location therefore provides us with this information if we assume it is linearly related to the creation and dispersal of elements. The situation for our galaxy is summarized in Figure 9.16.

9.6.2 Supernovae Are Bad?

Ironically, the very events that help to both produce and disperse the heaviest elements in the cosmos—thereby setting the scene for planet formation and the potential for life—are also sufficiently violent and energetic that they may serve to disrupt habitable zones. The question mark in the title of this section serves, however, as a reminder that when considering habitability there are many potential complications.

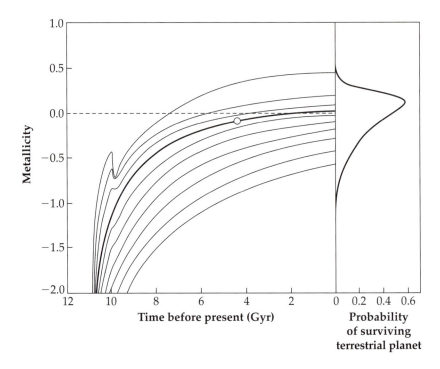

Figure 9.16. Plots of the local stellar metallicities used to derive an estimate of terrestrial planet likelihood. Metallicity ([Fe/H]) is plotted versus the time before the present (Gyr) for different distances from the galactic center. The uppermost curve corresponds to a galactic radius of 2.5 kpc, increasing in steps of 2 kpc to the lowermost curve at 20.5 kpc. The short-term peaks seen in the innermost regions at around 10 Gyr assume a 2-phase star formation model, where the inner Galaxy (bulge) has rapid early formation, before giving way to the slower, steadier star formation rate of the disk—seen in the lower curves. Heavy curve corresponds to the solar system, with the Sun indicated as a circle. To the right is a plot of the net probability of a surviving terrestrial planet (after giant planet migration) following Equation 9.33. This assumes "strong" disruption of terrestrial planets by giant planet migration (Chapter 3). Adapted from Lineweaver et al., 1004.

It is not inconceivable that past catastrophic events affecting the Earth, while undoubtedly causing mass extinctions of life and the destruction of habitats, might have also acted as catalysts for life to diversify into empty niches following such events (e.g., deep life, Chapter 5, and Chapter 8). While it is not pleasant to think that occasional destruction might be good for life, in the cosmic sense of life as an efficiency-seeking phenomenon, it may well be true.

Nonetheless, in the discussion below we will assume for now that supernovae are indeed bad for life. The reasons for this are as follows. The core collapse that can occur in stars of masses $\sim 8 M_\odot$ and up initiates a supernova explosion. The typical net energy release is some 10^{53} ergs over a short period. Of this energy, some 99% is carried away by **neutrinos**, some 1% by the physical blast wave of ejected material, and some 0.01% in the form of electromagnetic radiation. For the latter, most of the actual energy is in X-rays, gamma rays, and accelerated particles (e.g., electrons, protons), although of course the optical photon production is enormous. All three outputs have the potential to impact habitability of terrestrial-type planets. Even the neutrinos, with such a huge flux, represent a very significant source of damaging radiation. Although an electron neutrino cross section of interaction with normal matter is only $\sim 10^{-43}$, a supernova at a distance of 10 parsec would produce about 10^4 neutrino/nuclei recoil events *per kilogram of living tissue.*

The mean supernova rate in our Galaxy has been estimated at somewhere between 0.1 and 0.01 per year (out of a total of $\sim 10^{11}$ stars). Assuming that massive stars are made more or less in proportion to the formation of all stars, then the probability of a planet being impacted by a supernova is just linearly proportional to the neighboring space density of stars. We can then use a map of the space density of stars in the galaxy (specifically in the disk) as a proxy for supernova damage. Based on expected supernova rates, which are of course related to the star formation rate, a supernova "danger factor" can be constructed, normalized to the Earth (i.e., we have *not* been destroyed by a supernova during the past 5 billion years). This is illustrated in Figure 9.17.

9.6.3 Mapping the GHZ

Finally, combining all factors we can write a galactic habitability probability P_{GHZ} for *complex* (i.e., multi-generational, multi-cellular, eukaryotic like) life:

$$P_{GHZ}(t) \propto SFR \cdot P_{TP} \cdot P_{SN} \cdot P_{evol}, \qquad (9.34)$$

where SFR is the star formation rate at a given time, P_{TP} is the probability of a surviving terrestrial planet, P_{SN} is the probability of a world

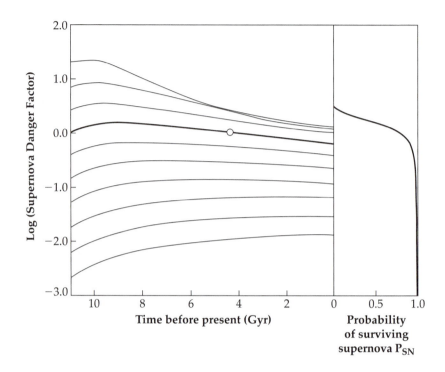

Figure 9.17. Plots of the "supernova danger factor" (Lineweaver et al. 2004) versus time before the present for different distances from the galactic center (uppermost curve corresponds to a galactic radius of 2.5 kpc, lowermost 20.5 kpc, in 2 kpc steps). Heavy curve corresponds to the solar system, with the Sun indicated as a circle. The danger factor is defined as an integral of the supernova *rate* (derived from stellar space densities) from the indicated time to 4 Gyr's earlier (this represents the required period to be supernova free) divided by the same quantity determined for the solar neighborhood. On the right-hand side the integrated probability of surviving a supernova for a given danger factor is given (P_{SN}).

remaining habitable after a supernova event, and P_{evol} is a rather broad probability based on the time taken to evolve complex life (in this case obtained by integrating a random (gaussian) distribution with a mean of 4 Gyr and dispersion of 1 Gyr, using the Earth as calibration).

Clearly this is a pretty audacious equation, and should be considered purely as an illustration of the reasoning involved rather than any type

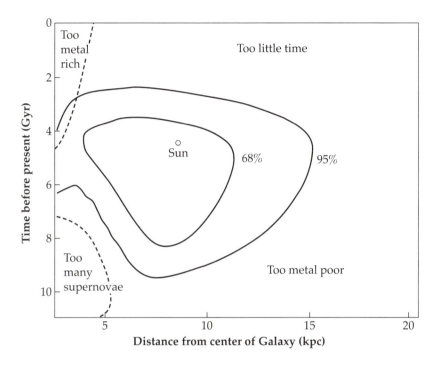

Figure 9.18. The galactic habitability probability (P_{GHZ}) map for *complex* life as a function of time (Gyr before the present) and distance from the Galactic center (in kpc). Solid curves indicate the regions encompassing 68% and 95% of the locations of the *highest* probability of habitable conditions according to Equation 9.34. Dashed curves indicate regions less likely to be habitable according to metal richness and supernova rates. Other less likely regions are labeled, due to a lack of time for complex life to evolve (based on an ad hoc assumption of 4 Gyr with a 1 Gyr gaussian spread), and due to too few metals (hence less likelihood of planet formation). The age and location of the Sun are marked. It is interesting to note that in this example the solar neighborhood is not quite at the peak of probability for complex life—this lies at slightly smaller galactic radius and with slightly older stars. Adapted from Lineweaver et al. 2004.

of definitive answer to these questions. In particular P_{evol} is a real shot in the dark, based solely—and crudely—on the apparent situation here on Earth. In Figures 9.18 and 9.19 the GHZ thus obtained is illustrated as a surface in time and galactic radial distance with, and without, P_{evol}.

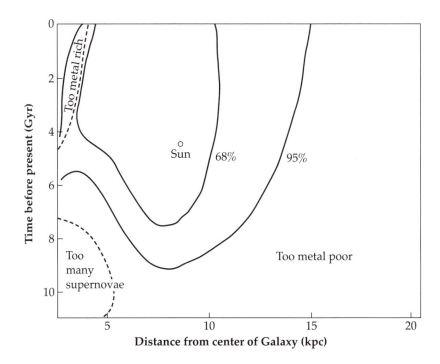

Figure 9.19. The galactic habitability probability (P_{GHZ}) map for life without any ad hoc criterion for complexity (as in Figure 9.18). In this case stars younger than the Sun and at smaller galactic radii are more likely to harbor habitable worlds. Adapted from Lineweaver et al. 2004.

Intriguingly both these maps suggest that the Galactic environment may be slightly more habitable than the solar neighborhood at radii somewhat smaller than the Sun's orbit and either (depending on P_{evol}) somwhat earlier or later than the present time, by a few billion years. The Galactic center is certainly not a good place for life, in part due to a much higher supernova rate owing to the higher stellar density.

9.6.4 Beyond the Galaxy

If we continue to expand the realm considered for life then we must consider the fact that the Milky Way is not an isolated galaxy. It is part of the Local Group, a gravitationally bound association of over

30 galaxies, ranging from the dominant spiral galaxies The Milky Way, and Andromeda (M31) to lesser galaxies such as M33 and many dwarf irregular galaxies occupying a region some 10^6 light years in diameter.

Our current understanding of the dynamics of the galaxies in the Local Group suggests that in approximately 5 billion years the Milky Way and Andromeda may actually collide. The collision of galaxies, or their interaction via gravity, is seen throughout the universe and undoubtedly plays a role in shaping the galaxies and the stellar populations they contain. The stars in both galaxies are effectively **collisionless**, in other words they present a sufficiently small cross section compared to the size of the galaxies and the typical interstellar separation that there will be almost no cases of direct stellar collision. The gravitational field of both galaxies will however dramatically alter these two spiral galaxies, strewing stars and gas off in long filaments and structures before the system (which is predicted to be gravitationally bound) settles into a single elliptical-type galaxy. In this process the *gaseous component* (the interstellar medium) of both galaxies *will* interact and experience significant torques and even shock heating. As a consequence we would expect there to be new star formation, triggered by the collision, which will produce what may amount to a new galactic habitable zone, albeit one likely of a more complex geometry as illustrated in Figure 9.20.

9.7 The Universe and Beyond

Although it is not a subject that we devote much space to (in part due to its potential complexity, but also due to its uncertainty, and our focus on dealing with the immediate technical and practical issues of astrobiology) it is worth making note of the questions surrounding the suitability of our universe for life such as that on Earth. There are many aspects of the fundamental laws of physics in our universe that are directly related to the phenomenon of life as we know it. We have already seen how the basic nature of gravity, momentum conservation, and energy minimization gives rise to stars and planets. On an equally deep level, the electronic structure of atoms—including the very particular structure of carbon - is of course a consequence of fundamental physical laws, as is the actual production of elements via nucleosynthesis. This list goes on, and has led to a variety of investigations into both the consequences

Figure 9.20. A Hubble Space Telescope image of the galaxy AM 0644-741 (NASA/STScI). This galaxy has experienced a "collision" with another, smaller galaxy (not shown). As a result of the dynamics of this close encounter a remarkable ring of new star formation (seen as the large bright ellipse-like structure) has been triggered. Already spanning over 46 kpc this ring is still expanding. As new stars are formed it seems likely that new planetary systems are also forming, with the potential to harbor habitable worlds.

for life if the laws of physics were slightly *different* than they are, and the value of **anthropic principles** for deducing facts about the universe (e.g., Barrow & Tipler 1986).

In terms of extending our discussion of habitable zones, we can continue to make similar arguments (avoiding any deep philosophical issues) applied to the universe as a whole. For example, prior to the first generation of stars (Chapter 1) there were no metals in the universe, and in the distant future (depending on the presumed cosmology) there may be no more star formation as normal matter becomes locked up in stellar remnants such as black holes, neutron stars, white dwarfs, and planets or brown dwarfs. Thus, there may certainly be a habitable zone in cosmic time—between the first stars, and the fading embers of all stellar activity.

To really go beyond, into total speculation, one can consider various models that suggest the possibility of multiple universes, or **multiverses**. These models range from an early effort to interpret the peculiarities (to us) of quantum mechanics via a "many-worlds" hypothesis (all possible quantum states actually exist, with a separate universe for each), to more modern efforts such as "bubble theory" (stemming from **chaotic inflation** and suggesting multiple "regions" of space–time each with different physical constants, dimensions, and fundamental particles). If we stick to our very mechanical approach to treating life and habitability then such models suggest that there could be other universes that meet the physical criteria for life (or complex, information carrying phenomena in general), and that there are others that don't. It is conceivable that there is a set of parameters that could be used to define a multiverse habitable zone. These parameters could be fundamental physical constants, or the higher level properties of a deeper theory that gives rise to such constants. Proponents of the "fine-tuning" required for life might take issue with this, but it is certainly a topic that can be discussed, and a possibility that cannot be readily discounted.

References

Barrow, J. D., & Tipler, F. J. (1986). *The Anthropic Cosmological Principle,* Oxford University Press.

Canup, R. M. (2004). Origin of terrestrial planets and the Earth–Moon system, *Physics Today,* **57**, 4, 56–62.

Chyba, C. (1987). The cometary contribution to the oceans of primitive earth, *Nature,* **330**, 632–635.

Gonzalez, G., Brownlee, D., & Ward, P., (2001) The Galactic Habitable Zone: Galactic chemical evolution, *Icarus,* **152**, 185.

Hart, M., (1979). Habitable Zones around Main Sequence stars, *Icarus,* **37**, 351.

Kasting, J. F., (1993). Earth's early atmosphere, *Science,* **259**, 920–926 .

Kasting, J. F. (2004). When methane made climate, *Scientific American,* 80–85 (July 2004).

Kasting, J. F., Whitmore, D. P., & Reynolds, R. T. (1993). Habitable zones around main sequence stars *Icarus*, **101**, 108–128 .

Laskar, J., & Robutel, P. (1993) The chaotic obliquity of the planets, *Nature*, **361**, 608–612.

Lineweaver, C., Fenner, Y., & Gibson, B., (2004). The Galactic Habitable Zone and the Age Distribution of complex life in the Milky Way, *Science*, **303**, 59.

Lissauer, J. J., & de Pater, I. (2004). *Planetary Sciences*, Cambridge University Press.

McGuffie, K., & Henderson-Sellers, A. (2005). *A Climate Modeling Primer*, 3rd ed., J. Wiley & Sons.

Murray, C. D., & Dermott, S. F. (1999). *Solar System Dynamics*, Cambridge University Press.

Nagasawa M., Lin, D., & Thommes, E. (2005). Dynamical shake-up of planetary systems. I. Embryo trapping and induced collisions by the sweeping secular resonance and embryo-disk tidal interaction, *Astrophysical Journal*, **635**, 578.

Peale, S. J. (1977). "Rotation histories of the natural satellites," In *Planetary satellites,* ed. Burns, J. A., University of Arizona Press, 87–111.

Raymond, S. N., Quinn, T., & Lunine, J. (2004). Making other earths: Dynamical simulations of terrestrial planet formation and water delivery, *Icarus*, **168**, 1.

Rigby, E., Symonds, M., & Ward-Thompson, D. (2004). A comet impact in AD 536? *Astronomy & Geophysics*, **45**, 23–26.

Tolstoy, M., Vernon, F. L., Orcutt, J. A., Wyatt, F. K. (2002). The breathing of the seafloor: Tidal correlations of seismicity on Axial volcano, *Geology*, **30**, 503–506.

Valley, J. W., et al. (2002). A cool early Earth, *Geology*, **30**, 351–354.

Walker, J. C. G., Hays, P. B., & Kasting, J. F., (1981) A negative feedback mechanism for the long-term stabilization of Earth's surface temperature, *Journal of Geophysical Research*, **86**, 1147–1158 .

Waltham, D. (2004). Anthropic selection for the Moon's mass, *Astrobiology*, **4**, 4, 460.

Wilde, S. A., Valley, J. W., Peck, W. H., & Graham, C. M. (2001). Evidence from detrital zircons for the existence of continental crust and oceans on Earth 4.4 Gyr ago, *Nature*, **409**, 175–178.

Problems

9.1   Describe the idea behind the classical Circumstellar Habitable Zone. Describe the *major* effect of a terrestrial-type planetary atmosphere on a planets' surface temperature and how this can operate as a negative feedback cycle. In what circumstances might this turn into a runaway effect?

Derive the range in equilibrium temperatures for an Earth-type planet (i.e., with a terrestrial-type atmosphere) on an eccentric orbit with $e = 0.3$ around a Solar luminosity star. You may assume that the planet instantly adjusts to the stellar flux at any location.

9.2 Imagine a terrestrial-type planet in an identical orbit, and with identical obliquity, to that of the Earth, about a solar mass star of the same age as the Sun. On this planet, unlike the Earth, the *mean* pressure and temperature at the mean surface elevation are exactly coincident with the primary triple point of water. Assuming that there is abundant water, then describe the conditions that one would encounter as a function of latitude, longitude, and elevation on this planet.

9.3 A probe lands on a hypothetical planet and measures a mean surface temperature of 373 K. The mean effective temperature of this planet (i.e., seen by an observer above the atmosphere) is 273 K. Compute the far infrared optical depth of this planet's atmosphere.

If the planet's atmosphere is identical in composition to that of the Earth and is approximated as a uniform sheet or slab, then by what factor should the atmospheric *density* be greater than that of the Earth? Given the surface temperature estimate the gas pressure on this planet (assume an ideal gas).

9.4 Describe how the carbon–silicate cycle works on Earth, and its role in climate stabilization. At some points in Earth's relatively recent geological history (e.g., during the Jurassic) the carbon dioxide content of the atmosphere was about 5 times its present concentration and the global mean temperature was higher by about 10 K. Using Equation 9.15, discuss the implications for rock weathering and how the climate might have been stabilized in this situation.

9.5 Compute the orbital radius for tidal locking of a planet in 4 Gyr around a $0.2M_\odot$ M-dwarf star. Assume first a Q value of 100. Re-compute the radius assuming $Q = 10$. Assuming that energy is efficiently distributed around the planet (e.g., by a thick atmosphere), and that the planet has an albedo of 0.3 and an Earth-like atmospheric infrared optical depth, compute the habitable zone for this star–planet system when $L_* = 0.005L_\odot$. Compare this to the tidal lock ranges previously calculated.

9.6 Describe the concepts of a circumstellar habitable zone, and a continuously habitable zone. With reference to Figure 9.6, discuss the nature of the habitable zone for a lower mass star (e.g., an M-dwarf). What other factors come into play for habitability in such a system compared to our own? By also making reference to earlier chapters, discuss the potential importance of extrasolar planetary systems around M-dwarf stars. [With a more extensive literature search this could be expanded to a larger project.]

9.7 Describe the potentially positive and potentially negative consequences of giant planets for the history of rocky, terrestrial-type planets and their habitability.

9.8 The Hill Sphere radius of a planet mass M orbiting a star of mass M_* at distance a is

$$R_H = a \left(\frac{M}{3M_*} \right)^{1/3} \qquad (9.35)$$

Extrasolar planet HD 150706a orbits a $1M_\odot$ star at a distance of 0.82 AU and has mass $1M_{Jupiter}$. Could the equivalent of any of the terrestrial planets from our own solar system exist in this system? (use a solar system data table).

9.9 Describe the origin and present nature of the Earth–Moon system. Discuss the various ways in which we think the Moon helps govern the habitable conditions, and phenomena, that life deals with on Earth. It is a popular notion that "the Moon makes Earth more suitable for life." Do you agree? Try to find counter arguments to this assertion—one approach may be to compare the modern Earth–Moon system to that a few Gyr ago. [Note: this question could evolve into a longer term project, involving both more detailed computation and literature research.]

9.10 Consider the idea of a "Goldilocks Zone" in the context of Galactic Habitability. In the text we assumed that the presence of a hot Jupiter implies the absence of terrestrial-type worlds. We know that this is likely to be an overstatement. We also assume that the probability of forming a terrestrial planet is linearly correlated with metallicity. This currently has no basis in observation and may be incorrect. Describe how you would modify the calculation of P_{TP} (the probability that a system harbors a surviving terrestrial planet) by assuming that hot Jupiters eliminate about 70% of inner terrestrial planets and that terrestrial planet formation is *independent* of metallicity above some minimum value.

CHAPTER 10

Alternate Habitable Zones and Beyond

10.1   Introduction

The classical definition of what constitutes a habitable environment may be overly restrictive. Indeed, from what we continue to learn about life on the Earth it would appear that there are many environments that, while unusual from our human perspective (for example, black smokers, Chapter 5), are *at least* as habitable as those hypothetical temperate planets we are eager to find around other stars. In this final chapter we discuss some of the possible *alternate* habitable zones that may exist in the universe. These include environments that are suggested by Earth itself, since the present-day terrestrial environment is by no means typical of the conditions throughout the past 4.5 Gyr. Other environments include those potentially offered by bodies such as moons or even planetesimals and asteroids. Some of these potential habitats for life are arguably less of an alternative to terrestrial worlds but possibly more of a primary environment. Just as our preconceptions about the dominance of multi-cellular life on Earth have turned out to be false, we may yet find that our preconceptions about acceptable conditions for life are appallingly ego-centric. We also discuss broader constraints on the suitability of a given location for life, such as those provided by considering the availability of critical elements. Much of this is highly speculative, but it is nonetheless a useful exercise, if only to demonstrate how to take what we have learnt in this book a little further—and as elsewhere in this text, how to begin to construct *quantitative* models.

Finally, we offer some speculative thoughts on the origins of life, and on the ways in which organisms "solve" problems of efficiency. We also consider some of the pitfalls in the ways in which the phenomenon of life is categorized, and provide some suggestions on how to avoid these.

10.2   Earth: But Not As We Know It

In preceding chapters we have touched upon the evolution of the Earth, its systems, and in particular its evolution as a habitable planet (according to a very limited set of criteria). If we investigate this a little further we find that for the majority of Earth's history it has really had little resemblance to its modern state. Not only that, but its global conditions have varied significantly at different times over the past 3.5–4 Gyr.

Among the most dramatic examples of this are the episodes of low global temperatures, and extensive surface water ice coverage that are generally termed **ice ages**. Strictly speaking we are currently within a long term ice age of lower global temperatures. 40 million years ago Antarctica did not have an ice cap, and it was really only about 3 million years ago that the Northern ice cap grew substantially. In addition, on much shorter timescales, there were periodic "surges" of **glaciation** (also often referred to as ice ages). The most recent of these began some 70,000 years ago, ended some 10,000 years ago, and involved extensive glaciation in the Northern Hemisphere. The present-day is therefore considered to be an **interglacial** period.

On longer timescales there are indications (although many aspects are still controversial) that the Earth has undergone far more radical episodes of global temperature change. In particular, there is evidence from the geological and isotopic record (e.g., Hoffman et al. 1998 and references therein, also Hoffman & Schrag 2000) that the Earth has suffered at least one, and possibly three or more *complete*, global glaciations. These events, lasting perhaps ∼100 Myr, have become known as **Snowball Earth** ice ages—where the mean global temperatures were well below 0 °C and the oceans literally froze over, from pole to pole. There is considerable uncertainty about whether these were either as extensive as suggested, or indeed whether the global climate could ever suffer such an enormous swing (e.g., see Chandler & Sohl 2000, Arnaud 2004). The location of the continental landmasses might have played some role in reorganizing the planet's energy circulation to a different equilibrium

(Chapters 5, 9). It is also unclear exactly how the Earth would "recover" from a snowball period—the carbon silicate cycle described in Chapter 9 might certainly play a role. Nonetheless, it is an intriguing possibility, and the Snowball Earth timing certainly corresponds to some of the major mass extinctions of families of terrestrial organisms (see also Chapter 8) and the growth of new ones, such as those responsible for oxygenating the atmosphere (Figure 10.1).

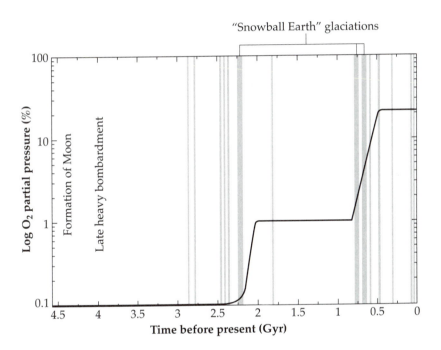

Figure 10.1. A schematic plot showing both the known major episodes of low global temperatures (ice ages) throughout Earth's history and the approximate variation in global oxygen content of the atmosphere. Time before present is plotted on the x-axis in Gyr. Filled vertical stripes indicate ice ages. Thin stripes are used to denote "regional" glaciations (e.g., extension of northern and southern ice caps to moderate latitudes). Thick stripes indicate possible global glaciation events, or Snowball Earth events, with ice extending to equatorial regions. The solid curve indicates the estimated variation on oxygen partial pressure with time, starting at less than 0.01% prior to ~2.3 Gyr, rising to ~1% between 2.3 Gyr and ~1 Gyr, and then increasing again to ~20% at the present.

Several interesting issues arise from this possibility. The first is that while life may have been severely depleted by such global change it nonetheless seems to have survived, and ultimately repossessed its environment once the climate returned to a temperate state. Indeed, it seems that the deep ocean and subsurface biospheres on the present-day Earth harbor an enormous genetic diversity. Much of this genetic capability is effectively dormant, in as much as many phyla are not particularly common. However, in the advent of a severe change in climate, and an associated reduction of the most populous phyla, many of these other species may find an opportunity. This is at least suggestive of a rather stochastic process, a terrestrial type world can harbor a wide variety of dominant organisms, but perhaps only a few major classes at a time (e.g., methanogenic or aerobic). The processes that determine which species are "in charge" at a given time in the planet's history may be essentially random. The second issue is that a Snowball Earth would present a very different type of world to any remote effort to classify it and to search for biosignatures. It would exhibit a high albeo, a potentially low atmospheric biosignature (i.e., little methane, little oxygen), and certainly no vegetation red edge (Chapter 6) or comparable biological surface markers. It would also present a very severe conundrum in the context of the circumstellar habitable zone (Chapter 9). It would appear as a frozen world that, on the basis of all other characteristics, should harbor liquid surface water, but did not.

As remarkable as the notion of a "frozen Earth" may seem, it is important to realize that between such periods the typical global mean temperature of the Earth was in fact significantly *higher* than it is today (remember, we are currently within an interglacial period). During the peak of the dinosaurs (from the early Triassic some 230 Myr ago up until the end of the Cretaceous 65 Myr ago, see also Chapter 8) *sea surface* temperatures seem to have had highs of $\sim 37\,°C$ at mid to low latitude ($\sim 25°$ N) (Steuber et al. 2005) (which are always significantly cooler than land). In addition, ocean levels were as much as 200 *meters* higher than they are today. Averaging out these extremes leads to estimated global mean temperatures as high as $22\,°C$. This may not sound like a great deal, but 8–10 degrees warmer globally than the present day is a *huge* difference. At these temperatures it is likely that the atmospheric water content would be significantly higher (Chapter 9). During the middle of this time (within the Jurassic period), some 180 Myr ago, the global concentration of atmospheric CO_2 was almost a factor 5 greater than it is today. At earlier epochs there is some evidence for even warmer global climates. At approximately 3–3.5 Gyr ago (during the Archean period) there are indications of *ocean* temperatures reaching 55–85 °C. Such high temperatures would result in enormous weathering rates (Chapter 9) and a high silicate content for the oceans (Knauth & Lowe 2003). Such conditions may seem hostile to us, but as we have seen, extremophilic organisms would be perfectly happy with such a planet (Chapter 5).

In fact, for most of Earth's history there is evidence that the global mean temperature was significantly higher than it is today (by at least 10 °C). While a warm terrestrial planet may not appear to be so much at variance with our concept of a habitable zone as a Snowball planet, it is nonetheless true that the *modern* Earth may be the worst template we could use in searching for life elsewhere. As we discussed at the outset (Chapter 1), one of the ultimate goals of astrobiology must be to place the Earth in full context. From Earth's own history we have the first indications of how to do this, by realizing that there are many faces to a habitable planet.

10.3   Moons

There is another class of "planet-like" body that we have largely ignored until now. Based on the example of our own Solar system, giant planets appear to play host to very substantial moon systems (Figure 10.2). These potentially represent an additional, and possibly numerous, harbor for life in the universe.

Moons are broadly classified as either "regular" or "irregular," the former implying moons with short periods, close-to-circular orbits, and a common orbital plane commensurate with the equatorial plane of the host planet. The latter typically includes many small satellites farther from the planet, often in orbital planes that are significantly offset in inclination and of high eccentricity. This classification is also generally used to assign a status to *captured* moons—satellites which did not form *in situ* with the planet. In this class of moons Jupiter has some 31 currently known irregulars, Saturn 13, Uranus 5, and Neptune one (Nereid). These moons are often "clumped" in orbital parameter space,

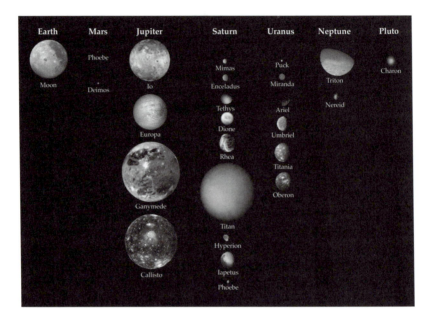

Figure 10.2. An illustration of some of the larger moons in our own solar system. All objects are scaled to the correct relative size. Note that both Ganymede and Titan have a larger radius than the planet Mercury (adapted from figure by NASA).

suggesting that many may arise from the disruption of a single, larger body, perhaps during their capture by the host planet. The most dramatic example of a captured moon (although one which is generally termed a regular moon) is Triton, Neptune's largest moon (2700 km diameter). Although in an almost perfectly circular orbit, Triton is in *retrograde* motion, and its orbital plane is inclined by 33 degrees compared to the other regular moons in the Neptunian system. As a consequence its orbit is slowly decaying due to tidal forces (Chapter 9), with a timescale of $\sim 10^9$ years. In fact, the tidal forces it experiences are likely responsible for driving internal heating, which has been inferred by the incredible observation of liquid nitrogen geysers (in 1989 by Voyager 2). Triton was the first example of active **cryo-volcanism** in the Solar System. Intriguingly, Triton's origin may have been as part of a *binary* Kuiper belt object (akin to Pluto and Charon), which was disrupted during an encounter with Neptune.

Considering regular satellites, and their potential as habitats, we have already discussed in Chapter 1 the possibility that the Jovian moon Europa has a liquid water ocean beneath a thick water ice crust. Europa has been the prototype for the notion that substantial liquid water environments can exist elsewhere in our Solar System. It does not appear to be alone, however. Saturn's small moon Enceladus has been seen to have active water geysers—suggesting the possible presence of substantial pockets of subsurface liquid water (and we note that the radiation environment around Saturn is much less extreme than that around Jupiter). The giant Jovian moon Ganymede (which is larger than Mercury) also has an icy outer crust and its own magnetic field, suggesting the possibility of a still molten inner core and therefore internal zones where hot rocks meet water ice. Even the ice-shrouded (but blackened) surfaces of the moons Miranda, Ariel, and Titania around Uranus show evidence for past melting and possible liquid flow on their surfaces. While this may have occured close to the time of formation it is nonetheless an intriguing reminder that at past periods the territories of the Solar System may have been quite different than they are today. In short, moons of giant planets may be excellent places to look for water-rich environments that may be currently active (in the geological sense) or have been active in the past.

Although it is likely some 50% water ice by composition, the Saturnian moon Titan presents a further class of object. Titan is larger (although less dense) than Mercury (but slightly smaller than Ganymede) but harbors a dense atmosphere. In fact this atmosphere, with a typical temperature of about 90–95 K, produces a surface pressure of about 1.5 Earth atmospheres on Titan. It is approximately 98.4% nitrogen in composition, with some 1.6% methane, other hydrocarbons, and traces of noble gases. The surface of Titan appears to exhibit a wide range of phenomena, from lake- and river-like structures (likely of liquid methane) to dunes and other features created by an active global climate and weather system. The presence of substantial amounts of atmospheric methane, but relatively small amounts of liquid methane on Titan's surface, indicates that it is likely stored in subsurface **clathrate hydrates** (water and methane in an ice-like structure) and undergoes episodic outgassing to resupply the atmosphere (Tobie et al. 2006). A further consequence of this internal arrangement is that Titan may actually harbor a subsurface liquid water ocean—maintained by the internal heat from the moon's formation and radiogenic activity. Such an ocean (surrounding the rocky core and a layer of high-pressure water ice and some 50 km below the surface) would have a high ammonia content, but would nonetheless represent a major zone of liquid water. A full discussion of Titan is beyond the scope of this text, but it is a fascinating object. Although ambient temperatures are very low, there is still the potential for an array of organic chemistry both on Titan's surface, subsurface, and within its atmosphere. It has been speculated that Titan represents some of the chemical conditions of a primordial Earth and might therefore provide insight to the processes that operated then.

10.3.1 Europa

As described in Chapter 1, if Europa has a sub-surface ocean (perhaps 10 km beneath the ice crust, Figure 10.3), possibly itself with a depth of some 60–100 km, then it is likely maintained by **tidal heating**, with some contribution from natural **radiogenic** heating due to the decay of elements in its rocky/metallic core. The Laplacian mean-motion orbital resonance between Io, Europa, and Ganymede results in eccentric orbits for these three moons—which are otherwise in *synchronous*, or spin-locked orbits. As the moons move between periapsis and apoapsis they experience significant differential tides in Jupiter's powerful gravitational field, and as a consequence are periodically stretched and relaxed—as energy is transferred from Jupiter's spin to the moons through the tidal bulges that are produced. Internal friction converts some of this translational motion into thermal energy. In the case of Io, closest to Jupiter and with an orbital eccentricity of 0.004, the heat flow has been enough to create silicate volcanism, which is active today, and has likely ensured the sublimation loss of surface volatiles such as water. In the case of Europa, further from Jupiter (but with the largest orbital eccentricity of these three moons; $e = 0.01$), this tidal heating is a factor of some six times or so less.

For small orbital eccentricities and synchronous orbits the surface heat flow due to simple tides can be expressed as (in units of energy per unit time per unit surface area)

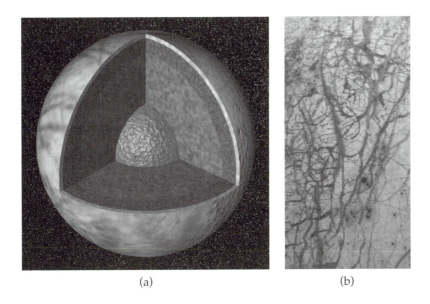

(a)   (b)

Figure 10.3. Left panel shows an illustration of Europa in cutaway, with a hypothetical subsurface ocean beneath an icy crust. The core is a rocky, silicate mass. Right panel shows an image of Europa's surface roughly 500 by 1000 km in size, with a resolution of 1.6 km taken by the Galileo mission. This can be compared with the images in Figure 1.10. What appear to be the equivalent of massive ice floes can be seen as the lighter toned features. Between them is what may be darker refrozen slush that has welled up between the fractures. (Adapted from NASA/JPL images.)

$$H_{\mathrm{T}} = \frac{\dot{E}}{4\pi R_{\mathrm{s}}^2} = \frac{21}{38} G^{5/2} \frac{\rho_{\mathrm{s}}^2 R_{\mathrm{s}}^5}{\mu Q} e_{\mathrm{s}}^2 \left(\frac{M_{\mathrm{p}}}{a_{\mathrm{s}}^3}\right)^{5/2}, \qquad (10.1)$$

where $\dot{E}$ is the total energy dissipation rate due to tides, R_{s} is the satellite radius, ρ_{s} is the mean density of the satellite, e_{s} is the satellite orbital eccentrcity, a_{s} is its orbital semi-major axis, and M_{p} is the host planet mass. The factor $1/\mu Q$ is a combination of the mean **elastic rigidity** μ of the material experiencing tidal flexure, and the **specific dissipation function** $1/Q$—a measure of the efficiency of energy dissipation (Chapter 9). For solid material such as that of the Earth $Q \approx 100$, and for water ice at 100 K $\mu \approx 4 \times 10^{10}$ dyne cm^{-2}. For rocky material μ is as much as a factor of ten larger, and consequently the heat flow is lower.

If we apply Equation 10.1 to the case of Europa, assuming an icy composition, then we find that $H_T \approx 50$ erg s^{-1} cm^{-2}, which corresponds to about 0.1 watts per square meter of heating. Detailed models of Europa, which include the hydrostatic equilibrium of a subsurface ocean and its heat capacity, suggest that this level of heating can indeed *sustain* a liquid ocean, but might not be able to *produce* one (Cassen et al. 1979). It is likely therefore that some earlier set of circumstances would be needed to put an ocean in place on Europa.

Nonetheless, if we extrapolate this outer surface heat flow down to a rocky core in Europa it suggests that on the ocean floor there could be situations akin to the hydrothermal vent systems (Chapter 5) on Earth. If this is the case then not only is there liquid water in abundance on Europa, but also the rich chemical and thermal environments driven by tectonic and/or volcanic activity necessary to complete the cycle in sustaining an ecosystem without direct solar energy.

That is, however, not to say that putative ecosystems on Europa need be non-photosynthetic. A body of literature exists that considers in great detail the possibility of ecosystems existing within the surface crack systems of that moon (e.g., Chyba & Phillips 2001). The tidal forces that heat Europa will also create a form of surface ice tectonics that may result in periodic opening and closing of fissures—potentially exposing the subsurface ocean to both space and to solar radiation. Although the upper few meters of Europa's crust are subject to intense particle radiation due to Jupiter's magnetospheric activity, beneath this the environment should be relatively benign, albeit also quite dark. It has been speculated that activity in these close-to-surface environments can allow, at least temporarily, for photo-synthesis, and the proximity to the surface can enable the exchange of useful elements and chemistry—such as oxygen from the radiation-driven dissociation of water on the surface.

Overall though, Europa might represent a habitat driven not by direct stellar input, but rather by the transfer of angular momentum from a gas giant planet to its moons. This is a radical shift in the standard paradigm that treats stellar energy as central to habitability. It is not without caveats however. The Laplacian resonance that may help sustain a liquid ocean in Europa will not last forever; it may be that such tidal heating lasts only a few 100 Myr, and that it comes and goes as the entire Jovian moon system evolves dynamically.

10.3.2 Exomoons

Since all of the giant worlds in our Solar System harbor quite extensive moon systems, which include both large and small moons, as well as moons with atmospheres (Titan) and clear signs of activity (Io, Enceladus), it seems a reasonable extrapolation to assume that giant worlds elsewhere may well harbor moons too. One has to proceed with caution however. We do not yet have a complete theory of planet formation, and for moons the situation is even worse. It does appear likely however that a forming giant planet will have its own **circumplanetary** disk of material, which in turn will interface with the proto-planetary disk of the system as a whole. This is illustrated in Figure 10.4. The circumplanetary disk gas seen in simulations appears to rotate in a prograde sense, and so would naturally give rise to the orbital configurations of moons such as those dominating the giant planets of our own system. Such a disk also appears to withstand the orbital migration of the host planet itself, thereby giving moons a chance to form in the presence of a gaseous phase. There is some debate about the stages and timing of moon formation, much of it is analogous to the debate on planet formation. One thing we can say with certainty is that moon systems are dynamically dense compared to planetary systems; relative to moon masses the orbital terrain (radii) is occupied more densely than it is for planets—and of course the orbital timescales are much smaller.

Despite these uncertainties, for a discussion of habitable zones we can speculate as to the broad features of moon environments. Foremost amongst the characteristics to consider is the orbital range, or terrain, within which a stable moon system can exist around a planet. Stable orbits exist for satellites of a planet within a range of radii defined by an inner radius within which tidal forces will rip bodies apart, and an outer radius beyond which the gravitational influence of the parent star (and even other planets in the system) results in unstable orbits.

We have already encountered this latter situation in our discussion of the **Hill Sphere radius** and planet formation (Chapter 3). The

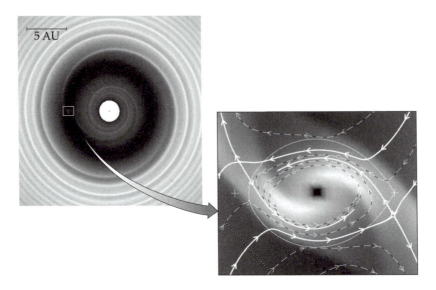

Figure 10.4. An illustration of the circumplanetary disk that may exist around a forming giant planet. The primary satellites of the planet may form from this circumplanetary disk in an analogous fashion to the planets themselves. Upper panel shows a face-on view of a numerical simulation of a proto-planetary disk with a giant planet at 5 AU from the central star (reproduced with permission from Wilhelm Kley, Univ. Tübingen). The density of the disk medium ranges from low (dark) to high (light). The planet has formed a disk gap (Chapter 3). The spiral density structure due to resonances is seen emanating from the location of the planet. Lower panel shows a close-up of the region around the planet, spanning some 0.1–0.2 AU. The oval shaped outline represents the Hill sphere region for stable versus unstable orbits around the planet. Arrowed dashed curves illustrate the **flow lines** of disk material, which circulates the planet as well as flowing in and out of the Hill region (Lubow et al. 1999).

Hill Sphere radius is a solution of the restricted 3-body problem—considering a small mass in the gravitational field defined by two orbiting large masses. Previously we had defined it as $R_{\mathrm{H}} = a_{\mathrm{p}}(M_{\mathrm{p}}/3M_*)^{1/3}$, where a_{p} and M_{p} are the planet orbital radius (for a circular orbit) and mass respectively. Numerical simulation of satellite orbital stabilities has been used to show that long-lived stable orbits in reality extend up to the so-called *critical semi-major axis*, which is some $\approx 0.36R_{\mathrm{H}}$. Furthermore, as we have seen (Chapter 4), exo-planets often have highly eccentric orbits. Thus, it is more appropriate to consider the *periastron* distance of the planet from the star—where the pull of the star is at a maximum. We then arrive at an outer stable orbital semi-major axis (e_s) of

$$a_s < 0.36(1 - e_p)a_p \left(\frac{M_p}{3M_*}\right)^{1/3}, \qquad (10.2)$$

where e_p is the planet's orbital eccentricity.

For the inner stable orbit we can use a rather simplified argument to derive the solution. Consider the situation shown in Figure 10.5. A small piece (of mass m) of the satellite on its surface experiences a force due to the satellite's self-gravity of

$$F_s = \frac{GM_sm}{r_s^2}. \qquad (10.3)$$

Now, the *tidal field force gradient* between the planet and the mass m is

$$\frac{dF}{dR} = \frac{-2GM_pm}{R^3}. \qquad (10.4)$$

Thus, at the location of m on the satellite surface the actual tidal force is just

$$F_p = \frac{2GM_pm}{R^3}r_s. \qquad (10.5)$$

When the tidal force and the satellite self-gravity force are equal then this is the limit we seek. Equating Equations 10.3 and 10.5 we obtain an expression for the **Roche limit**, or innermost satellite orbit

$$a_s > R = \left(\frac{3M_p}{2\pi\rho_s}\right)^{1/3}, \qquad (10.6)$$

where ρ_s is again the satellite mean density. This limit is strictly true only for a body that cannot deform without falling apart (e.g., a rubble pile). In reality a solid satellite will have additional forces resisting its

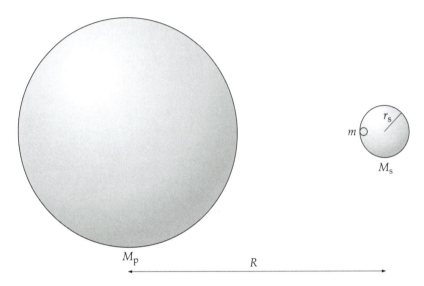

Figure 10.5. Illustration of the configuration adopted in deriving the inner stable satellite orbit around a giant planet—within which tidal forces will begin to disrupt a satellite (assuming a rigid composition to first order). The forces due to the satellite's gravity and due to the tidal field of the planet on an arbitrary surface mass m on the satellite are considered.

tidal disruption due to molecular and atomic forces in its material. We also note that of course this inner radius does not (unlike the outer) correspond to *unstable* orbits, but rather the limit at which a body of any size can remain intact.

Armed with the expressions in Equations 10.2 and 10.6 we can determine the possible extent of moon systems around any planet, as long as we have basic information about the planet mass and orbit, the mass of its parent star, and speculation about the mean density of a satellite. In Figure 10.6 the results of applying Equations 10.2 and 10.6 to a sample of known exoplanets is shown (these planets have been detected using the radial velocity technique). Within orbits of about 0.6 AU radius around solar mass stars more detailed calculations show that moon systems may not be stable over long periods due to stellar tides (Barnes & O'Brien 2002). The sample of exoplanets shown in Figure 10.6 is therefore limited to those with semi-major axes greater than 0.6 AU. What is immediately apparent is that the size of the stable orbital terrain for

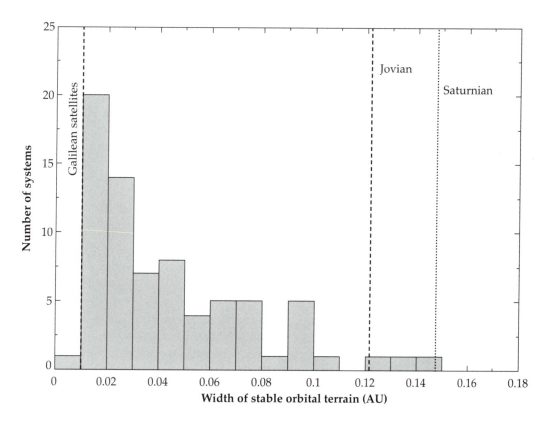

Figure 10.6. The number of exoplanet systems is plotted as a function of the net width of the orbital terrain around each planet which can harbor stable satellites (i.e., Equation 10.2 minus Equation 10.6). A subset of known exoplanets is used (detected using radial velocity measurements, Chapter 4) with semi-major axes greater than 0.6 AU to ensure the possibility of long-lived satellite systems. The widths of the stable orbital terrains for Jupiter and Saturn are indicated. The major moons of these planets however lie within the inner 10% or so of this total terrain—as indicated by the vertical dashed line labeled for the Galilean satellites.

moons around these exoplanets is significantly less than it is for Jupiter and Saturn. This is almost entirely due to an observational bias, first discussed in Chapter 4, which results in preferential detection of planets *closer* to their parent stars than either Jupiter or Saturn are to the Sun. From Equation 10.2 we can see that this will shrink the outer stable orbit for moon systems. However, if we take a look at the Galilean satellites (e.g., Io, Europa, Ganymede, and Callisto) we find that they *all* inhabit the *inner 10%* of the stable orbital terrain (as indicated in Figure 10.6). Thus, the majority of the exoplanets in this sample could still quite happily hold on to an equivalent set of moons; it is just the outermost smaller regular and irregular satellites that might no longer be viable.

We can also apply our knowledge of the exoplanet orbits and the luminosities of their parent stars to estimate the stellar insolation experienced by any hypothetical moons and, with some assumptions about albedo and atmosphere, estimate their equilibrium surface temperatures. Specifically, as we have done before (Chapter 9) the surface temperature can be written as

$$T_{\text{surface}} = \left[\frac{(1 - A_{\text{B}})L_*}{16\pi \epsilon \sigma d^2} \right]^{1/4}, \tag{10.7}$$

where ϵ is a factor that approximates an atmospheric greenhouse effect and is $\epsilon \approx 0.62$ for a terrestrial type atmosphere or $\epsilon = 1$ for no atmosphere. We can compute the *time averaged* surface temperature T_{surface} for hypothetical moons by integrating this expression around the (typically non-circular) exoplanet orbits. An example of this is shown in Figure 10.7. One of the results from this exercise is that some 15–28% of the planets in this sample could harbor moons below the water–ice sublimation temperature of 170 K. Thus, ice-rich moons, such as Europa, could certainly exist around such worlds without losing their water over time.

Intriguingly, some work has shown (Williams et al. 1997) that as moons approach masses of $\sim 0.1 M_{\oplus}$ (compared to $0.008 M_{\oplus}$ for Europa) they should be capable of retaining a full (warm) atmosphere, even within the intense radiation environment that may exist around gas giant planets. In this case, large moons could exhibit many of the characteristics of terrestrial planets such as the Earth or Mars. If, in addition, such moons are subject to tidal heating as described above, they could exhibit sustained geological activity (beyond that just due to primordial and radiogenic heating) and be subject to the same kind of feedback systems that we see on Earth (e.g., the carbon–silicate cycle). This possibility is made even more intriguing by the indication in Figure 10.7 that there may be a respectable fraction of moon systems experiencing stellar irradiation that results in surface temperatures at, or above, 273 K.

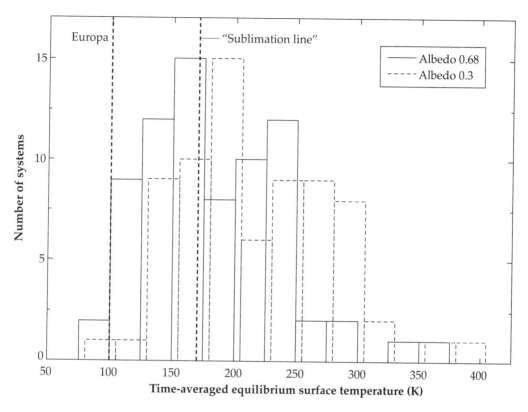

Figure 10.7. A plot showing the distribution of surface temperatures for hypothetical moons around the exoplanet subset chosen in Figure 10.6. Observed and estimated stellar luminosities, orbital parameters, and two different albedos are used to integrate over the orbital period to obtain a time-averaged equilibrium surface temperature (assuming instantaneous re-equilibration as the stellar flux varies with orbital position). Two different surface albedos are used: 0.68, corresponding to a water ice surface such as that of Europa, and 0.3 corresponding to a dry terrestrial surface composition. No atmospheric greenhouse effect is included—which would otherwise shift all temperatures upwards. While a significant fraction (15–28%) of the systems would harbor moons below the water-ice sublimation temperature of 170 K (indicated by vertical dashed line), the majority would be warmer. Many could exist in with temperatures close to 273 K.

Table 10.1. The fundamental elements critical for life on Earth. In this case the terminology we use to describe metals versus non-metals is in the chemical sense, not in the astronomical sense. Note that certain elements such as silicon exhibit significantly *different* abundances between plants and animals (a factor 1000 more abundant in plants).

	Elements in all terrestrial organisms								
Non-metals	H	C	N	O	P		S	Cl	Se	Si
Metals	Na	K	Mg	Ca	Fe and/or Mn	Zn			
Occasional additions	Co	Mo							

10.4 Elemental Constraints on Life

The fundamental elements that are considered critical for life on Earth are shown in Table 10.1. The absence of any of these from an environment on Earth will essentially render it incapable of supporting life. This raises several interesting questions. First, are there places in the Galaxy where some of these critical elements are absent? This could be due to a lack of stellar processing or any mechanism that sequesters the elements into forms inaccessible by organisms (e.g., hot interstellar gas). It seems possible that (for example) the Galactic bulge and halo stars, which generally have a very low abundance of heavier elements, might have little or none available (Chapter 9). However, in these locations the expectation is also that insufficient heavy elements exist to build planetary bodies—and so it may be a moot point. Second, are there situations in places that are capable of building planets where nonetheless some of the key elements involved in life are somehow in short supply?

We can begin to address this second question by looking in a little more detail at the relative abundances of some of these key elements with respect to the actual solar abundances in our own system. In Table 10.2 some of the representative biogenic elements are listed, together with their abundances relative to hydrogen in the Sun, in terrestrial animals, and in the terrestrial lithosphere (crust, see Chapters 1 & 5). As might be expected, heavier elements are far more abundant in planets and life than they are in the Sun, and by extension the proto-stellar nebula.

Table 10.2. Abundances of key biogenic elements with respect to hydrogen. Solar abundances are considered to reflect the undifferentiated abundances in planets.

Element	Solar abundance (relative to H)	Abundance in animals (relative to H)	Abundance in lithosphere (relative to H)
H	1	1	1
C	0.00042	0.15	0.18
N	0.000087	0.021	0.26
O	0.00069	0.41	209.6
P	0.00000024	0.004	0.28
S	0.000016	0.0009	0.12
Fe	0.000032	0.0000071	6.44

In fact the abundance ratio between two environments (or "concentration parameter") can be used to give us an idea of the extent to which an object, or organism, sequesters or "harvests" key elements from its surroundings (Table 10.3). For terrestrial organisms (and plants or prokaryotic life have very similar values to those shown here) this is some indication of what needs to occur for them to be viable. Not surprisingly, the heavy elements are highly concentrated in the Earth's lithosphere compared to solar values, as are those in terrestrial life.

There are several features worth noting in Table 10.3. The animal/lithosphere concentration of elements is low—whereas the animal/solar concentration is high—indicating that the planet has already done the work of sequestering critical biogenic elements. Furthermore the Earth is of course highly differentiated (Chapter 3), which also acts to concentrate certain elements in its lithosphere.

A particularly interesting element is phosphorus. This shows a very high animal/solar concentration of approximately 17,000, two orders of magnitude greater than the other listed elements. Phosphorus and phosphates are key parts of the double helix "backbone" of DNA as well as in many other critical functions (such as the energy transport molecule ATP, Chapter 5). However, in terms of the proto-stellar disk, phosphorous must have been in relatively short supply. This is further indicated by the lithosphere/solar concentration of phosphorus (1.2 ×

Table 10.3. The ratio of the abundances, or the "concentration parameter,"
gives an indication of the degree to which elements are sequestered,
or accumulated, relative to another environment.

Element	Abundance ratio (concentration) Lithosphere/Solar	Abundance ratio (concentration) Animal/Solar	Abundance ratio (concentration) Animal/Lithosphere
H	1	1	1
C	429	357	0.83
N	2989	241	0.08
O	3.0×10^5	594	0.002
P	1.2×10^6	17000	0.014
S	7188	56	0.008
Fe	2.0×10^5	0.2	0.000001

10^6), which is almost an order of magnitude greater than the most concentrated of the other selected elements. To put this another way: life needs a lot of phosphorus compared to the solar abundance (and by extension the cosmic abundance), and the lithosphere of the Earth has managed to concentrate phosphorus to a much greater extent than other biogenic elements.

What does this imply? This is suggestive, at least, of a possible bottleneck in biotic chemistry. Phosphorous is manufactured only in moderately massive stars ($M > 15 M_{\odot}$ via Ne and C fusion), and its presence in interstellar molecular species appears to be quite sensitive to dust grain mantle chemistry and fragility. It is conceivable then that planetary systems could exist that have low phosphorus contents and are then perhaps less suitable for Earth-type biochemistry. It is also conceivable that the processes that have resulted in the Earth concentrating phosphorus to a high degree are not ubiquitous in terrestrial worlds. Furthermore, much of the phosphorus in the terrestrial lithosphere (Tables 10.2 and 10.3) is in the form of phosphates (particularly the **apatite** minerals), which are not particularly soluble and require further processing in order to be available to organisms. Thus, the abundance of phosphorus in solution (or easily soluble form) is significantly lower than is suggested from the pure value in the lithosphere. In addition, if much of the initial pre-biotic chemistry of a planet takes place in interstellar space and the circumstellar environment of a proto-star (Chapters 7 and 8), then the animal/solar concentration may be a better indicator of the degree to which phosphorus must be harvested in order for the eventual success of entire biospheres.

Conversely, we see that an element such as iron is present in significant excess of biological need (i.e., an animal/lithosphere concentration of 0.000001). While iron may not be a huge threat in terms of toxicity to life, there are certainly heavy elements that are. Although life can deal with a great deal of toxicity (i.e., chemistry that interferes with biologically critical processes) and evolve to cope with many situations (Chapter 5), there may be a limit to this. Again, there could be situations where either the local stellar abundance, or the formation pathway of planets, results in a high concentration of toxic elements (or compounds) that bio-chemistry simply cannot deal with.

This is all quite speculative. Nonetheless, we tend to take it for granted that life is relatively insensitive to the abundances of elements surrounding it. After all, life has not converted *all* material on the surface of the planet into biological structures—there appears to be plenty of raw material lying around (an observation supported by Table 10.3). On the other hand, we do not know how large a buffer of this material is actually required for a biosphere to succeed—perhaps, for some reasons of long-term bio-geochemical processing, life on Earth is actually at capacity (if not today, then at points in the past).

10.5 Speculations on the Origins of Life

Throughout this text we have avoided any real confrontation with the vexing questions of **biogenesis**: the origins of life. It has been our general and rather vague contention (Chapter 1) that part of our quest is to see whether life, or at minimum the conditions for life, arises as a "simple" consequence of the nature of the universe—in terms of cosmochemistry, planet formation, and the like. By way of a suitable bookend to this quite quantitative exploration it seems valid to discuss some rather more qualitative ideas.

In particular, can we do better than simply stating that the investigations of the preceding chapters suggest that many of the steps towards life come about as a direct consequence of the nature of gravity and thermodynamics?

At the risk of oversimplifying a great deal of debate on the subject (see some of the references in Chapter 1), it can probably be stated that origin-of-life arguments currently tend to fall into two categories:

1. Life will emerge more or less wherever "Earth-like" conditions exist in the universe.

2. The emergence of life is incredibly unlikely, but it only has to happen occasionally and once it does it is dispersed throughout the cosmos.

The latter is generally labeled as **panspermia** (Chapter 1). Both arguments try to help avoid the unappealing notion that life on Earth is somehow unique in the universe.

Based upon what we have learnt in the rest of this text, then at first glance either theory seems within the bounds of plausibility. This is especially true with some modifications. For example, perhaps we can relax the Earth-like conditions requirement of 1 to include environments like that of Europa, or perhaps life is a bit more likely than in 2, but is only dispersed more locally, within the planetary system, or within the local stellar concentration.

What neither argument yields, of course, is any true insight to the actual first (chemical) steps that set in motion the development of life as we know it. Such mechanisms have also been the subject of extensive debate for a long time, and include general notions such as "emergent phenomena," self-organization, and more specific ones, such as inorganic chemical templates (e.g., clay lattices), polymerization, and many more. Slightly further along the evolutionary tree are ideas including those of an **"RNA-world,"** where RNA reigns as the information carrying molecule, and where there exist deep connections to the seemingly ancient phyla of archaea and bacteria. This stems from a long running debate about whether DNA or proteins came first. Since the discovery of **ribozyme** molecules, a potential solution has been apparent. In brief, RNA molecules can exhibit *catalytic* activity (something previously considered the province of proteins). Within cells the **ribosome**, a complex assemblage of protein and RNA, is a site of protein *synthesis*.

Certain RNA enzyme (ribozyme) molecules catalyze such activity *in addition* to being capable of "self-splicing" (i.e., replicating and expressing gene function). This stunning capability then suggests the possibility of what has become known as an "RNA world" prior to the current "DNA world." In such a bio-molecular world, information is carried, and function is performed, by the same family of molecules—without the "permanent" storage mechanism of DNA. This would be a much less specialized world, awaiting the arrival of the DNA mechanism for evolution to move along new pathways. It would however allow for rapid experimentation and exploration of "survival space"—the ability to try out bio-chemical strategies without great cost.

Embedded in all such discussions (sometimes implicitly) is the connection to environment—was it some temperate wet soup, or was it some ancient low-temperature chemistry in the harsh vacuum of interstellar space? In order to provoke further thought we consider here what might be a *third* category to the list of models for the origin of life. Namely

3. Life can emerge where circumstellar and proto-planetary chemistry and planet formation yield a diverse set of chemical and physical environments.

Rather than shunt the question of the actual mechanism of biogenesis off to either the distant universe (category 2) or to some "temperate soup" (category 1), we can speculate that it is an *integral part* of the remarkable working and re-working of material during the entire process of planet formation, from nebula collapse to the processing and rearrangement of planetesimals, comets and planetary embryos and the eventual formation of a stable planetary architecture.

Consider the following picture of a planetary system like our own in the mid-stages of its early formation. Millions (if not trillions) of planetesimals exist throughout the system, some are rocky, some are icy bodies from a kilometer in size to larger. In amongst these are a few hundred larger planetary embryos. At the center of the system is a still-shrinking proto-star, producing a huge infrared flux, together with vigorous particle flares and X-ray flux. The natural radioactivity of elements is greater than it is today, giving rise to significant radiogenic heating in any large solid bodies. Over a timescale of at least ten million years this melting pot of material undergoes continuous dynamical processing, and as a consequence is chemically and physically altered. A planetesimal (rocky and/or icy) may be heated and cooled many times, it may then be incorporated into a planetary embryo, where gravity and heat begin the differentiation of materials, only to be shattered by a later collision, and eventually reformed with the addition of new material. The conditions experienced will even encompass those which are quite temperate. Indeed, the recent detection of clay-like minerals in comet Tempel 1 (via the Deep Impact mission) has added support for the notion that within both planetesimals and planet embryos, conditions allowing the presence of liquid water must exist at certain times—albeit possibly in very thin layers (e.g., Lisse et al. 2006). Furthermore, the rich organic chemistry seen in carbonaceous chondrites and comets (Chapter 8) is likely to be only the tip of the iceberg in terms of the organic mix in the proto-planetary disk system.

Here then is a somewhat alternative route to take in the quest for the origins of life. It can be suggested that the conditions in a proto-planetary system are such that *extensive* organic chemistry must take place. This is above and beyond that which takes place in interstellar space, or around aged massive stars, or indeed that which takes place on the surface of a fully formed terrestrial world. Not only that, but the conditions will be incredibly diverse, from relatively temperate to extremes of heat and cold, together with potentially strong and fluctuating radiation fields. The possible chemical pathways thus explored by the forming planetary system may far exceed those accessible either on a temperate planetary body or in deep space.

What is further intriguing about taking the investigation for pre-biotic chemistry and possibly biogenesis itself, away from the environment of a planetary surface, is that a very different set of possibilities for the *combination* of chemically processed environments exists. For example, two planetesimals with wildly different formation pathways and chemistry may end up merging. The new body may in turn experience a wide range of changing environments and further agglomeration or fragmentation. The merger of two distinct sets of organic inventories may result in a whole new chain of products and byproducts. Similar events are harder to envisage on a single planetary surface—except through the impact of material, and even that must limit certain products due to the energetics of collision with a massive body.

The hypothesis we make then is that there is indeed something *special* about the processing of material in a proto-planetary environment—a fact that is really self-evident. Furthermore, we suggest that the incredible mixing and range of environments that must occur allow for a vastly wider chemical diversity. This doesn't solve the origin of life problem, but it moves the focus away from either the planetary surface origin or the seeding origin (which again requires a planetary surface). Suppose that all the pre-biotic, and even the first rudimentary biotic molecules, are really produced "off-world," but only in the window of opportunity afforded by planet formation itself.

In such a picture, life on Earth or its *immediate* precursors will either be planted following the final formation of the planet through the merger of multiple planetary embryos, or, in a variation of the usual impact delivery of raw materials, over some extended period thereafter. This model still leaves room for the notion that a terrestrial-like environment is necessary for life to develop further, but makes it necessary *only* for the final combination of ingredients, rather than as a primary incubator.

How could we test this idea? Finding increasingly complex organic chemistry in comets and former planetesimals would certainly be one way. Similarly, finding evidence for life on Mars or places such as Europa would also support this picture—it would however not uniquely confirm this idea. It is also worth pointing out that, unlike the dispersal of life either cosmically or within a given planetary system, which would suggest a common root ancestor, the proto-planetary origin of life does *not* necessarily imply the same root ancestor on every world or body with life. Perhaps the ultimate way to test this idea is through observation of proto-planetary systems themselves. If, as we have argued, the formation period of a planetary system creates a unique chemical factory, which may disperse as the system evolves, then finding complex organic chemistry at the right time in a system would provide considerable support for the idea. Identifying complex molecules via spectral information is however a tricky business. Big molecules exhibit band-like absorption and emission characteristics rather than the discrete spectral lines of atoms or small molecules. Nonetheless, it seems likely that it would be possible to determine something about the mix of organic compounds in a proto-planetary system.

Finally, in the realm of pure speculation, suppose one day we can drill into a large cometary body (for example, a short-period object, likely processed in the inner solar system)—we might pull up a core sample to find that it is swarming with RNA life, no cell structures, but reproducing, protein-building, complex molecules. Or, suppose that on drilling into a Kuiper-belt type object we come across a deep pocket of liquid water, heated by the radiogenic volcanism of the equivalent of a terrestrial hydrothermal vent system. Living around this vent are familiar extremophiles, bacteria, tube-worms, and the like. Eons past these organisms had found their way to the safe haven of Earth's deep mid-ocean ridges when a similar planetary embryo had crunched into the forming world. At that point we might consider that we had found our true ancestors.

10.5.1 The Universality of Bio-mechanics

Somewhere between the discussions of whether the carbon-based chemistry of terrestrial life is likely to be universal (Chapter 5) and discussions of the "special" nature of life (see below) is a gray area revolving around the particular mechanics of life "functions." To give an example: all of what we term life on Earth (i.e., prokaryotes and eukaryotes) operates with **cells** (Chapter 5). Cells are essentially just membrane-like structures that encapsulate the critical machinery for an organism—allowing for the input and output of raw materials and waste products by being semi-permeable.

Clearly, at some point in the evolving biochemistry of the planet, such structures came into use. There are excellent logical reasons for the need to have this physical packaging—at the crudest level it just prevents stuff from floating apart. But it raises an intriguing question—would life on other worlds also utilize such a mechanism? Arguments for the origin of terrestrial cell-structures focus on this need for coherence, in order for the products of DNA and RNA molecules to support their continued existence. It is hard to envisage another way of doing things. The question then becomes this: at what point, what level of complexity, might we really expect divergence between the bio-mechanics utilized on Earth and that utilized on other distant planets?

It is extremely hard to know whether the phenomenon of life actually has a relatively limited number of bio-mechanical options available to it—or whether there are possibilities that we have never imagined. It is tempting to argue that over the 3–4 Gyr that life has been present on Earth it must have had ample opportunity to "explore" (through evolution, natural selection, and random selection) most options. In this vein the concept of **convergent evolution** suggests that there may actually be a rather limited number of successful evolutionary strategies and bio-mechanical structures. Hence both birds (reptiles) and insects have converged on the mechanism of flight using wings, and numerous other—otherwise distantly related—organisms have found similar approaches to dealing with environmental pressures. How deep this convergence really goes is a matter of debate, and it is hard to quantify such things as the limits of complex molecular "solutions" to any particular biological or energetic need. The cautionary note to consider though is that life on Earth appears very much interrelated, and that the ubiquitous prokaryotes are likely the grandparents of everything else, so if they chose one route then it follows that everything else more or less took this path.

To use another example: multicellular life on Earth tends to exhibit strong bilateral symmetry. This symmetry offers an organism the ability to "centralize" critical functions (e.g., digestive systems, nervous systems), as well as being a potential consequence of the most basic cell division processes. Is this going to be a universal function of macroscopic life? Not all multicellular terrestrial organisms have bilateral symmetry. The most notable example is the phyla of *Porifera*, or **sea sponges**. These have no symmetry—they also lack internal organs, nervous systems, and so on. It has been speculated that they represent one of the intermediate evolutionary classes between prokaryotes and eukaryotes. Nonetheless, these organisms have continued to occupy their particular niche over geological timescales, and so should be considered remarkably successful. Thus, bilateral symmetry may be advantageous in some situations, but not all. While we may not expect to find a planet-of-the-sponges, it would certainly be wrong to assume that bi-lateral symmetry is going to be a universal mechanism. As with our understanding of global paleoclimate (see above), the template offered by the Earth should be considered only as a starting point.

10.6   Concluding Commentary and Cautions

At the start of this book we posited that life is a phenomenon that emerges in this Universe as naturally as physical "laws," such as Newtonian gravity. It certainly seems that many of the pieces that go together to enable life as we know it are indeed inevitable. Star and planet formation, and complex carbon chemistry, are generic features of the cosmos, and these appear to be critical for life. We are still, however, looking for the steps that take us from a mix of chemistry to the self-organizing phenomenon of life as we recognize it on Earth (see above).

In earlier chapters (e.g., §5.3) we have alluded to the question of what life really is. We have chosen to not fully address this, and have instead considered life as an intricate phenomenon to understand without imbuing it with any special significance. This can be a hard thing to do—after all we have a very personal involvement, and it feels confusing (if not a little depressing) to imagine that there is nothing special about living, self-aware organisms. The problem with starting down the path of considering "life" as distinct from the rest of the material universe is one that has been around since ancient Greece. This concept is known as **vitalism**, which, put simply, states that life cannot be explained entirely by the laws of physics and chemistry. In other words, life is to some extent self-determining, or a "vital spark." Now, most self-respecting modern scientists would not dream of being advocates for vitalism in any form—however, it can sneak in through innocent discussion of what distinguishes life from non-life and how life can "originate." Indeed, it is implicit in such considerations that there is something fundamentally different between, for example, a bacterium and a piece of rock, even though both can incorporate highly complex structures.

If one looks at life on Earth from a very simple viewpoint—for example, considering diversity versus scale—then it becomes much less "obvious" where life ends and chemistry begins. In Figure 10.8 a rather general measure of diversity (or combinatorial multiplicity) is plotted versus the physical size of organisms, molecules, and atoms. For organisms this is a measure of the estimated terrestrial *genetic diversity* (number of distinct phyla). For molecules it is a measure of the number of potential combinations of atomic constituents, and for atoms it is simply the number of known elements. This should be treated purely as an illustrative guide, but it does point towards a notional trend that the

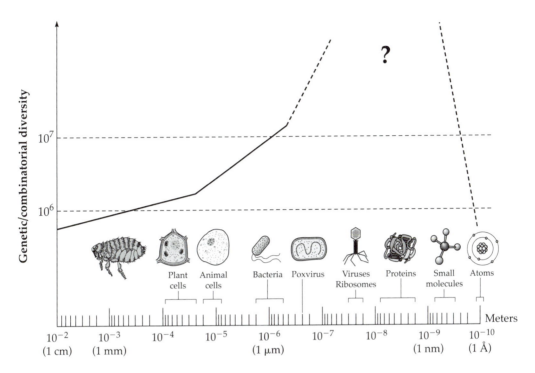

Figure 10.8. A schematic diagram indicating the broad relationship between the scale of organisms, molecules, and atoms, and the potential genetic diversity or combinatorial multiplicity of a given structure. The genetic diversity of prokaryotes is understood to be significantly higher than that of eukaryotic organisms. The potential numbers of different molecular structures at the scale of proteins and other large molecules is hard to compute, but readily exceeds levels of 10^7. While genetic diversity and pure combinatorial numbers are clearly different measures of complexity, this diagram still provides an indication of the idea that it is in the range of the smaller biochemical structures that the most functional diversity can exist.

peak of diversity in organic structures lies somewhere in the regime of proteins to viruses, and that it diminishes in more complex structures.

In fact it is not clear that life is anything more than a continuation of the combinations and permutations of increasingly large molecules. The concept of an RNA-world (see above) appears to fit in this notional scheme—it is simply of intermediate diversity (or functional complexity) between quite complex carbon chemistry and full DNA-based chemistry. Atoms and molecules have the capacity to combine, with varying routes to minimizing energy (which is fundamental to all physical phenomena). These routes get more and more complex but the atoms and molecules just keep doing what they do—exploring the possible combinations and permutations. More complex structures can do this in ways that may be more efficient than one would suspect from just a list of the basic atomic ingredients. Thus, where one places the boundary for "life" is entirely ad hoc, there is simply a continuum of possible form and structure, all the way from a single atom up to a multi-cellular organism. While this may seem obvious, it may also be a bit shocking because it is really the same as saying that a bacterium and a rock *are* the same, except for their degree of complexity or particular arrangement of atoms. In practical terms this may cause panic.[1] because if life is not strictly distinct from non-life then how can we search for it successfully? The solution is most likely one that has served modern science very well. Just as the study of exoplanets gives rise to the field of comparative planetology, the study of life elsewhere gives rise to the field of **comparative biology**. In other words, given a phenomenon too complex to be described by simple parameters, we must first seek ways to group and classify data for comparison. That data may be atmospheric molecules, signatures of photosynthetic pigments, DNA fragments from soils, mineral depositions in rocks, and even walking, talking organisms. To put this another way; rather than setting out to find a specific characteristic indicative of "life," we should try to cast a wide net. This net should include tests based on terrestrial life (amino acids, chirality, cell structures, etc.), but it should also include the means to characterize and parameterize planet-wide systems, local environments, and even stellar systems. There is good reason to do this. Even if life walks by our microscope or telescope and waves at us, as fantastic as that would be, it alone may not allow us to determine the true pathways to life in the universe. We need to follow the example of Darwin and those before him; we must canvas nature, record it, and characterize it. Eventually a more fundamental pattern may emerge, and from there a fundamental theory for the phenomenon of which we are a part.

---

1. Neil Gaiman (2005). *Don't Panic*: Douglas Adams & The hitchhikers guide to the galaxy, Titan Books.

References/Suggested Reading

Arnaud, E. (2004). Giant cross-beds in the Neoproterozoic Port Askaig Formation, Scotland: Implications for snowball Earth, *Sedimentary Geology,* **165**, 155–174.

Barnes, J. W., & O'Brien, D. P. (2002). Stability of satellites around close-in extrasolar giant planets, *Astrophysical Journal,* **575**, 1087.

Cassen, P. M., Reynolds, R. T., & Peale, S. J. (1979). Is there liquid water on Europa? *Geophysics Research Letters,* **6**, 731–734.

Chandler, M. A., & Sohl, L. E. (2000). Climate forcings and the initiation of low-latitude ice sheets during the Neoproterozoic Varanger glacial interval, *Journal of Geophysical Research,* **105**, 20737–20756.

Chyba, C., & Phillips, C. (2001). Surface-subsurface exchange and the prospects for life on Europa, *Lunar and Planetary Sciences,* **32**, 2140.

Hoffman, P. F., Kaufman, A. J., Halverson, G. P., & Schrag, D. P. (1998). A neoproterozoic snowball Earth, *Science,* **281**, 1342-46.

Hoffman, P. F., & Schrag, D. P. (2000). Snowball Earth, *Scientific American,* **282**, 68–75.

Kasting, J., & Ono, S. (2006). Paleoclimates: The first 2 billion years, *Philosophical Transactions of the Royal Society,* **361**, 917.

Knauth, L. P., & Lowe, D. R. (2003). High Archean climatic temperature inferred from oxygen isotope geochemistry of cherts in the 3.5 Ga Swaziland Supergroup, South Africa, *Geological Society of America Bulletin,* **115**, 566–580.

Lisse, C. M., et al. (2006). Spitzer spectral observations of the deep impact ejecta, *Science,* **313**, 635.

Lubow, S., Seibert, M., & Artymowicz, P. (1999). Disk accretion onto high-mass planets, *Astrophysical Journal,* **526**, 1001.

Steuber, T., et al. (2005). Low-latitude seasonality of Cretaceous temperatures in warm and cold episodes, *Nature,* **437**, 1341–1344.

Tobie, G., Lunine, J. I., & Sotin, C. (2006). Episodic outgassing as the origin of atmospheric methane on Titan, *Nature,* **440**, 61.

Williams, D. M., Kasting, J. F., & Wade, R. A. (1997). Habitable moons around extrasolar giant planets, *Nature,* **385**, 234.

Problems

10.1 Using the information in this chapter, and other sources, discuss the changing nature of the Earth's global climate during the past 3.5 Gyr. In particular, discuss the apparent relationship of climate changes to the biological diversity and biomechanical strategies before, during, and after such episodes. Can you comment on the timescale of some of these changes?

10.2 Using the review article by Kasting & Ono (2006), discuss the various possible mechanisms considered for causing glaciations in the early (2–3 Gyr) Earth. Could any such mechanisms operate today? Speculate as to how likely it might be for another Earth-like planet to experience similar episodes of global change—is there anything about climate change and evolution that seems inevitable for such types of planet?

10.3 If the present-day Earth suddenly experienced a glaciation extending to $\pm 30°$ in latitude what would be its effective temperature (recall Chapter 9)? You may assume the same atmospheric content and solar insolation as today, an initial mean albedo of 0.3, and a surface ice albedo of 0.5.

10.4 Compute the surface heat flow on Io due to tides. You may assume a mean density for Io of 3.53 g cm^{-3} and rigidity and specific dissipation the same as for Europa. Convert this into a total heat flow and then compute the total energy lost (assuming all other things remain the same) over 1 Gyr. How does this compare to the total rotational energy of Jupiter? (You will need to research the necessary information for this, but you may approximate Jupiter as a sphere of uniform density.)

10.5 Review the paper by Chyba & Philips (2001) on the prospects for life on Europa. In particular, discuss how the exchange of material at the outer ice surface could play a central role in driving certain types of biosystems.

10.6 Keeping all other parameters the same, how small could Jupiter's orbital semi-major axis be if it were to retain all four Galilean moons ? Assuming an albedo of 0.68 and no atmosphere, compute the effective surface temperature of Europa at this location.

10.7 Describe the concept of convergent evolution. Provide four examples of apparently convergent evolution in the species inhabiting the Earth—these may be biochemical or biophysical examples. Some scientists have used the idea of convergent evolution (e.g., Conway-Morris) to argue that successful life anywhere in the universe really has only a limited number of options available to it in terms of biochemical or biophysical strategies. Do you agree? Given what you have learnt about habitable zones, can you speculate as to whether there might be environments where organisms could exist, but be radically different to that on Earth?

Constants and Data

Astronomical and Physical Constants

Solar mass (1 $M_\odot$)	1.989×10^{33} g
Jupiter mass (1 M_J)	1.8986×10^{30} g
Earth mass (1 $M_\oplus$)	5.974×10^{27} g
Solar equatorial radius (1 $R_\odot$)	6.9599×10^{10} cm
Jupiter equatorial radius (1 $R_\odot$)	7.149×10^{9} cm
Earth equatorial radius (1 $R_\oplus$)	6.378×10^{8} cm
Solar luminosity (bolometric) ($L_\odot$)	3.826×10^{33} erg s^{-1}
Solar effective temperature	5770 K
Jupiter effective temperature	112 K
Earth effective temperature	255 K
Astronomical unit (AU)	1.4960×10^{13} cm
Light year (ly)	9.4607×10^{17} cm
Parsec (pc)	3.0857×10^{18} cm
Sidereal year	3.155815×10^{7} s
Gravitational constant (G)	6.67259×10^{-8} dyne cm^2 g^{-2}
Speed of light (c)	$2.99792458 \times 10^{10}$ cm s^{-1}
Planck's constant (h)	$6.6260755 \times 10^{-27}$ erg s
Boltzmann's constant (k)	1.380658×10^{-16} erg K^{-1}
Stefan–Boltzmann constant (σ)	5.67051×10^{-5} erg cm^{-2} s^{-1} K^{-4}
Proton mass (m_p)	$1.6726216 \times 10^{-24}$ g
Electron mass (m_e)	$9.1093819 \times 10^{-28}$ g
Hydrogen mass (m_H)	1.673532×10^{-24} g

Other Data

DNA helix diameter	2×10^{-7} cm (2 nm)
Ribosome size	1.1×10^{-6} cm (11 nm)
Prokaryotic cell diameter (typical)	0.0001–0.001 cm (1–10 μm)
Eukaryotic animal cell diameter (typical)	0.001–0.003 cm (10–30 μm)
Eukaryotic plant cell diameter (typical)	0.001–0.01 cm (10–100 μm)
Prokaryotic cell mass (typical, wet)	1×10^{-12} g (1 pg)
Eukaryotic cell mass (typical, wet)	1×10^{-9} g (1 ng)
Water ice density (1 atmos pressure, 273 K, Ice-one phase)	0.99984 g cm^{-3}
Water ice density (3000 atmos pressure, 100 K, Ice-nine phase)	1.16 g cm^{-3}
Silicate rock density	3 g cm^{-3}
Iron–nickel density	8 g cm^{-3}

Selected Stellar Data

Main Sequence stellar classes

Class	Mass ($M_\odot$)	Radius ($R_\odot$)	Temperature (K)	Luminosity ($L_\odot$)
O5–O9	60–20	15–10	45,000–33,000	7.9–0.97×10^5
B0–B9	17.5–3.0	8.4–3.0	30,000–10,500	52,000–95
A0–A8	2.9–1.8	2.7–1.7	9,520–7,580	54–8.6
F0–F8	1.6–1.1	1.6–1.1	7,200–6,200	6.5–2.1
G0–G8	1.05–0.85	1.1–0.85	6,030–5,570	1.5–0.66
K0–K7	0.79–0.6	0.79–0.66	5,250–4,060	0.42–0.10
M0–M8	0.51–0.08	0.63–0.17	3,850–2,640	0.077–0.0012

Stellar to brown dwarf regime

Class	Mass ($M_\odot$)	Radius ($R_\odot$)	Temperature (K)	Luminosity ($L_\odot$)
L	0.07–0.012	~ 0.1	1,300–2,000	0.0003–0.00005
T	0.01–0.005	~ 0.1	770–1000	~ 0.00001

Note: parameter ranges are approximate and should be used only as a guide. Stellar (and substellar) classes are based on spectroscopic data, and therefore the mass, radii, temperature, and luminosity ranges indicate an approximate range and may not always overlap with adjacent classes. For objects in the L and T class the observed temperature and luminosity is also a strong function of age, see Chapter 3.

Selected Planet Data

Name	Mass (M$_\oplus$)	Radius (R$_\oplus$)	Density (g cm^{-3})	Semi-major axis (AU)	Eccentricity	Orbital inclination[a] (°)
Solar system bodies						
Mercury	0.06	0.38	5.4	0.39	0.206	3.38
Venus	0.82	0.95	5.3	0.72	0.007	3.86
Earth	1	1	5.5	1	0.017	7.25
Mars	0.11	0.53	3.9	1.52	0.093	5.65
Jupiter	317.8	11.19	1.3	5.20	0.048	6.09
Saturn	95.2	9.46	0.7	9.54	0.054	5.51
Uranus	14.6	4.01	1.3	19.22	0.047	6.48
Neptune	17.2	3.81	1.6	30.06	0.009	6.43
Dwarf planets						
Ceres	0.0002	0.04	1.98	2.76	0.08	10.59
Pluto	0.0022	0.90	2.0	39.48	0.249	17.14
Eris	0.0025	0.10	2.3	67.67	0.442	44.19
Selected moons						
Luna (The Moon)	0.0124	0.272	3.3	0.0026	0.055	18.29-28.58 [b]
Europa	0.008	0.246	3.0	0.0045	0.0101	0.464[b]
Ganymede	0.025	0.413	1.9	0.0072	0.0011	0.2[b]
Callisto	0.018	0.376	1.8	0.0126	0.0074	0.192[b]
Titan	0.023	0.404	1.9	0.0082	0.0288	0.349[b]
Enceladus	0.000018	0.0395	1.6	0.0016	0.0047	0.019[b]
Triton	0.0036	0.212	2.1	0.0024	0.0000	157.34[b]

a. relative to solar equator
b. relative to host planet equator

Index